VCF 9000

STATI
FOR
QU

STATISTICAL METHODS FOR DETECTION AND QUANTIFICATION OF ENVIRONMENTAL CONTAMINATION

Robert D. Gibbons
University of Illinois at Chicago

David E. Coleman
Alcoa Technical Center

JOHN WILEY & SONS, INC.

New York / Chichester / Weinheim / Brisbane / Singapore / Toronto

This book is printed on acid-free paper. ∞

Copyright © 2001 by John Wiley & Sons. All rights reserved.

Published simultaneously in Canada.

No part of this publication may be reproduced, stored in a retrieval system or transmitted in any form or by any means, electronic, mechanical, photocopying, recording, scanning or otherwise, except as permitted under Sections 107 or 108 of the 1976 United States Copyright Act, without either the prior written permission of the Publisher, or authorization through payment of the appropriate per-copy fee to the Copyright Clearance Center, 222 Rosewood Drive, Danvers, MA 01923, (978) 750-8400, fax (978) 750-4744. Requests to the Publisher for permission should be addressed to the Permissions Department, John Wiley & Sons, Inc., 605 Third Avenue, New York, NY 10158-0012, (212) 850-6011, fax (212) 850-6008, E-Mail: PERMREQ @ WILEY.COM.

This publication is designed to provide accurate and authoritative information in regard to the subject matter covered. It is sold with the understanding that the publisher is not engaged in rendering professional services. If professional advice or other expert assistance is required, the services of a competent professional person should be sought.

Library of Congress Cataloging-in-Publication Data:

Gibbons, Robert D., 1955–
    Statistical methods for detection and quantification of environmental contamination / Robert D. Gibbons, David E. Coleman.
      p. cm.
    Includes index.
    ISBN 0-471-25532-7 (cloth: alk. paper)
      1. Pollution—Measurement—Statistical methods. 2. Environmental monitoring—Statistical methods. I. Coleman, David E. II. Title.
    TD193.G547 2001
    628.5′0287—dc21   2001017855

Printed in the United States of America.

10 9 8 7 6 5 4 3 2 1

To Carol, my wonderful wife of 22 years, and to our children, Julie and Jason

Robert Gibbons

To Sally, my wonderful wife of 25 years, and to our four blessings from the Lord: Elizabeth, Brittany, Rebecca, and Timothy

David Coleman

In memory of Dorothy Homa

CONTENTS

ACKNOWLEDGMENTS — xv

1 Introduction — 1

PART I DETECTION AND QUANTIFICATION IN THE LABORATORY — 7

2 Conceptual Foundations — 9

   2.1 Chemical Measurement Process / 9
   2.2 Calibration and Evaluation Functions / 11
   2.3 Sensitivity / 12
   2.4 Accuracy and Precision / 12
   2.5 Detection and Quantification (Quantitation) / 14
      2.5.1 Conceptual Definitions of the Three Limits / 14
      2.5.2 Interpretation of the Intervals Defined by the Critical Level, Detection Limit, and Quantitation Limit / 16
   2.6 Between-Laboratory Environment / 17
      2.6.1 Bias and Variation Within a Single Laboratory / 17
      2.6.2 Bias and Variation Between Laboratories / 18
      2.6.3 Currie's Schematic / 18

3 Statistical Foundations and Review — 21

   3.1 Hypothesis Testing / 21
      3.1.1 Truth and Certainty of Belief / 22
      3.1.2 Statistical Yardstick for Hypothesis Testing / 24
      3.1.3 Statistical Significance and Practical Importance / 26

3.2 Interval Estimation / 27
    3.2.1 Confidence Intervals / 27
    3.2.2 Prediction Intervals / 30
    3.2.3 Statistical Tolerance Intervals / 32

4 Calibration-Based Regression Models   34

4.1 Calibration Designs / 34
4.2 Ordinary Least-Squares Estimation / 34
4.3 Weighted Least-Squares Estimation / 35
4.4 Estimating the Weights / 36
    4.4.1 Rocke and Lorenzato Model / 36
    4.4.2 Exponential Model / 38
    4.4.3 Linear Model / 39
4.5 Iteratively Reweighted Least-Squares Estimation / 39
4.6 OLS Prediction Intervals / 40
4.7 OLS Tolerance Intervals / 40
4.8 WLS Prediction Intervals / 41
4.9 WLS Tolerance Intervals / 42
4.10 Illustration / 42

5 Single-Concentration-Based Detection Limit Methods   48

5.1 Introduction / 48
5.2 Kaiser–Currie Method / 49
5.3 U.S. EPA 40CFR, Part 136 (GLASER et al.) / 50
5.4 Prediction Limits / 52
5.5 Tolerance Limits / 52
5.6 Simultaneous Control of False Positive and False Negative Rates / 53
5.7 Limitations / 54
5.8 Illustration / 56

6 Single-Concentration-Based Quantification Limit Methods   57

6.1 Introduction / 57
6.2 Currie's Determination Limit / 58
6.3 ACS Limit of Quantitation / 58
6.4 EPA Practical Quantification Limit / 58
6.5 EPA Minimum Level / 59
6.6 Illustration / 59
6.7 Discussion / 60

CONTENTS

7 Calibration-Based Detection Limit Methods — 61

7.1 Introduction / 61
7.2 Models for Constant Variance (Hubaux and Vos) / 62
7.3 Procedure Due to Clayton and Co-workers / 64
7.4 Generalization to Multiple Future Detection Decisions / 65
7.5 Nonconstant Variance / 66
7.6 Variance Proportional to Concentration / 66
7.7 Modeling Observed Variance / 67
7.8 Iteratively Reweighted Least Squares / 68
7.9 Illustration / 69
7.10 Between-Laboratory Detection Estimate / 70
    7.10.1 Interlaboratory Detection Estimate and the ASTM / 72
    7.10.2 IDE Approach / 73

8 Calibration-Based Quantification Limit Methods — 79

8.1 Introduction / 79
8.2 Alternative Minimum Level / 79
8.3 Alternative Formulations / 82
8.4 Estimator Due to Gibbons and Co-workers / 84
8.5 Interlaboratory Quantitation Estimate / 85
    8.5.1 IQE Approach / 86

9 Significant Digits — 90

9.1 Introduction / 90
9.2 Measurement Uncertainty Intervals from Calibration / 90
9.3 Common Use of Significant Digits / 92
9.4 Pair of Problems Due to Having Ten Fingers and Their Solutions / 92
    9.4.1 Order-of-Magnitude Uncertainty Range: Fractions of Significant Digits / 93
    9.4.2 Order-of-Magnitude Relative Uncertainty: Bounding Relationship / 94
9.5 Definition of Detection: At Least Zero Significant Digits / 96
9.6 Definition of Quantitation: At Least One Significant Digit / 98
    9.6.1 Use of QL_{1D} / 99
    9.6.2 Development of QL_{1D} / 99
    9.6.3 Graphical Example of Computing QL_{1D} / 100

9.7 Reporting Measurement and Uncertainty / 101
    9.7.1 Adorning a Measurement with Uncertainty / 101
    9.7.2 Measurement, Standard Deviation, Radical, and Degrees of Freedom: MRDS Format / 102

10 Experimental Design of Detection and Quantification Limit Studies and Related Studies   104

10.1 Introduction / 104
10.2 Basic Considerations and Recommendations / 105
10.3 Variance Components / 108
10.4 Youden Pairs / 111
10.5 Matrix Effects and Blanks / 113
    10.5.1 Matrix Effects / 113
    10.5.2 Blanks / 115
10.6 Blind Studies / 116

11 Between-Laboratory Detection and Quantification Limit Estimators   117

11.1 Introduction / 117
11.2 Naive Estimators / 118
11.3 Approximate Estimator / 118
11.4 More Sophisticated Approach / 118
    11.4.1 Random-Effects Model for Homogeneous Errors / 119
    11.4.2 Random-Effects Model for Heterogenous Errors / 120
11.5 Applications / 122
11.6 Illustration / 123

PART II DETECTION AND QUANTIFICATION IN THE FIELD   129

12 Comparison of a Single Measurement to a Regulatory Standard   131

12.1 Introduction / 131
12.2 Determining Regulatory Compliance / 132
    12.2.1 Case I: STD = 0 / 132
    12.2.2 Case II: $0 < \text{STD} < L_Q$ / 132
    12.2.3 Case IIa: $0 < \text{STD} \leq L_C$ / 133

  12.2.4 Case IIb: $0 < L_C <$ STD $< L_Q$ / 133
  12.2.5 Case III: $L_Q \leq$ STD / 133
 12.3 Confidence Interval for True Concentration / 134
 12.4 Illustration / 134

13 Censored Data   136

 13.1 Introduction / 136
 13.2 Conceptual Foundation / 137
 13.3 Simple Substitution Methods / 139
 13.4 Maximum Likelihood Estimators / 139
 13.5 Restricted Maximum Likelihood Estimators / 143
 13.6 Linear Estimators / 144
 13.7 Alternative Linear Estimators / 145
 13.8 Delta Distributions / 153
 13.9 Regression Methods / 155
 13.10 Substitution of Expected Values of Order Statistics / 157
 13.11 Comparison of Estimators / 158
 13.12 Further Simulation Results / 161
 13.13 Summary / 162

14 Testing Distributional Assumptions   164

 14.1 Introduction / 164
 14.2 Simple Graphical Approach / 165
 14.3 Shapiro–Wilk Test / 166
 14.4 Shapiro–Francia Test / 171
 14.5 D'Agostino's Test / 171
 14.6 Methods Based on Moments of a Normal Distribution / 173
 14.7 Multiple Independent Samples / 175
 14.8 Testing Normality in Censored Samples / 177
 14.9 Kolmogorov–Smirnov Test / 181
 14.10 Summary / 182

15 Testing for Outliers   183

 15.1 Introduction / 183
 15.2 Rosner's Test / 184
 15.3 Skewness Test / 187
 15.4 Kurtosis Test / 188
 15.5 Shapiro–Wilk Test / 188
 15.6 E_m-statistic / 188
 15.7 Dixon's Test / 190

15.8 Illustration / 191
15.9 Summary / 193

16 Detecting Trend — 194

16.1 Introduction / 194
16.2 Sen's Test / 195
16.3 Mann–Kendall Test / 197
16.4 Seasonal Kendall Test / 200
16.5 Statistical Properties / 203
16.6 Summary / 203

17 Detection Monitoring — 204

17.1 Introduction / 204
17.2 Groundwater Detection Monitoring / 204
17.3 Statistical Prediction Intervals / 206
    17.3.1 Single Location and Constituent / 206
    17.3.2 Multiple Locations / 207
    17.3.3 Verification Resampling / 208
    17.3.4 Multiple Constituents / 209
    17.3.5 Problem of Nondetects / 209
    17.3.6 Nonparametric Prediction Limits / 210
17.4 Intrawell Comparisons / 211
17.5 Illustration / 212
17.6 Methods to be Avoided / 213
    17.6.1 Analysis of Variance / 213
    17.6.2 Cochran's Approximation to the Behrens–Fisher t-Test / 215
17.7 Summary / 217

18 Assessment and Corrective Action Monitoring: Overview — 218

18.1 Introduction / 218
18.2 Strategy / 219
18.3 Application to Specific Media / 223
    18.3.1 Soils: Evaluation of Individual Source Areas (PAOCs) / 223
    18.3.2 Soils: Area- or Site-wide Evaluations / 224
    18.3.3 Groundwater: Aquifer / 224
    18.3.4 Groundwater: Groundwater–Surface Water Interface / 226

        18.3.5   Groundwater: Long-Term Monitoring / 226
        18.3.6   Groundwater: Natural Attenuation Evaluation / 227
        18.3.7   Waste Stream Sampling / 228

   18.4 Summary / 228

19 Assessment and Corrective Action Monitoring: Comparison to a Standard — **229**

   19.1 Introduction / 229
   19.2 LCL or UCL? / 230
   19.3 Normal Confidence Limits for the Mean / 231
   19.4 Lognormal Confidence Limits for the Median / 232
   19.5 Lognormal Confidence Limits for the Mean / 232
        19.5.1   Exact Method / 232
        19.5.2   Approximating Land's Coefficients / 233
        19.5.3   Approximate Lognormal Confidence Limit Methods / 238
   19.6 Nonparametric Confidence Limits for the Median / 242
   19.7 Confidence Limits for Other Percentiles of the Distribution / 243
        19.7.1   Normal Confidence Limits for a Percentile / 243
        19.7.2   Lognormal Confidence Limits for a Percentile / 244
        19.7.3   Nonparametric Confidence Limits for a Percentile / 244

20 Assessment and Corrective Action Monitoring: Comparison to Background — **252**

   20.1 Introduction / 252
   20.2 Normal Prediction Limits for $m = 1$ Future Measurements at Each of k Locations / 253
   20.3 Normal Prediction Limits for the Mean(s) of $m > 1$ Future Measurements at Each of k Locations / 258
   20.4 Lognormal Prediction Limits for $m = 1$ Future Measurements at Each of k Locations / 262
   20.5 Lognormal Prediction Limits for the Median of $m > 1$ Future Measurements at Each of k Locations / 264
   20.6 Lognormal Prediction Limits for the Mean of $m > 1$ Future Measurements at Each of k Locations / 265
   20.7 Nonparametric Prediction Limits for $m = 1$ Future Measurements in Each of k Locations / 267
   20.8 Nonparametric Prediction Limits for the Median of $m > 1$ Future Measurements at Each of k Locations / 267

21 Assessment and Corrective Action Monitoring: Case Studies   273

21.1 Case Study 1: Long-Term Monitoring / 273
21.2 Case Study 2: Soil PAOC and Soil Phase III Evaluation / 274
   21.2.1 PAOC 2 / 274
   21.2.2 PAOC 18 / 276
   21.2.3 Results / 276
21.3 Case Study 3: Site-wide Groundwater Evaluation / 277

22 Review of Available Computer Software   280
23 Summary   281

APPENDIX: LAND'S TABLES   283

GLOSSARY OF MEASUREMENT TERMINOLOGY   323

MATHEMATICAL SYMBOLS   339

WEB REFERENCES   341

ANNOTATED BIBLIOGRAPHY   343

INDEX   381

ACKNOWLEDGMENTS

We would like to thank Michael Zorn for his tremendous assistance in preparing the bibliography, and Paul Hewett and Gary Ganzer for sharing their algorithms for the computer approximation of Land's coefficients. We would also like to thank Fred Blickle, Mark Napolitan, and John Baker for help with the assessment and corrective action monitoring examples and formulation of the problems and strategies. Special thanks to Jay Auses, Nancy Grams, Stu Hunter, Steve Koorse, Larry LaFleur, Ray Maddalone, Phil Ramsey, Jim Rice, Phil Stein, and Lynn Vanatta. Each has played a substantial role in our work in this area and has advanced the science of metrology. Thanks also to other ASTM colleagues: Paul Britton, Naomi Goodman, John Hubbling, Babu Nott, John Phillips, Marty Sara, Judy Scott, and Nina Whiddon; to Alcoans: Cheryl Begandy, Lee Blayden, Jay Goodman, Brad Novic, John Smith, and Bill Snee; and to others: Lloyd Currie, Dulal Bhaumik, Don Hedeker, Darrell Bock, Dave Patterson, Tony Gray, Kirk Cameron, and Charles Davis. Special thanks to Dave Burt, Mark Verwiel, and Louis Bull for continuing to bring interesting environmental problems to our attention after so many years and for providing a test track for our solutions. Additional thanks to Mike Caldwell, Steve Clarke, Terry Johnson, Jeff Shanks, Lori Tagawa, and Eric Wallis for providing valuable feedback on the implementation of environmental monitoring statistics across the nation. Finally, we would like to thank Joe Flaherty and Boris Astrachan of the University of Illinois at Chicago for their unending support of our statistical research efforts.

1

INTRODUCTION

In recent years, statistical methods have played a major role in environmental monitoring programs. With the development of a modern statistical approach to environmental regulatory statistics (Davis, 1994; Davis and McNichols, 1987; Gibbons, 1987a,b, 1994, 1996; Gilbert, 1987) there has been a major evolution in the way in which environmental impact decisions are made. This early work has focused largely on the earliest possible detection of a release, often termed environmental *detection monitoring*. As these new methods have become incorporated into state and federal regulation and guidance (U.S. EPA, 1987, 1988, 1989, 1992), the need for improved statistical approaches to related problems of assessment, compliance, and corrective-action monitoring has grown as well. Unfortunately, far less statistical work has been done in this area, and corresponding environmental impact decisions are still often based on a comparison of individual measurements to fixed standards, or at best, simple normal confidence bounds. Often, a facility or property is declared as environmentally impacted if a single measured concentration exceeds an environmental standard. First, we should be concerned about such a practice because we should be interested in comparison of the true concentration to the standard, not simply the measured concentration. Of course, without infinite sampling we can never really know the true concentration; however, statistical analysis provides a means of drawing inference to the true concentration distribution from a series of measured concentrations. Second, we should be concerned about such a practice because it treats all environmental problems as being equal. For example, exceeding an environmental standard in 1 of 5 samples is a very different problem than exceeding an environmental standard in 1 of 500 samples. Again, the statistical approach to this problem incorporates our uncertainty in the true concentration distribution rather than simply assuming that the measurements are made without error and represent truth in and of themselves. In fact, nothing could be further from the truth.

Since much of the threat of pollution comes from human beings and manufactured products, the constituents of concern are often anthropogenic and do not exist naturally in the environment. Here, the variability in the system is based in large part on the practice of the laboratory and the preparation and analysis of individual samples. In almost all cases, analytical measurements are accepted as true concentrations without regard to their uncertainty. Confidence bounds on measured concentrations are rarely reported, despite the availability of historical data that could routinely be used for this purpose. In the absence of such uncertainty estimates, laboratories most often rely on limits of detection and quantification to screen analytical measurements. The limit of detection, which we denote as L_D following the pioneering work of Currie (1968), allows us to make the binary decision of "detected" with specified levels of confidence for errors of both the first (false positive) and second (false negative) kinds. The limit of quantification, L_Q (Currie, 1968), is the concentration at which the true concentration can reliably be measured. In many ways, these two types of limits are simply points along a continuum that describes the relationship between true concentration and uncertainty. As we will see, the statistical modeling of such a relationship is complicated by the fact that uncertainty is rarely, if ever, constant, even for small intervals on this continuum. Of course, the role of this uncertainty must be incorporated in making environmental monitoring decisions, although in practice, it rarely is.

The purpose of this book is to describe the statistical theory that underlies the detection and quantification of environmental pollution in both the laboratory and the field. In the laboratory, we present the foundation of relating measured concentrations to true concentrations and the development of intervals of uncertainty for true concentrations given a new measured concentration. Related to this problem is the problem of estimating thresholds on this curve that define concentrations at which detection and quantification decisions can reliably be made. In this book we present a comprehensive review of this topic with directions for future research.

In the field, we discuss how analytical measurements can be used in making environmental impact decisions and more broadly, how environmental data can be compared to regulatory standards, naturally occurring background concentrations, or both. Again, we present a comprehensive review of this problem and directions for future research.

Overview of the Book In most chapters a general introduction to the problem is presented, followed by increasingly complex solutions. In some cases, statistical theory is presented that may not be accessible to all readers; however, it is included for completeness and the hope that this book may provide a foundation for further statistical research in this area. Despite complexity, for each solution or statistical approach to a particular problem, a relevant example is provided with computational details and/or tables that can be used for routine application of the statistical results. Attention is paid to statistical properties of alternative approaches, including false positive and false negative rates associated with each test and factors related to these error rates where possible. Recommendations are provided for specific problems based on characteristics such as number of monitoring wells, number of constituents, distributional form of measurements, and detection frequency. The reader may use the book to help craft an assessment, compliance, or corrective-action monitoring program for most environmental media that have been or potentially have been affected by one or more pollutants. Although discussed to some degree for completeness, the reader interested in routine environmental detection monitoring is referred to the previous book by Gibbons (1994), where this topic is dealt with in considerable detail. Similarly, the reader interested in statistical aspects of environmental sampling is referred to the excellent book by Gilbert (1987).

Part I contains 10 chapters in which conceptual and statistical issues of detection and quantification in the laboratory are discussed. Although much of the work and illustrations involve problems in the environmental sciences, the problems and solutions apply to all aspects of analytical chemistry and to calibration problems in other fields as well. For example, much of the work done in analysis of the contents of the foods that we eat are directly amenable to solution using the methods described in Part I. Chapter 2 begins with a discussion of the *conceptual foundations* underlying the chemical measurement process, calibration function, sensitivity, precision, and accuracy. In addition, in Chapter 2 we discuss the conceptual basis for detection and quantification decisions both within and across laboratories.

In Chapter 3 we review the *statistical foundations* that underlie the methods described in the book. They are divided into the areas of *hypothesis testing* and *interval estimation*. Under hypothesis testing, we describe issues of confidence and statistical power. Under interval estimation we discuss confidence limits, prediction limits, and tolerance limits.

Chapter 4 provides a detailed overview of *calibration-based regression models* that relate measured concentrations or instrument responses to the underlying true concentration. Here we present two general types of estimation procedures: ordinary least squares (OLS) and weighted least squares (WLS). OLS assumes that the measurements are independent and have equal uncertainty or variability regardless of concentration, although the latter assumption is rarely true in practice. WLS assumes that the measurements are independent; however, their uncertainty or variability is nonconstant and often is some function of the true concentration in the sample. For both OLS and WLS estimators, we provide details for the computation of confidence, prediction, and tolerance intervals for the calibration function itself (i.e., a confidence interval) and for individual deviations from the calibration function (i.e., prediction and tolerance intervals).

In Chapter 5 we introduce the concept of the *detection limit* for single-concentration-based designs. The detection limit is used of make the binary decision as to whether or not an analyte is present in the sample. The various approaches to this problem differ primarily as to whether and how false positive and false negative error rates are specified and at what concentration, if any, the samples are "spiked." Although the single-concentration design is the least statistically rigorous in that it provides no way in which to estimate the variance function (i.e., the relationship between variability and true concentration), it is by far the most widely used approach to estimating the detection limit and the easiest to compute. We present it here as a review and as a way of fixing ideas for the presentation of calibration-based (i.e., more than one concentration) detection limit estimators, which are the primary theme of this part of the book.

In Chapter 6 we introduce the concept of the *quantification limit* for single-concentration designs. The quantification limit describes the point on the calibration curve at which the signal-to-noise ratio is sufficiently large to have reasonable confidence in the true concentration based on measured value(s). As one might expect, this conceptual definition can lead to a multitude of operational definitions. Again, the single-concentration-design estimators are presented here as a means of fixing ideas and preparing the reader for the more statistically advanced calibration-based estimators.

In Chapters 7 and 8 we extend the concepts of detection and quantification limits to calibration-based designs in which multiple concentration points are used to model simultaneously the relationships between measured and true concentration and variability and true concentration. A variety of approaches are described, based largely on the statistical foundation provided in Chapters 3 and 4. As in all cases, the methods are illustrated using a number of relevant examples.

In Chapter 9 we extend the methods developed in Chapters 4, 7, and 8 to the important problem of determining the number of significant digits for an estimated concentration. In this chapter we also present new guidelines for reporting measurements and corresponding uncertainty estimates.

In Chapter 10 we present experimental design issues that underlie detection and quantification limit studies. In this chapter we consider problems of variance component estimation, between- and within-laboratory studies, selecting calibration standards, and various traditional experimental designs that can be used to provide representative conditions.

In Chapter 11 we extend the previous developments to the case of multiple laboratories. When the same data can be analyzed by multiple laboratories, both within- and between-laboratory variance components must be incorporated into the interval estimates for the calibration function. This is often the case when split samples are obtained and analyzed by different laboratories for regulatory or quality control purposes. In this chapter we introduce new work on generalized mixed-effects models (i.e., the laboratory is random and the concentration is fixed) and corresponding interval estimates.

Part II deals with the impact that use of detection and quantification limits have on traditional environmental statistics. As described in these chapters, the presence of nondetected values have a profound effect on the usual complete data case for testing hypotheses and constructing interval estimates.

In Chapter 12 we discuss how laboratory uncertainty estimates can be used to determine if the underlying true concentration for an individual measured concentration exceeds a regulatory standard. In Chapters 13 to 16 we discuss methods for dealing with the resulting data, which comprise a mixture of detected and nondetected concentrations. In Chapter 13 we review the literature on methods for dealing with censored data. In this chapter we consider imputation methods, linear estimators, regression methods, methods based on normal order statistics, maximum likelihood estimators, restricted maximum likelihood estimators, and delta distributions. In

Chapter 14 we present methods for testing for distributional form, including cases in which the data are censored. In Chapter 15 we provide a general overview of testing for outliers in environmental data. In Chapter 16 we provide a general overview of nonparametric methods for testing trend that are all suitable for data that comprise a mixture of quantifiable and nonquantifiable data.

Chapter 17 is a broad overview of statistical methods for environmental detection monitoring, including problems of statistical prediction, multiple comparisons, treatment of nondetects, nonparametric alternatives, intrawell comparisons, applied to problems in groundwater, surface water, and air monitoring.

In Chapters 18 to 21 we present a unified treatment of statistical methods for analysis of data collected as part of compliance, assessment, and corrective-action programs. Although statistical methods for detecting an environmental impact (Chapter 17) have been well studied, considerably less attention has been paid to the characterization of the rate and extent of a release and the effects of remediation. In Chapter 18 we provide a general overview of the problem and a sketch of the general approaches that are available. In Chapter 19 we discuss comparisons to regulatory standards and in Chapter 20, statistical methods for comparison to background concentrations. The methods are then illustrated in Chapter 21 using a series of three case studies. Much of the material presented in these four chapters will be new to practitioners in this area. We consider methods for normal and lognormal distributions, as well as nonparametric alternatives. For the comparison to regulatory standards, we present normal, lognormal, and nonparametric confidence bounds for the mean, median, and other percentiles of the concentration distribution. As in all cases in this book, methods for the treatment of censored laboratory data are emphasized.

The book concludes with a review of currently available software (Chapter 22), a summary (Chapter 23), and an annotated bibliography. The bibliography covers most papers published in the area of detection and quantification and a brief discussion of each, as a well as a list of references used in the text.

I

DETECTION AND QUANTIFICATION IN THE LABORATORY

2

CONCEPTUAL FOUNDATIONS

2.1 CHEMICAL MEASUREMENT PROCESS

Analytical chemistry is a tremendously complex field which has as its principal objective the process of making chemical measurements. The complexity of the field of analytical chemistry has grown to the point where there are tens or hundreds of subfields defined by such aspects as the analyte being measured, the medium (*matrix*) in which the analyte is found, the physical principle used (more specifically, the analytical method used), the required instrumentation, and perhaps even the sampling technique involved. To get some idea of the complexity, one need only attend one of the large national conferences for analytical chemists, such as PittCon, or visit the conference Web page, *www.pittcon.org*. On display at such conventions are hundreds of complex measurement systems. It is almost certainly true that no one fully understands even 1% of the equipment that is on display.

Following Currie (1995), this book is concerned with a much higher level and abstract view of the chemical measurement process. At this high level, all chemical measurement processes (indeed, all measurement processes) look essentially the same and can be divided into three phases: (1) initialization phase, (2) routine measurement phase, and (3) quality assurance/quality control phase. We consider each of these phases briefly, in turn.

Initialization Phase Prior to making routine measurements, the following must be selected:

- An analyte to measure (typically, this will be the concentration of a chemical constituent, such as free cyanide, or Aroclor 1428, but it could

instead be a property or physical state, such as turbidity or total solids in solution)
- A matrix within which one measures the analyte (e.g., tap water, stack emissions, river sediment)
- A suitable sample (e.g., weight, volume, environmental conditions)
- A physical principle on which to base the measurement (e.g., effect of molecular weight on gas-phase elution time in a column, $F = MA$)
- Standard reference materials (SRMs) to define what are considered true levels of the analyte [e.g., NIST (National Institute of Standards and Technology) standards, internal standards]
- Standard test data to be used to transform measurement signals into usable measurement values via the calibration function

Routine Measurement Phase The routine measurement phase is the *up-and-running phase* of measurement, the phase in which measurement customers take greatest interest, and which laboratory managers seek to maximize (as proportion of total time). One can identify different steps in the routine measurement phase, not all of which will be present for all measurement systems and which may be followed in slightly different order:

1. Obtain the sample (time and space may be randomly or deliberately selected).
2. Homogenize the sample.
3. Split the sample (into two or more samples), or obtain subsamples.
4. Preserve the sample (may involve chemical additives).
5. Store the sample (may involve a special container, and special environmental conditions).
6. Prepare the sample (may involve dilution, concentration, extraction, centrifuging, etc.).
7. Initialize the measurement system.
8. Adjust the measurement system parameters for this particular type of analyte, matrix, solvent, sample, and measurement.
9. Conduct the measurement process.
10. Ensure that no error condition was encountered during the measurement process (respond, as necessary).
11. Record measurement results and supporting information (e.g., solvent used, time and date, analyst name).
12. Safely dispose of waste and by-products.

Quality Assurance/Quality Control (QA/QC) Phase The QA/QC phase of measurement is a "necessary evil" of good measurement practice. It can help ensure reliable measurements, at the cost of the loss of some productive time for the measurement system. Generally, QA/QC checks are interlaced with routine measurement.

For any individual measurement, there may be "sufficiency" requirements that must be satisfied for the measurement to be declared valid, such as the recovery rate of a related compound spiked into the sample. For any set of measurements there are contemporaneous "before" and "after" measurements of standard reference materials (SRMs) that pass the test of no unusual bias or variation. This "test" is typically one or more control charts of such measurements (or combinations of the measurements), such as x-bar and s charts.

2.2 CALIBRATION AND EVALUATION FUNCTIONS

The calibration function is, in general, an algorithm that converts a desired measurement value into an expected response:

$$E(y) = F(x)$$

where y is the response, $E(y)$ the expected value of the response, x the true quantity being measured (typically, the concentration or amount of an analyte), and $F(\cdot)$ the transformation algorithm, typically a function, such as $A + bx$, or Ae^{bx}. Because all measurements are made with error, and because we do not know $F(\cdot)$ exactly, we must accept the fact that for any given measurement with true value x, the response, y, is obtained with error:

$$y = F(x) + \varepsilon$$

where ε is random, possibly normally distributed.

For example, a simple spring scale has hash marks scribed on a metal plate. These hash marks convert a linear displacement (say, in millimeters) to mass (say, in grams). A chromatographic system may use a straight line or a polynomial to convert peak area (or log peak area) into a mass, then a concentration. In a flowmeter, fluid may turn a propeller, which turns a magnet in a coil of wire, producing a voltage, which is then converted by a spline function to liters per second.

We consider here primarily the simplest case, a straight-line relationship between x and y. This is also the most common case (at least after natural transformation of x or y, such as log or square):

$$F(x) = A + Bx$$

A and B are unknown constants, which are estimated, perhaps using least-squares regression, by a and b. Note that different calibration functions may be defined for different ranges of measurement. Collectively, they may be "optimally" chosen and estimated using spline functions, perhaps with attractive properties of connectivity and smoothness. Alternatively, someone might simply examine residuals and somewhat arbitrarily select different regions over which different functions will be estimated and used. A piecewise linear calibration function is not uncommon.

12 CONCEPTUAL FOUNDATIONS

There are a few desirable properties of the calibration function, $F(x)$:

- It satisfies the definition of a mathematical function; that is, for any given x, there is a single value, y (e.g., straight line, log, or sine, but *not* arcsine).
- Its inverse (called the *evaluation function*) satisfies the definition of a mathematical function (e.g., $F(x)$ is a straight line, log, or arcsine, but *not* sine).
- The two properties above, if satisfied, imply that $F(x)$ is strictly monotonic in x; that is, given any value, $\delta \neq 0$, for any x, either

$$F(x) > F(x + \delta) \quad \text{or} \quad F(x + \delta) > F(x)$$

- $F(x)$ has a mathematical inverse, $G(y) = F^{-1}(y) = x$.
- The form of $F(x)$ is given by first-principles laws or relationships, such as $F = MA$, or chemical balance equations.

2.3 SENSITIVITY

Sensitivity is a term that suffers from having a variety of definitions, depending on the technical discipline. Here we use the term as it is used in analytical chemistry: The sensitivity is the slope of the calibration function. In the case of straight-line calibrations, sensitivity is a scalar value. Otherwise, sensitivity is the first derivative of the calibration function, wherever that is defined within the range of measurement. Generally speaking, the higher the sensitivity, the better, because higher sensitivity generally means better ability to discriminate between different true values. Higher sensitivity generally results in lower detection limits and lower quantitation limits (provided that measurement imprecision is not correspondingly higher). However, one must be careful to only compare sensitivities expressed in the same units; 800 mm/kg is *not* better than 1 m/kg. As Currie notes, sensitivity is defined differently for clinical chemistry: Of all test subjects with a particular condition, the sensitivity of a test is defined as the proportion of those subjects that are correctly identified as having the condition.

2.4 ACCURACY AND PRECISION

Accuracy and precision are to metrologists what high employment and low mortgage rates are to politicians: Everyone is "for" them! That is, everyone wants accuracy and precision. Unfortunately, the terms are used loosely—sometimes deliberately so, for commercial advantage.

Accuracy, in particular, is a misleading term because there are two different definitions. They overlap but are not identical:

1. *Accuracy* is absence of systematic error (i.e., absence of average bias, where bias represents the difference between the measurement and true values; see the definition below).

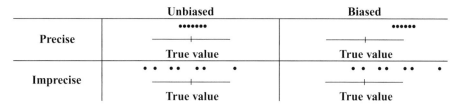

Figure 2.1 Bias and precision.

2. *Accuracy* is absence of error of any form, systematic and random. In this book we follow a recommendation from U.S. Environmental Protection Agency (EPA) scientists: To avoid confusion, the term *accuracy* should not be used. Instead, the terms *precision* and *bias* should be used, to convey separately the information about systematic and random error, respectively (see Figure 2.1).

Revising the EPA definitions, we thus define *bias* as systematic or persistent error due to distortion of a measurement process, which deprives the result of representativeness (i.e., the expected sample measurement is different than the sample's true value). Note that for a given analyte, matrix, and method, bias may or may not be constant with respect to the true value, x. This may be due to faulty or obsolete calibration. For straight-line calibration, an error in the estimate of the slope (i.e., $b - B \neq 0$) leads to nonconstant bias. An error in the estimate of the intercept (i.e., $a - A \neq 0$), alone, leads to constant bias.

The complement of systematic error is random error (i.e., precision, often quantified by standard deviation). We define *precision* as the degree to which a set of observations or measurements of the same property, usually obtained under similar conditions, conform to themselves (i.e., are self-consistent). Precision is usually quantified by standard deviation, variance, or range, in either absolute or relative terms. Precision may not be constant over the entire measurement range. For analytical measurements, measurement standard deviation typically increases with concentration, so that relative standard deviation (i.e., standard deviation divided by true concentration) may be roughly constant over most of the working range of the measurement system. Note that for a measurement system, we want *low* bias and *low im*precision (i.e., small standard deviation, representing good precision)

The difference between bias and imprecision is more than semantic; the difference has practical implications:

- To ensure low bias, one must have access to traceable standard reference materials, and one must be willing and able to conduct potentially large calibration studies.
- To ensure low imprecision (i.e., good precision), one might have to replace routine single measurements with routine multiple measurements, and report averages instead of single values.

14 CONCEPTUAL FOUNDATIONS

- Bias can be tolerated if the only use to which the measurements are put is monitoring a process for change, provided that precision is good relative to the size of change to be detected and the sampling frequency. A constant bias of 5 g may be immaterial if all one is interested in is a shift by 100 g, or a trend where the mass slowly increases by 100 g, or a single outlier measurement that is lower than the average by 100 g, or a quadrupling in the process standard deviation by a factor of 4, from 20 g to 80 g.
- Bias that changes with true concentration, due to an error in estimating the calibration function slope, can usually be corrected simply by recalibrating, often by using ordinary least squares. However, precision that changes with true concentration may force the routine use of weighted least squares for calibration, a technique that is somewhat more complicated than ordinary least squares.

2.5 DETECTION AND QUANTIFICATION (QUANTITATION)

Over the years, there have been numerous definitions and interpretations of *detection limits* and *quantitation limits* for trace-level measurement (see Coleman et al., 1997; Currie, 1968; or Oresic and Grdinic, 1990). *Critical limits* (also called *critical levels*) have been little discussed in the literature but have often been confused with detection limits. The basic concepts of all three types of limits are simple. In this section we once again use the foundation laid by Currie (1968, 1995). Referring to Figure 2.2, we first define each of the limits, from lowest to highest, then we discuss the technically sound interpretation of the *intervals* between the limits. In later chapters we deal with the statistical procedures used to compute these limits.

2.5.1 Conceptual Definitions of the Three Limits

L_C The lowest of these three limits is the *critical level* (L_C). L_C is the lowest measured concentration above which one can confidently assert that the analyte has been detected. It is the lowest measurement that is unlikely to have been obtained from a blank sample. We reserve the right to choose a confidence level (e.g., 99%) to define what *is* and what *is not* considered unlikely. Hence any measurement above L_C should be considered strong evidence that the analyte is present (at least one molecule), where "convincing" is only to the degree of the confidence level chosen. Because of this implicit decision that the analyte is present (when a measurement exceeds L_C), L_C is sometimes referred to as the *detection threshold*.

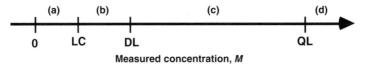

Figure 2.2 Metrological partitioning of the low end of the real number line. Distance not to scale.

Unfortunately, L_C has sometimes been confused with a detection limit, as in the EPA's *'method detection limit'* (MDL, as described in 40 CFR, Part 136, App. B). Without developing a comprehensive critique of the MDL in this chapter, let it be observed that the EPA MDL is defined basically as

$$\text{MDL} = t_{[\text{df}=6, 1-\alpha=0.99]} s = 3.14s$$

where t is a Student's t critical value and s is a sample standard deviation from seven replicate samples of spiked reagent-grade water. Thus the MDL is a critical level that has been set up to be exceeded approximately 1% of the time when a blank sample is measured (actually, based on its construction, it is technically more correct to say that a measurement will be negative approximately 1% of the time when measuring a sample with concentration = MDL).

Note that there is little or no assurance that samples with real concentrations at or below L_C will be detected. For that, we need to consider a higher concentration: the detection limit.

DL (Also Called L_D) The middle limit among these three is the *detection limit* (DL). The DL is the lowest concentration at or above which an analyte can confidently be detected (i.e., distinguished from zero). Thus the detection limit defines the lowest concentration at which the measurement signal consistently emerges from the noise. That is, when the true concentration is at or above the DL, the reported measurement will, with high confidence, exceed L_C. At the DL, a measurement provides the most elementary form of information, a binary value indicating (at some level of confidence) whether the analyte is detected or *not* detected. See Chapter 9 for a discussion of how this basic information content is consistent with an alternative definition of the DL as "the minimum concentration at which one can be sure that a measurement will have at least zero significant digits."

There is tremendous diversity in proposed DLs. Some of this diversity is due to semantic confusion between L_C and DL. However, the diversity in proposed DLs is due primarily to differences in DL applications, differences in required confidence, and different statistical approaches to calculating DLs. Some of the key issues are:

- For which sources of bias and variation should the DL account (i.e., include in its estimate of measurement variation, e.g., σ)? Does the limit account for laboratory-to-laboratory biases, preparation variation, calibration error (lack of fit, coefficient error, error in the calibration standard concentrations), matrix effects, analyte impurity, differences in the identification or computation algorithm used by the measurement system, bias, and variation in the instrument or analytical method?
- Does the limit define detection for a single future determination, a week's worth of future determinations, or all future determinations—and with respect to the *current* or *all future* calibrations?
- What are required rates of correct detection when measuring samples at the DL? Correct nondetection when measuring blanks?

- Is standard deviation of measurement error assumed known? Is it assumed constant within the range of concentrations of interest?

QL (Also Called L_Q) *Quantitation limits* (QLs) are also defined in diverse ways. In concept, a QL is the lowest concentration at which there is some confidence that the reported measurement is relatively close to the true value. This is what is truly meant by "to quantitate" or "to quantify." Some of the diverse definitions are given below.

The quantitation limit (QL) is the lowest concentration at or above which:

- *One can quantitate.* Amazingly, this definition has been proposed by some chemists in all seriousness, but the definition merely shifts the burden to defining *quantitate*.
- *One can have "assurance" of detection.* However, this is provided by the DL. It is appropriate for detection but is inconsistent with the common usage of the related words: *quantify*, *quantity*, *quantitative*, and is inconsistent with the historical use of the term, introduced by Currie (1968).
- *Measurements have a low, prescribed standard deviation* (e.g., 5 ppb). This is a reasonable definition for ensuring a certain number of significant digits of measurement for a known range of measurement values, say 100 to 999 ppb.
- *Measurements have limited relative standard deviation* (e.g., RSD < 10%). This also is a reasonable definition and is used by the American Chemical Society to define the *limit of quantitation* (LOQ): LOQ = 10σ, the solution to requiring that the 99%+ confidence interval about a measurement be within span $\pm 30\%$: $\pm 3\sigma = \pm 30\%$. Additionally, it has been required to have RSD < 10%, in Gibbons (1994). Two weaknesses of the "10σ" approach are: (1) there is no indication of degrees of freedom in the estimate of the RSD, so it is not possible to determine the multiplier required to base the QL on a valid statistical interval; and (2) the percentage is arbitrary and typically has no relationship to significant digits.
- *Measurements have limited relative measurement uncertainty* (RMU) (e.g., RMU 5%) or some other prescribed proportion at some level of confidence (e.g., 95%). The RMU is the RSD times a multiple (>1) which is a function of concentration, and depends on the confidence levels and the calibration design. See Chapter 9 for a complete development of the QL as a guarantor of limited RMU, which can provide assurance of at least one significant digit in a measurement.

2.5.2 Interpretation of the Intervals Defined by the Critical Level, Detection Limit, and Quantitation Limit

As discussed in Section 2.5.1, conceptually, the critical level, detection level, and quantitation level partition the real number line into four intervals into which a trace-level measurement, *M*, can fall. Figure 2.2 provides a graphical representation of this partitioning of the number line. The intervals of the number line (a) to (d) as divided by the L_C, DL, and QL are listed below, with interpretation:

(a) The measurement M is less than the critical level ($M < L_C$). By definition of L_C, M is indistinguishable from zero concentration and hence should be considered a nondetect. M might even be less than zero. If $M < 0$, M can still be useful—if it is one of several measurements. It should not be discarded, labeled "nondetect," nor set to a prescribed nonnegative value (such as 0, DL/2, or DL).
(b) M is at or above the critical level but below the detection limit ($L_C \leq M < $ DL). By definition of L_C, M is treated as a detection. However, by definition of DL, it should be realized that there is low confidence of detection in this interval. Note that the label "low confidence" is not a value judgment; it is a factual statement given that a level of confidence has been selected.
(c) M is at or above the detection limit but below the quantitation limit (DL $\leq M <$ QL). By definition of L_C, M is treated as a detection, and by definition of DL, any true concentration in this interval is *likely* to be detected. By definition of QL, M is a very noisy measurement value which should only be reported with an error interval, and used only with extreme caution in comparison or computation.
(d) M is at or above the quantitation limit (QL $\leq M$). By definition of QL, M can be reported as a measurement (preferably with an error interval) and can generally be used for comparison and computation.

2.6 BETWEEN-LABORATORY ENVIRONMENT

Once again, we draw on the Currie (1995) insightful treatment of this topic, but we start at a more basic level and examine sources of bias and variation from several perspectives.

2.6.1 Bias and Variation Within a Single Laboratory

Chemical measurement is complex, even when one considers measuring a single analyte in a single type of matrix by a single method. Under these constraints, any given measurement still has several sources of bias and variation that may be nontrivial: intrinsic instrument noise, some "typical" amount of carryover error, plus differences in analysts, sample preparation, instrumentation, and even data-processing algorithms (thresholds, signal filters, etc.). Here we ignore sampling variation, which results in different true values for different samples.

A thorough approach to characterizing measurements within a laboratory would involve developing a plausible statistical model of measurement components (probably a complicated *mixed-effects model*, a model containing factors with random effects and factors with fixed effects), design a study to collect the necessary data, conduct the study, then fit one or more models to the results. Unfortunately, although it may appeal to the purist, this approach is intractably complex and expensive. Instead, a simpler approach is followed: The effects of the various sources are split into two categories, bias and variation. It is generally accepted that within a laboratory there are ("fairly independent") sources of bias that conspire to create a consistent lab bias relative to the true value (here, we are simplifying by assuming that bias is constant over the measurement range, but even if this is not true, the concept can be generalized). Additionally, it is generally accepted that within a laboratory there are (fairly independent) sources of variation that add up to *within-laboratory variation* (Currie uses the label *repeatability variation*, quantified by the estimated repeatability standard deviation).

One of the responsibilities of a laboratory expert in a particular measurement area is to minimize the within-laboratory bias and minimize repeatability standard deviation.

2.6.2 Bias and Variation Between Laboratories

Taking a bird's-eye view, many laboratories make the same types of measurements. Within any particular laboratory, there will be measurement bias and variation, but a hard truth is that the problem of measurement error is more serious than one can discern by analyzing data from only one laboratory. If one were to send identical replicate samples to a random selection of laboratories, it is inescapable (given the right study design and sufficient measurements) that there will be evidence of laboratory-to-laboratory bias, plus overall laboratory bias (comparing grand mean to true value). The latter component may be labeled—perhaps correctly—*method bias*.

Why does method bias occur? The speculative reasons are legion, including standard reference materials (SRMs) at different laboratories that are nominally the same value, but in fact, are different (e.g., PCB standards), contamination problems that affect some laboratories more than others, differences between instruments, data-processing systems or versions, and laboratory standard operating procedures (SOPs). At this level it is impractical to try to study and understand what contributes to the differences in measurements from different laboratories. Pragmatically, all that one can do is ensure that between-laboratory bias and variation are *not* ignored. Particularly, between-laboratory bias and variation should not be confused with within-laboratory bias and variation.

2.6.3 Currie's Schematic

In a 1995 missive, Lloyd Currie adapted one of his own graphics, a schematic of the partitioning of method, interlab, and within-laboratory error and uncertainty. We reproduce this excellent graphic and add a discussion in Figure 2.3. Understanding the concepts in Currie's schematic is essential to understanding the two basic levels at which modeling and analysis of between-laboratory data take place. There are two parts: the single-laboratory perspective (top half) and the group-of-labs

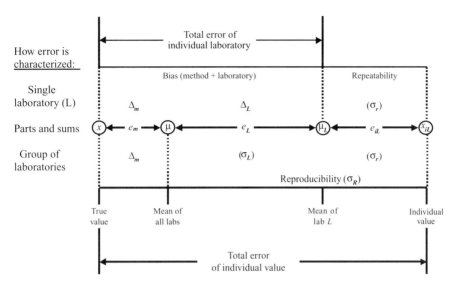

Figure 2.3 Currie's schematic of measurement error components in the interlaboratory context.

(i.e., between-laboratory) perspective (bottom half). Common to both halves is x_{iL}, an individual measurement made at lab L (representing the population of such measurements). That measurement consists of several components:

$$x_{iL} = x + e_m + e_L + e_{iL} \qquad (2.1)$$

where x_{iL} is the ith measurement made at lab L, x the true value, e_m the error due to the bias of the method, e_L the error due to the bias of lab L, and e_{iL} the random error of the ith measurement made at lab L.

Within lab L it is recognized that there are two types of bias, method and lab. These are characterized by fixed effects, Δ_m and Δ_L, and their sum is the total bias of lab L. The total bias ($\Delta_m + \Delta_L$) of any lab, L, can be estimated by conducting a *within-laboratory* study. The two fixed effects, Δ_m and Δ_L, can be estimated using a *between-laboratory* study. Within lab L it is also recognized that there is random error associated with each measurement, called *repeatability variation*. This random error cannot be estimated for any individual measurement, but it is accepted that such errors are distributed according to a normal distribution, with mean equal to zero, and standard deviation σ_r. Although σ_r is unknown, it should also be possible to estimate it using the results of a within- or between-laboratory study. From the lab L perspective, the method improvement goal is to reduce bias (get the total bias of lab L, μ_L, closer to x), and improve repeatability (reduce σ_r, i.e., get x_{iL} closer to μ_L).

From the more comprehensive, less detailed perspective of the group (population) of laboratories in the study, the same error component equation (2.1) holds, but no longer is the lab L bias treated as a fixed effect to be added to the method bias (for any lab, L), producing the total bias of lab L. Instead, e_L is regarded as a random realization of a (normal) random variable representing the distribution of all laboratory biases. This random variable has mean equal to zero and standard deviation labeled σ_L—which is unknown but can be estimated from the interlaboratory study. It is then convenient to add the random variable that characterizes laboratory biases and the random variable that characterizes repeatability variation. Assuming that they are independent, their variances add, and we get σ_R, reproducibility standard deviation:

$$\sigma_R^2 = \sigma_L^2 + \sigma_r^2$$

For the perspective of the group of laboratories, the method improvement goal is to reduce method bias (i.e., get the laboratory consensus mean, μ, closer to x) and improve reproducibility (reduce σ_R, i.e., get x_{iL} closer to μ).

3

STATISTICAL FOUNDATIONS AND REVIEW

3.1 HYPOTHESIS TESTING

Hypothesis testing is a bedrock of classical statistics, and it has particular relevance to detection, although not so much relevance to quantification. The essential idea in hypothesis testing is that one proposes a plausible assertion (called a *null hypothesis,* labeled H_0), an assertion that may or may not be true. Then one makes a decision: Is the assertion true, or not? The decision is based on the analysis of relevant data and comparing the likelihood of the analysis results to an acceptable statistical significance level. For example, H_0 might be: There is no cyanide in the glass of tap water sitting on the desk before me. Before I take a sip, I want to decide whether or not I accept H_0. (Note that whether or not there is any cyanide is a matter of fact—but possibly an *unknowable* fact.)

To test H_0, I could take one or more samples from the glass or the faucet and have them tested for cyanide by gas chromatography/mass spectroscopy (GC/MS). Or, I could call the local water utility and ask them for recent test results. Or, I could ask my doctor (or my mother) whether the water was safe. Suppose I take the first option and get 10 measurements: 970, 1010, 1020, 990, 1010, 1000, 990, 980, 1000, 1000 (all ppb). These values would seem to be well above zero, and let us suppose that they are well above the maximum nontoxic allowable level (which is true; the maximum contaminant level for cyanide is 200 ppb according to the U.S. EPA National Primary Drinking Water Regulations as of January 2000).

22 STATISTICAL FOUNDATIONS AND REVIEW

Let us now formalize the null hypothesis, as follows:

$$H_0: c = 0$$

where c is the concentration of cyanide in the water. The next logical step is to consider what it means for H_0 to be true or not true. As stated above, H_0 states that there is no cyanide in the water in the glass. For the time being, we ignore detection issues and hair-splitting considerations such as the near certainty that there is at least *one* cyanide molecule in the glass. If H_0 is *not* true, then apart from the existence of antimatter, the only possibility is that $c > 0$. We call this the *alternative hypothesis*,

$$H_1: c > 0.$$

Note that H_0 and H_1 are mutually exclusive (i.e., they are disjoint hypotheses) and they are complete (i.e., one of them must be true). The objective of hypothesis testing is to determine, to some degree of likelihood, whether the null hypothesis is true or the alternative hypothesis is true.

We use our measurement data to decide between the hypotheses. Our best estimate of c is the sample mean, $\bar{x}$. Therefore, as common sense would dictate, we use $\bar{x}$ to determine whether we can accept hypothesis H_0 ($\bar{x}$ is close to 0) or reject hypothesis H_0 in favor of hypothesis H_1 ($\bar{x}$ is far from 0). The outstanding question is: What does it mean to be close and what does it mean to be far? Additionally, how certain should we be of our decision to accept or reject hypothesis H_0?

3.1.1 Truth and Certainty of Belief

We tackle the second question first: How certain should we be of our decision to accept or reject hypothesis H_0? The best way to answer this question is to consider a simple truth table, shown as Table 3.1. There are four possible outcomes, labeled R00, R01, R10, and R11, depending on the true condition and the belief about the condition. They are shown and discussed below.

Table 3.1 Truth table for hypothesis testing

Belief or Decision (Based on Data)	True Condition	
	$c = 0$	$c > 0$
$c = 0$	R00: H_0 correctly recognized as true $\Pr\{\text{accept } H_0 \mid H_0\} = 1 - \alpha$	R10: H_0 incorrectly believed to be true $\Pr\{\text{accept } H_0 \mid H_1\} = \beta$
$c > 0$	R01: H_1 incorrectly believed to be true $\Pr\{\text{reject } H_0 \mid H_0\} = \alpha$	R11: H_1 correctly recognized as true $\Pr\{\text{reject } H_0 \mid H_1\} = 1 - \beta$

R00 This favorable outcome is that the null hypothesis is true, and it is believed to be true. In our drinking water example, R00 would correspond to cyanide-free water that we decide is perfectly safe to drink. If the null hypothesis were true (the $c = 0$ column of the table), there are only two possibilities: H_0 is correctly believed to be true (R00) or is incorrectly *not* believed to be true (outcome R01). There is some probability associated with each of these beliefs, given that the water is free of cyanide. These two probabilities must sum to 1 since there are no other outcomes, given that $c = 0$. A common convention is to define α implicitly by stating that the R00 (given that the true condition is $c = 0$) is $1 - \alpha$. Assuming that we set up our statistical tests properly, $1 - \alpha$ is called the *confidence level*. Clearly, we want $1 - \alpha$ to be as close to 1 as is practically possible. Typically, people choose confidence levels of 95% or 99%; rarely would it be below 90% or 80%.

R01 This unfavorable outcome is that the null hypothesis is true, but it is believed to be false, so it is rejected. In our drinking water example, R01 would correspond to cyanide-free water that we decide is contaminated. Keeping the same notational convention and taking as a given that $c = 0$, there is a probability of α that R01 will be the outcome. Clearly, we want α (called the *significance level* or *Type I error rate*) to be small; identically, we want $1 - \alpha$ to be near 1.

R10 This unfavorable outcome is that the null hypothesis is false (the $c > 0$ column of the table), but it is believed to be true. In our drinking water example, R10 would correspond to contaminated water that we decide is safe to drink (in which case the result may be *very* unfavorable). As with the $c = 0$ column, if we take as a given that the null hypothesis, H_0, is false, and the alternative hypothesis, H_1 ($c > 0$), is true, there are only two possibilities: H_0 is incorrectly believed to be true (outcome R10) or H_0 is correctly rejected as false (outcome R11). These two probabilities must sum to 1 since there are no other outcomes given $c > 0$. A common convention is to label the probability of R10 as β (this is an implicit definition of β). Assuming that we set up our statistical tests properly, β is called the *Type II error rate*. Clearly, we want β to be small, especially if it involves people ingesting cyanide.

R11 This outcome is "statistically" favorable, albeit undesirable for cyanide. In this case, the null hypothesis is false, and it is believed to be false. In our drinking water example, R11 would correspond to contaminated water that is recognized to be contaminated. Consistent with the notational convention used for R01 and taking as a given that $c > 0$, R11 will be the outcome with probability $1 - \beta$. Assuming that we set up our statistical tests properly, $1 - \beta$ is called the *statistical power* of the hypothesis test. Clearly, we want $1 - \beta$ to be as close to 1 as possible; identically, we want β to be small.

In the scientific community, and even in the public at large, there is wide appreciation for high confidence (i.e., high $1 - \alpha$). Unfortunately, there is poor recognition of the importance of statistical power and the consequences of inadequate power. As we discuss later, the decision to accept or reject H_0 for a situation such as cyanide in water is based on comparing $\bar{x}$ to a threshold (or critical value, c'). We reject H_0

24 STATISTICAL FOUNDATIONS AND REVIEW

if $\bar{x} > c'$, where c' depends directly on the number of measurements (i.e., sample size) and is computed to ensure an acceptable level of confidence $(1 - \alpha)$ and/or an acceptable level of statistical power $(1 - \beta)$. The difficulty can be in the trade-off, because for a given sample size and a given amount of variability in the measurement, the only way to increase confidence is to give up power, and vice versa. To ensure both high confidence and high power, one needs an adequate sample size.

See Table 3.2 for some examples of hypothesis testing in real life.

3.1.2 Statistical Yardstick for Hypothesis Testing

We now return to the question posed in Section 3.1.1. For the example of cyanide in drinking water, suppose that we use $\bar{x}$ to determine whether we can accept hypothesis H_0: $c = 0$ ($\bar{x}$ is close to 0) or reject hypothesis H_0 in favor of hypothesis H_1: $c > 0$ ($\bar{x}$ is far from 0). What does it mean to be close, and what does it mean to be far? We propose to reject H_0 if $\bar{x} > c'$, where c' depends directly on the number of measurements (i.e., sample size) and is computed to ensure an acceptable level of confidence $(1 - \alpha)$ and/or an acceptable level of statistical power. How do we compute c'?

Elementary statistics textbooks generally provide an adequate explanation of the mechanics of hypothesis testing of a mean compared to a standard, and develop other tests for hypothesis testing. Therefore, in this section we develop only the simplest and most classical approach to hypothesis testing, using our example of cyanide in water.

The first fact to be understood about the statistics of hypothesis testing is that there is no *single* way to do it. However, there are preferred approaches, and there is a means by which one can make quantitative comparisons of one approach versus another. For example, in the case of cyanide in water, and assuming that negative measurements can be reported (as is usually the case), one could reject the null hypothesis of no cyanide

(H_0: $c = 0$) if one of the following holds:

1. The mean of a set of n measurements exceeds zero.
2. Any of n measurements exceeds zero.
3. The average of the maximum and the (negative) minimum of n measurements is greater than zero.

Although each of the tests above is logical, in the sense that each one is more likely to reject H_0 as the true concentration increases, none of the tests explicitly controls α or β. Furthermore, the level of α and/or β will depend on n, and for a given α and n there is no assurance that any one of the tests is optimal (i.e., achieves the minimum possible β).

We present the optimal (and classical) test, given the following simplifying assumptions:

1. The measurement process through which we obtained the cyanide measurements is unbiased.

Table 3.2 Hypothesis testing and risk analysis in everyday life

Type of Test or Tester	Description of Decision to Be Made	Typical Null Hypothesis, H_0	Possible Consequences of "False Alarm" (Falsely Reject H_0; probability = α)	Possible Consequences of "False AOK" (Falsely Accept H_0; probability = β)	Focus to Reduce:
Metal detector	Is the airline passenger carrying a weapon or bomb?	No weapon or bomb	Passenger is delayed and will miss flight	Many people could be injured or killed	β
ELISA blood test	Does this blood donor have AIDS?	No infection	Donor is put through temporary trauma and must get more tests	Innocent blood recipients could contract AIDS	β
Crash sensor	Should the airbag be deployed?	No collision; do not deploy	Serious accident	Injuries are unnecessarily serious	Both
PCR with sequence and match software	Does this DNA match?	Perfect match	Criminal remains at large	Innocent person arrested/convicted	α
Breathalyzer	Is this driver drunk?	Not drunk	Driver detained, although innocent	Drunk driver waved on; may cause accident	α
Ultrasonic test	Is there a crack in this elevator beam?	No crack	Lost time to verify; may be added cost of scrapping good metal	Dangerous crack may cause part failure/airplane crash	β
X-ray reading	Does this lung have cancer?	No tumor	Added expense, time, and stress of unneeded tests	Deadly cancer that could have been treated successfully	β

26   STATISTICAL FOUNDATIONS AND REVIEW

2. The cyanide measurements are statistically independent of one another.
3. The cyanide measurements are identically distributed according to the normal distribution (we can make this assumption even though we know it cannot be exact; measurements cannot be arbitrarily negative).

To test H_0, the following statistics are computed, where we let C represent the value of interest in the null hypothesis (H_0: $c = C = 0$ for the cyanide example).

$$\text{Sample mean: } \bar{x} = \frac{1}{n}\sum_{i=1}^{n} x_i = 997$$

$$\text{Sample standard deviation: } s = \sqrt{\frac{1}{n-1}\sum_{i=1}^{n}(x_i - \bar{x})^2} = 14.94$$

$$\text{Student's } t \text{ statistic: } t_{n-1} = \frac{\bar{x} - C}{s/\sqrt{n}} = 211 \text{ (for } C = 0\text{)}$$

$$p\text{-value: } T(t_{n-1}, n-1) < 0.0001$$

where $T(\cdot)$ is the cumulative Student's t (found in the back of most elementary statistics textbooks), and the second argument ($n - 1$) is the number of degrees of freedom for s. To decide whether to reject H_0, the p-value is compared to α (selected a priori). In this case, $p < 0.0001 < 1\%$, so we would reject H_0.

For any assumed values of s, $\bar{x}$, and n, and any selected values of C and α, the value of β can be calculated using the noncentral t distribution (see Johnson and Kotz, 1970). Some important facts to remember are as follows:

- To decrease β, one must increase n, α, or Δ (= $\bar{x} - C$), *or* one must decrease s.
- One cannot change any one of the values, n, α, Δ, s, or β without changing one or more of the others.
- Typically, one can make Δ arbitrarily small (for reasonable values of α and β, and for any given s) by making n sufficiently large, although this might be expensive. In other words, as long as one does not care how *small* it is, one can almost always find a difference that is statistically significant if one has a large enough budget.

3.1.3   Statistical Significance and Practical Importance

A error common among practitioners of applied statistics is to confuse statistical significance with practical importance. The two are *not* the same. Consider the implications of the last statement of the preceding section. With enough samples, one can prove that there are statistically significant differences between cancer rates of people with or without mustaches, or between up- and downgradient cyanide concentrations, or between average popcorn consumption of left- and right-handers.

Statistical significance is an assertion that a difference is "unlikely to have happened by chance." The needed yardstick is a "by chance" yardstick, typically a multiple of a

Table 3.3 Combinations of statistical significance and practical importance

Practically Important? (Choose Δ)	Statistically Significant? (Choose α)	
	Yes	No
Yes	Take action (easy decision)	Get more data if cost/benefit warrants (hard decision)
No	Record for future; may want to reassess importance (hard decision)	No action needed (easy decision)

standard deviation. If such a yardstick shows that two values (e.g., $\bar{x}$ and C) are widely separated, we reject the implausible notion that the underlying ("population") values are actually equal.

Practical importance is an assertion that a difference matters to us, apart from any statistical/uncertainty considerations. A doubled cancer rate matters because life is important. A mean cyanide level above the EPA's MCL matters for the same reason. It is a little harder to concoct a reason why popcorn consumption levels that differ by handedness might be important!

Decision making in the face of uncertainty is made easier when statistical significance coincides with practical importance (see Table 3.3). Complex judgment is required when there is no such coincidence. In fact, if Table 3.2 is reexamined, it is obvious that every reference to β has an implicit Δ that *is practically important*.

3.2 INTERVAL ESTIMATION

Statistical intervals are technically equivalent to hypothesis tests but often provide a more useful way to communicate uncertainty. Unfortunately, it is not broadly appreciated outside the statistical community that there are three main types of statistical intervals: confidence, prediction, and tolerance intervals. These different types of intervals have different applications, different widths, and different interpretations. A summary of these and other differences is provided for the simplest cases in Table 3.4. Here we provide only an introduction, focusing on the concepts. For a much more complete treatment, see Hahn and Meeker (1991).

3.2.1 Confidence Intervals

A confidence interval is an interval on the real number line that is computed so as to contain an unknown characteristic of a population, such as the true mean. Thus a confidence interval is descriptive, not predictive. Typically, a confidence interval is *two-sided* and of finite length, such as (A, B). It can be *one-sided* and semi-infinite, such as values $< B$ *or* values $> A$, which can also be expressed as $(-\infty, B)$ or (A, ∞).

Table 3.4 Comparison of statistical intervals[a]

Characteristic	Type of Interval		
	Confidence	Prediction	Statistical Tolerance
Contains:	μ	Next observation	Central proportion, P, of population
Comparative width	Narrow	Medium width	Wide
Parameters to specify	α = nominal significance (Type I error rate)	α	α, P
Basic formula (example)	$\bar{x} \pm t_{[n-1,1-\alpha/2]} \dfrac{s}{\sqrt{n}}$	$\bar{x} \pm t_{[n-1,1-\alpha/2]} s \sqrt{1 + \dfrac{1}{n}}$	$\bar{x} \pm K_{P,\alpha} s$
Needed factors	$t_{[n-1,1-\alpha/2]}$ = Student's t	$t_{[n-1,1-\alpha/2]}$ = Student's t	$K_{P,\alpha}$ = STI factor (tabulated)
Example of use	"With 95% confidence, the mean is 7 ± 1"[b]	"With 95% confidence, the next observation will be within 7 ± 2"[c]	"With 95% confidence, at least 90% of the population is within 7 ± 5"
Variants	• One- vs. two-sided • Can be created for σ or other population parameters • Asymmetric • Nonnormal distributions • Nonparametric	• One- vs. two-sided • Asymmetric • Nonnormal distributions • Nonparametric • r of m future observations • r of m observations from k sites	• One- vs. two-sided • Asymmetric • Nonnormal distributions • Nonparametric
Extension to straight-line regression (calibration)	Joint confidence region for slope and intercept; useful for calibration QA/QC, but not for prediction of new measurement uncertainty	Single-observation prediction interval for regression line is almost useless; instead, use r of m version	Regression STI is best interval for making statement re uncertainty of arbitrary number of future observations

[a] Single concentration; two-sided; normal distribution; $\bar{x}$ = sample mean; s = sample standard deviation.
[b] Technically, this should be restated as follows: "95% of all such intervals—computed independently—will contain the true mean."
[c] An alternative interpretation is that the average proportion of all future observations falling outside of many such intervals is 5% = 100% − 95%.

Technically, a 95% confidence interval is an interval about a population parameter for which a declaration can be made, such as the following: *On average, 95% of such computed intervals will contain the true value of the parameter of interest.* A similar declaration can be made for any desired confidence level. This declaration is commonly transformed to the pragmatic (although technically incorrect) declaration for any given interval: *With 95% confidence, the true parameter lies within this interval.* These declarations are made given that certain standard assumptions hold—the same as listed above—for hypothesis testing. In this book we avoid discussion of the distinction between these two declarations.

A confidence interval is the complement of a hypothesis test for a population parameter. In a hypothesis test, one wants to make a binary decision (i.e., decide yes or no) at a selected significance level, whether a hypothesis (such as $C = 0$) is true, given a relevant sample value (e.g., $\bar{x} = 997$) and other values (e.g., sample estimate of σ). A test statistic is computed (e.g., Student's t is used for the mean). Based on the presumed known and tabulated distribution of the test statistic, a p-value is computed, then compared to the desired significance level, α. If the p-value exceeds α, the hypothesis is not rejected. If the p-value does not exceed α, the hypothesis is rejected at significance level α.

Plausible Values For comparison, the same types of assumptions and calculations can be used to construct an interval of "plausible values" about the sample value, at the same desired significance level, α. This interval of plausible values is a confidence interval. Let $\alpha = 1\%$ and consider constructing a $100(1 - \alpha) = 99\%$ confidence interval about the mean for the cyanide example. Recall that it was known that

$$\bar{x} = \frac{1}{n}\sum_{i=1}^{n} x_i \text{ (sample mean)} = 997$$

$$s = \sqrt{\frac{1}{n-1}\sum_{i=1}^{n}(x_i - \bar{x})^2} \text{ (sample standard deviation)} = 14.94$$

From a table of Student's t critical values, we find that

$$t_{[n-1, 1-\alpha/2]} = t_{[9, 0.995]} = 3.25$$

A symmetric 99% confidence interval for the mean is then

$$\bar{x} \pm t_{[n-1, 1-\alpha/2]} \frac{s}{\sqrt{n}}$$

which is

$$997 \pm 3.25 \frac{14.94}{\sqrt{9}}$$

which simplifies to 997 ± 16.2; this can also be expressed as (980.8, 1013.2) ppt.

30 STATISTICAL FOUNDATIONS AND REVIEW

Since we are concerned, a priori, about values that exceed 0, and since true concentrations less than zero are physically impossible, we would actually prefer to use the one-sided interval defined by its lower bound,

$$\bar{x} - t_{3n-1, 1-\alpha 4} \frac{s}{\sqrt{n}}$$

which is

$$997 - 2.82 \frac{14.94}{\sqrt{9}} = 982.9 \text{ ppt}$$

The one-sided 99%-confidence interval is thus (982.9, ∞).

Since this interval does not include 0, we see that at the 1% significance level, the mean concentration cannot be zero. Or, if we had hypothesized a priori that the concentration was 200 ppt, we can see that 200 is not contained in the interval, so the "concentration = 200 ppt" hypothesis also would have been rejected. Note that neither of the confidence intervals permits a statement regarding individual, future measurements. One cannot say "with 99% confidence, the next measurement will fall in the interval (980.8, 1013.2) or (982.9, ∞)." The reason, as stated previously, is that the confidence interval is a descriptive interval for a population parameter. It is not a predictive interval (see the next section). This fact should be made obvious by studying the form of the confidence interval for the mean:

$$\text{CI} = \bar{x} \pm t_{[n-1, 1-\alpha/2]} \frac{s}{\sqrt{n}}$$

which implies that as the sample size, n, increases, the interval shrinks inversely with the square root of n, hence becomes vanishingly small as n tends to infinity. Clearly, this makes sense for increased knowledge of the mean, but it violates common sense for the next measurement. Analogously, having an ever-larger number of IQ scores in a database does *not* make it increasingly likely that the next score (or scores) will be arbitrarily close to 100!

See Table 3.4 for other characteristics of confidence intervals.

3.2.2 Prediction Intervals

A *prediction interval* is an interval used to make a statement about one or more future "like" measurements. As with a confidence interval, a prediction interval is calculated at a selected confidence level (or equivalently, at a selected significance level). *Unlike* a confidence interval, a prediction interval reflects the intrinsic variability of an individual measurement (apart from degree of uncertainty about the mean). See the basic formula in Table 3.4. Thus, unlike a confidence interval, there is a lower limit on the width of a prediction interval as the sample size, n, goes to infinity.

3.2 INTERVAL ESTIMATION

Consider the contrast between a confidence interval and a prediction interval for the cyanide example. The symmetric 99% confidence interval for the mean is computed in the preceding section to be

$$\bar{x} \pm t_{[n-1, 1-\alpha/2]} \frac{s}{\sqrt{n}}$$

which is

$$997 \pm 16.2 \text{ or } (980.8, 1013.2) \text{ ppt}$$

Recall the interpretation of this 99% confidence interval: 99% of all such intervals actually contain the true value (given needed assumptions are met). Informally, we assert that we are 99% sure that the true mean is within this interval (in reality, the true mean either *is* or *is not* in the interval we are simply ignorant of which is true).

For comparison, the symmetric 99% confident prediction interval is

$$\bar{x} \pm t_{[n-1, 1-\alpha/2]} s_B \sqrt{1 + \frac{1}{n}}$$

which is

$$997 \pm 3.25 \times 14.94 \sqrt{1 + \tfrac{1}{10}}$$

which simplifies to 997 ± 50.9, also expressed as (946.1, 1047.9) ppt. This is somewhat wider than the confidence interval for the mean, due to the contribution of the variability of the future observation, reflected by the presence of the "1" in the radical. The interpretation of this prediction interval is as follows: With 99% certainty, the next cyanide measurement will fall within (946.1, 1047.9), with the usual caveats regarding assumptions being met.

Note that the normality assumption is more critical for prediction intervals than with the confidence interval constructed for the mean. The reason is that the confidence interval benefits from application of the *central limit theorem* (CLT), which states that the distribution of the mean of n independent, identically distributed values tends toward the normal as n increases (assuming that the underlying distribution has finite variance, which almost always hold in practice). The prediction interval does *not* benefit from the CLT or any similar theorem; the Student's t critical value in the formula holds only if the normal distribution is true for individual measurements.

An alternative interpretation of the prediction interval is possible: On average, over all such 99% intervals, only 1% of observations will fall outside the intervals. This interpretation is useful for the rationale behind a Hubaux–Vos (H-V)-based detection limit, as follows: On average, all such DLs will have the desired (nominal) rates for correct detection at the DL and correct nondetection when measuring blanks. Thus, if a DL based on a statistical tolerance interval about the calibration line is unacceptably high, one can use the H-V DL, with the weaker assurance that it will provide 99% confidence on average.

3.2.3 Statistical Tolerance Intervals

A statistical tolerance interval is a confidence interval for a proportion of a population. We use the term *statistical tolerance interval* to distinguish this type of interval from a tolerance value or tolerance interval that might be generated as a result of an engineering study. As an example of the latter, a dimension on a part might be 5 mm, with a tolerance interval (i.e., maximum acceptable deviation from target) of ±0.1 mm.

Whereas a confidence interval is descriptive of a population parameter and a prediction interval is predictive of one or more future samples, a statistical tolerance interval is both descriptive *and* predictive. It makes an assertion about a population, and that can be used to define both rare and ordinary future samples from that population. For example, a statistical tolerance interval can allow one to make statements such as "With 95% confidence, at least 90% of the population is within 7 ± 5" and "With 90% confidence, at least 99% of future measurements will fall below 3—in the long term, assuming many measurements."

Statistical tolerance intervals are the least understood of the three intervals, have the least accessible tables, appear in asymmetric form far more than the other intervals, and when generalized to the calibration line, are by far the most complex of the intervals. For these and other reasons, statistical tolerance intervals are often improperly replaced by other, simpler intervals. To compute a statistical tolerance interval, one needs access either to tabulated values of statistical tolerance interval factors, $K_{P,\alpha}$ (where P is the population proportion and α is the significance level) or access to software that can generate such factors (see Hahn and Meeker, 1991). Furthermore, $K_{P,\alpha}$ factors are extremely sensitive to the underlying assumptions of independence, and one must be careful to ensure that the correct choice is made between a one- and two-sided interval. There is also great sensitivity to the normal distribution assumption, although nonparametric statistical tolerance intervals can be computed instead (they are based on order statistics and are comparatively very wide); other K factors are available for a few other distributions (Hahn and Meeker, 1991).

For the cyanide example, it would be highly presumptuous to assert, based on only 10 measurements, that the cyanide measurement distribution is normal and the measurements are independent. Let us assume nonetheless that these assumptions are true. The symmetric 99% confidence statistical tolerance interval for the central 90% of the population is

$$\bar{x} \pm K_{90\%,1\%} s$$

which is

$$997 \pm 3.62 \times 14.94 = 997 \pm 54.1$$

which can be rewritten (942.9, 1051.1) ppt.

Note that this interval is only slightly wider than is the 99% prediction interval for a single future observation. We could make the following interpretation of this interval: Making the assumptions listed above, and assuming that nothing changes in the process by which cyanide measurements are generated:

- In the future, over the long term, we can be 99% certain that at least 9 out of 10 cyanide observations will fall within the interval (942.9, 1051.1). (The reason for the "at least" is that the statistical tolerance interval is not a point estimate that provides an expected proportion; it guarantees that proportion at some confidence level.)
- With 99% confidence, the next k measurements will fall within the interval (942.9, 1051.1) with a probability of at least 0.9.

A one-sided statistical tolerance interval is more suitable for the cyanide case, as with many environmental measurements. For example, we may wish to know at what concentration c' we can be 99% certain that we will have 10% or fewer exceedences. To determine that we would compute the one-sided 99% confidence statistical tolerance interval for the lower 90% of the population:

$$\bar{x} + K'_{90\%,1\%} s$$

which is

$$997 + 3.05 \times 14.94 = 1042.6 \text{ ppt}$$

Or, the question might be inverted: If we know that a regulatory limit is 1053 ppt, what is the proportion of values that we can say, with 99% confidence, will *not* exceed 1053 ppt? The calculation is as follows:

$$997 + K'_{?\%,1\%} \times 14.94 = 1053 \text{ ppt}$$

Hence $K'_{?\%,1\%} = 3.74$, which (by statistical tolerance interval table look-up) corresponds to $P = 95\%$ of the population.

4

CALIBRATION-BASED REGRESSION MODELS

4.1 CALIBRATION DESIGNS

By far the most rigorous approach to modeling measurement uncertainty and related detection and quantification limits involves analysis of the calibration function. To estimate the parameters of the calibration function and its related properties, a series of samples are spiked at known concentrations in the range of the hypothesized detection and quantification limit, and variability is determined by examining the deviations of the actual response signals (or measured concentrations) from the fitted regression line of response signal on known concentration. In these designs it is generally assumed that the deviations from the fitted regression line are normally distributed. As discussed in this chapter, the concepts of tolerance limits and prediction limits apply to the individual measurement deviations from the fitted calibration function (and not the calibration function itself) and in general, there is considerable confusion in the literature regarding the choice of the appropriate interval (Hubaux and Vos, 1970). In the following, we review ordinary least-squares (OLS) and weighted least-squares (WLS) approaches to the problem of estimating the calibration function and related interval estimates.

4.2 ORDINARY LEAST-SQUARES ESTIMATION

As preparation for the following discussion, we generally conceive of the relationship between response signal y and spiking concentration x in the region of the detection and quantification limits as a linear function of the form

$$y = \beta_0 + \beta_1 x + \varepsilon \qquad (4.1)$$

where ε is a random variable that describes the deviations from the regression line, distributed with mean 0 and constant variance $\sigma_{y \cdot x}^2$. The assumption of constant variance is not critical to this approach and will be relaxed in a later section; however, it is useful to simplify the initial exposition.

The sample regression coefficient

$$b_1 = \frac{\sum_{i=1}^{n}[(x_i - \bar{x})y_i]}{\sum_{i=1}^{n}[(x_i - \bar{x})^2]} \quad (4.2)$$

provides an estimate of the population parameter β_1 (i.e., the slope of the calibration function). The sample intercept

$$b_0 = \bar{y} - b_1 \bar{x} \quad (4.3)$$

provides an estimate of the population parameter β_0 (i.e., the intercept of the calibration function) which describes the mean instrument response or measured concentration when the true concentration $x = 0$. An unbiased sample estimate of $\sigma_{y \cdot x}^2$ (i.e., the variance of deviations from the population regression line) is given by

$$s_{y \cdot x}^2 = \sum_{i=1}^{n} \frac{(y_i - \hat{y}_i)^2}{n - 2} \quad (4.4)$$

where $\hat{y}_i = b_0 + b_1 x_i$.

4.3 WEIGHTED LEAST-SQUARES ESTIMATION

When variance is not constant, as is typically the case in the calibration setting, the previous OLS solution for constant or *homoscedastic* errors no longer applies. There are several approaches to this problem, but in general, the most widely accepted approach is to model the variance as a function of true concentration x and then to use the estimated variances as weights in estimating the calibration parameters, which are now denoted as β_{0w} and β_{1w}.

The weighted least squares regression of measured concentration or instrument response (y) on true concentration (x) is denoted by

$$\hat{y}_{wi} = b_{0w} + b_{1w} x_i \quad (4.5)$$

where

$$b_{1w} = \frac{\sum_{i=1}^{n}[(x_i - \bar{x}_w) y_i / k_i]}{\sum_{i=1}^{n}[(x_i - \bar{x}_w)^2 / k_i]} \quad (4.6)$$

$$b_{0w} = \bar{y}_w - b_{1w} \bar{x}_w \quad (4.7)$$

$$\bar{y}_w = \frac{\sum_{i=1}^{n}[y_i/k_i]}{\sum_{i=1}^{n}[1/k_i]} \tag{4.8}$$

$$\bar{x}_w = \frac{\sum_{i=1}^{n}[x_i/k_i]}{\sum_{i=1}^{n}[1/k_i]} \tag{4.9}$$

and the weight $k_i = s_{x_i}^2$ is the variance for sample i, which is computed from those samples with true concentration $x_i = x$. The weighted residual variance is

$$s_w^2 = \sum_{i=1}^{n} \frac{(y_i - \hat{y}_{wi})^2/k_i}{n-2} \tag{4.10}$$

4.4 ESTIMATING THE WEIGHTS

When the number of replicates at each concentration is small, as is typically the case, the observed variance at each concentration provides a poor estimate of the true population variance. Two better alternatives are to (1) model the observed variance or standard deviation as a function of true concentration, or (2) model the sum of squared residuals as a function of concentration. The latter approach can also be performed iteratively in which improved estimates of β_0 and β_1 are obtained from weights computed from the current sum of squared residuals on each iteration. These new estimates of β_0 and β_1 are in turn used to obtain a new set of estimated weights and so on until convergence. This algorithm is commonly termed *iteratively reweighted least squares* (see Section 4.5). An essential element of either approach is to identify a plausible model for the variance function. In the following sections we consider a few models that are particularly well suited to this problem.

4.4.1 Rocke and Lorenzato Model

To measure the true concentration of an analyte (x) the traditional simple linear calibration model, $y = \beta_0 + \beta_1 x + e$ with the standard normality assumption on errors is not appropriate, as it fails to explain increasing measurement variation with increasing analyte concentration, which is commonly observed in analytical data. To overcome this situation, one may propose a log linear model, for example, $y = xe^\eta$, where η is a normal variable with mean 0 and standard deviation σ_η. This model also fails to explain near-constant measurement variation of y for low true concentration level x (Rocke and Lorenzato, 1995). To better model the calibration curve, Rocke and Lorenzato (1995) proposed a combined model that has both types of errors:

$$y_{jk} = \beta_0 + \beta_1 x_{jk} e^\eta + e_{jk} \tag{4.11}$$

where y is the kth measurement at the jth concentration level, x_{jk} the corresponding true concentration, and β_0 and β_1 the fixed calibration parameters. In this model, η represents proportional error at higher true concentrations and the e_{jk}'s are the additive errors that are present primarily at low-level concentrations. We assume that η and the e_{jk}'s are independent and follow normal distributions with means 0's and variances σ_η^2 and σ_e^2, respectively. Data near zero (i.e, $x \approx 0$) determine σ_e^2, and data for large concentrations determine σ_η^2. The model specification also indicates that errors at larger concentrations are lognormally distributed and at low concentrations errors are normally distributed, which agrees with common experience.

In their original paper, Rocke and Lorenzato (1995) derived the maximum likelihood estimators for their model based on maximizing the likelihood function

$$\prod_{i=1}^{n} \frac{1}{2\pi\sigma_e\sigma_\eta} \int_{-\infty}^{\infty} \exp\left(\frac{-\eta^2}{2\sigma_\eta^2}\right) \exp\left(\frac{(y_i - \beta_0 - \beta_1 x_i e^\eta)^2}{2\sigma_e^2}\right) d\eta \quad (4.12)$$

These computations require complex numerical evaluation of the required integrals. Alternatively, Gibbons et al. (1997a) and Rocke and Durbin (1998) have described a weighted least-squares (WLS) solution which involves the following algorithm:

1. Use OLS regression to find initial estimates of β_0 and β_1 by fitting the linear model

$$y = \beta_0 + \beta_1 x + e \quad (4.13)$$

2. Using the sample standard deviation of the lowest concentration as an estimate for σ_e and the standard deviation of the log of the replicates at the highest concentration as an initial estimate for σ_η, refit the model in step 1 using WLS with weights equal to

$$w(x) = \frac{1}{\sigma_e^2 + \beta_1^2 x^2 e^{\sigma_\eta^2}(e^{\sigma_\eta^2} - 1)} \quad (4.14)$$

3. Using the new estimates of β_0 and β_1, compute the predicted response $\hat{y} = \hat{\beta}_0 + \hat{\beta}_1 x$ and standard error of the calibration curve at each concentration x:

$$s^2(x) = \frac{\sum_{i=1}^{m(x)}(\hat{y} - y_i)^2}{m(x)} \quad (4.15)$$

where $m(x)$ is the number of replicates for concentration x.

4. Using WLS, fit the variance function

$$s^2(x) = \gamma + \delta x^2 + e \quad (4.16)$$

where

$$\gamma = \sigma_e^2 \tag{4.17}$$

and

$$\delta = \beta_1^2 e^{\sigma_\eta^2}(e^{\sigma_\eta^2} - 1) \tag{4.18}$$

using weights

$$w(x) = \frac{m(x)}{s^2(x)} \tag{4.19}$$

5. Compute the new estimates of $\sigma_e^2 = \gamma$ and

$$\sigma_\eta^2 = \log_e \frac{1 + \sqrt{1 + 4\delta/\beta_1^2}}{2} \tag{4.20}$$

6. Iterate until convergence.

In general, this algorithm will converge to positive values of γ and δ. Note that this algorithm uses WLS to compute the parameters of the calibration curve (β_0 and β_1) as well as the parameters of the variance function (γ and δ). In this way, the lowest concentrations with the smallest variances provide the greatest weight in the estimation. The net result is that we do not sacrifice precision in estimating the calibration function and corresponding interval estimates at low levels by including higher concentrations in the analysis. This is quite useful if our interest is in low-level detection and quantification. In particular, it is generally possible to obtain lower detection limits and quantification limits than by application of OLS.

4.4.2 Exponential Model

An alternative parameterization of the variance function involves modeling the relationship between σ and x as an exponential function of the form

$$\sigma_x = a_0 e^{a_1(x)} \tag{4.21}$$

Although less well theoretically motivated than the Rocke and Lorenzato model, the exponential model provides excellent fit to a wide variety of analytical data (see Gibbons et al., 1997a). The model can be applied either to the observed standard deviations at each concentration or applied iteratively to the sum of squared residuals. In terms of estimating a_0 and a_1, the traditional approach involves substituting s_x for σ_x and using nonlinear least squares (e.g., Gauss–Newton) or using OLS regression of natural log–transformed observed standard deviation on true concentration

(Snedecor and Cochran, 1980). Similarly, WLS can also be used on the regression of $\log_e(s)$ on x using weights:

$$w(x) = \frac{m(x)}{s^2(x)} \quad (4.22)$$

4.4.3 Linear Model

The linear model has also been used to model the variance function (Currie, 1995). The linear model is of the form

$$\sigma_x = a_0 + a_1(x) \quad (4.23)$$

The primary disadvantage of the linear model is that the small sampling fluctuations in the sample variance observed at each concentration can lead to a negative intercept (i. e., $a_0 < 0$) and negative variance estimates. This can lead to improper detection and quantification limit estimates and corresponding interval estimates. As such, the linear model is generally not recommended for routine use. This is not a problem for either of the two preceding models, which can mimic a linear model if required.

4.5 ITERATIVELY REWEIGHTED LEAST-SQUARES ESTIMATION

An alternative to modeling the observed variance at each concentration is to model the squared residuals as a function of x and then to use this estimated variance function to obtain weights which are then used in reestimating the regression coefficients. This process is iterated until convergence, hence the term *iteratively reweighted least squares* (see Carol and Rupert, 1988). As noted by Neter et al. (1990), the methods of maximum likelihood and weighted least squares lead to the same estimators for linear regression models of the form considered here. The previous example of the WLS estimator for the Rocke and Lorenzato model is an example of iteratively reweighted least squares.

The general algorithm is as follows:

1. Use OLS estimation to find initial estimates of β_0 and β_1 by fitting the linear model

$$y = \beta_0 + \beta_1 x + \varepsilon \quad (4.24)$$

2. Using the OLS estimates of β_0 and β_1, compute the predicted response $\hat{y} = \hat{\beta}_0 + \hat{\beta}_1 x$ and standard error of the calibration curve at each concentration x:

$$s^2(x) = \frac{\sum_{i=1}^{m(x)}(\hat{y} - y_i)^2}{m(x)} \quad (4.25)$$

where $m(x)$ is the number of replicates for concentration x.

3. Using an appropriate model for the variance function, fit the variance function to the sum of squared residuals:

$$s^2(x) = f(x) \tag{4.26}$$

4. Using the provisional weights

$$w(x) = \frac{m(x)}{s^2(x)} \tag{4.27}$$

recompute β_0 and β_1 using WLS.

5. Iterate until convergence.

4.6 OLS PREDICTION INTERVALS

In the regression case, prediction limits for a single new measurement parallel limits for the case of a fixed concentration design. In this case the estimated standard error of the prediction $\hat{y}_x$ for a new value of y at point x is

$$s(\hat{y}_x) = s_{y \cdot x}\sqrt{1 + \frac{1}{n} + \frac{(x - \bar{x})^2}{\sum_{i=1}^{n}(x_i - \bar{x})^2}} \tag{4.28}$$

The prediction interval for a response signal obtained from a new test sample given a concentration x (e.g., $x = 0$) is

$$\hat{y}_x \pm t_{[1-\alpha/2, n-2]} s(\hat{y}_x) \tag{4.29}$$

For example, at a concentration of $x = 0$ the prediction limit is

$$\bar{y} - b\bar{x} \pm t_{[1-\alpha/2, n-2]} s(\hat{y}_x) \tag{4.30}$$

The one-sided limit can be obtained by substituting α in place of $\alpha/2$.

4.7 OLS TOLERANCE INTERVALS

The notion of a tolerance interval for a random sample of measurements can also be extended to the calibration setting in which the intervals are simultaneous in each possible value of the independent variable x (e.g., spiking concentration levels). Lieberman and Miller (1963) give four techniques for deriving such intervals, the simplest and most robust is based on the Bonferroni inequality.

For the response signal $\hat{y}$ predicted at concentration x, the interval is

$$\hat{y}_x \pm s_{y \cdot x} \left\{ (2F_{2,n-2}^{1-\alpha/2})^{1/2} \left[\frac{1}{n} + \frac{(x_i - \bar{x})^2}{\sum_{i=1}^{n}(x_i - \bar{x})^2} \right]^{1/2} + \Phi(P)\left(\frac{n-2}{\alpha/2 \chi_{n-2}^2} \right)^{1/2} \right\} \quad (4.31)$$

where $F_{2,n-2}^{1-\alpha/2}$ is the upper $(1 - \alpha/2)$ percentile point of the F distribution on 2 and $n - 2$ degrees of freedom (df), $^{\alpha/2}\chi_{n-2}^2$ is the lower $\alpha/2$ percentile point of the χ^2 distribution with $n - 2$ df, and $\Phi(P)$ is the two-sided P percentile point of the unit normal distribution.

For the case of $x = 0$, the upper tolerance limit is

$$\hat{y}_0 = \bar{y} - b\bar{x} + s_{y \cdot x} \left\{ (2F_{2,n-2}^{1-\alpha/2})^{1/2} \left[\frac{1}{n} + \frac{\bar{x}^2}{\sum_{i=1}^{n}(x_i - \bar{x})^2} \right]^{1/2} + \Phi(P)\left(\frac{n-2}{\alpha/2 \chi_{n-2}^2} \right)^{1/2} \right\} \quad (4.32)$$

The value predicted for $\hat{y}_0$ in equation (4.32) specifies a proportion P of the population of response signals that is possible when the true concentration is 0, given a $(1 - \alpha)100\%$ confidence level. This interval estimate corresponds to the concept of a decision limit defined by Currie (1968) for the case in which the data arise from a calibration experiment, μ and σ at $x = 0$ are unknown, and we wish to provide coverage of a proportion of the population of possible test samples and not just the next single test sample. This approach is well suited to the case of detection limits used as regulatory thresholds, or the case when detection limits are applied to a large or unknown number of future sample determinations.

4.8 WLS PREDICTION INTERVALS

For WLS estimates of β_0 and β_1, the variance estimated for a predicted value $\hat{y}_{wj}$ is

$$V(\hat{y}_{wj}) = s_w^2 \left[k_j + \frac{1}{\sum_{i=1}^{n}(1/k_i)} + \frac{(x_j - \bar{x}_w)^2}{\sum_{i=1}^{n}(x_i - \bar{x}_w)^2/k_i} \right] \quad (4.33)$$

where k_j is the estimated variance at concentration x_j. An upper $(1 - \alpha)100\%$ confidence interval for $\hat{y}_{wj}$ (i.e., an upper prediction limit for a new measured concentration or instrument response at true concentration x_j) is

$$\hat{y}_{wj} + t\sqrt{V(\hat{y}_{wj})} \quad (4.34)$$

where t is the upper $(1 - \alpha)100$ percentage point of Student's t-distribution on $n - 2$ degrees of freedom.

For example, at $x = 0$, the upper prediction limit is

$$\hat{y}_{0w} = \bar{y}_w - b\bar{x}_w + ts_w \sqrt{s_{0w}^2 + \frac{1}{\sum_{i=1}^n (1/k_i)} + \frac{\bar{x}_w^2}{\sum_{i=1}^n (x_i - \bar{x}_w)^2 / k_i}} \quad (4.35)$$

where s_{0w}^2 is the variance of the measured concentrations or instrument responses for a sample that does not contain the analyte.

4.9 WLS TOLERANCE INTERVALS

For the predicted response signal $\hat{y}$ at concentration x, the interval is

$$\hat{y}_{wx} \pm s_w \left\{ (2F_{2,n-2}^{1-\alpha/2})^{1/2} \left[\frac{1}{\sum_{i=1}^n w_i} + \frac{(x_i - \bar{x}_w)^2}{\sum_{i=1}^n (x_i - \bar{x}_w)^2} \right]^{1/2} \right.$$

$$\left. + \left(\frac{1}{w_x} \right)^{1/2} \Phi(P) \left(\frac{n-p-2}{\alpha/2 \chi_{n-p-2}^2} \right)^{1/2} \right\} \quad (4.36)$$

where F, χ, and $\Phi(P)$ are as defined earlier for the OLS case.

For the case of $x = 0$, the upper WLS tolerance limit is

$$\hat{y}_{0w} = \bar{y}_w - b\bar{x}_w + s_w \left\{ (2F_{2,n-2-p}^{1-\alpha/2})^{1/2} \left[\frac{1}{\sum_{i=1}^n w_i} + \frac{\bar{x}_w^2}{\sum_{i=1}^n (x_i - \bar{x}_w)^2} \right]^{1/2} \right.$$

$$\left. + \left(\frac{1}{w_x} \right)^{1/2} \Phi(P) \left(\frac{n-p-2}{\alpha/2 \chi_{n-p-2}^2} \right)^{1/2} \right\} \quad (4.37)$$

The WLS predicted value $\hat{y}_{0w}$ in the previous equation (4.37) specifies a proportion P of the population of response signals that is possible when the true concentration is 0, given a $(1 - \alpha)100\%$ confidence level. As in the OLS case, this interval estimate corresponds to the concept of a decision limit defined by Currie (1968) for the case in which the data arise from a calibration experiment, μ and σ at $x = 0$ are unknown and we wish to provide coverage of a proportion of the population of possible test samples and not just the next single test sample. This approach is well suited to the case of detection limits used as regulatory thresholds, or the case when detection limits are applied to a large or unknown number of future sample determinations.

4.10 ILLUSTRATION

To illustrate the general approach for estimating the calibration function and corresponding interval estimates, these methods were applied to the data in Table 4.1.

Table 4.1 Benzene data

Sample	Concentration in Water (mg/L)					
	1.0	4.0	10.0	20.0	30.0	40.0
1	1.07	3.66	10.13	20.22	31.04	41.14
2	1.01	3.75	10.20	19.46	31.40	40.96
3	0.99	3.77	9.77	19.02	31.84	42.77
4	1.09	3.92	10.28	20.06	30.77	38.36
5	0.94	4.08	8.71	20.94	32.13	39.71
6	1.00	3.99	9.79	19.92	31.17	39.61
7	1.01	3.95	9.82	19.96	30.36	40.73
8	0.98	4.10	9.85	20.04	29.95	39.42
9	1.00	3.88	10.07	19.67	30.74	40.72
10	1.03	4.04	9.63	19.96	30.67	40.24
11	1.00	3.68	10.02	20.16	31.28	41.26
12	1.11	3.98	9.90	19.44	31.03	40.46
13	0.96	3.64	9.61	20.21	31.28	42.78
14	0.97	3.82	9.32	19.67	32.64	41.63
15	1.03	3.99	10.20	19.13	28.87	42.50
16	1.00	4.08	9.93	19.75	31.02	38.78
17	1.00	4.07	10.09	20.14	29.69	39.17
18	1.00	4.45	9.67	19.51	29.16	39.08
19	0.94	4.04	9.97	19.97	31.24	38.11
20	1.10	3.83	9.63	19.92	31.28	41.67
21	—	3.94	10.29	20.07	30.11	39.21
22	—	3.75	9.72	20.71	31.07	40.95

Table 4.1 presents typical calibration data for benzene spiked into distilled water at concentrations of 1, 4 10, 20, 30, and 40 µg/L and analyzed using GC/MS Method 8260. The results of the unweighted analysis (OLS regression and corresponding uniform prediction bands) are displayed in Figure 4.1. Figure 4.1 reveals that the slope of the OLS regression is close to 1 (b_1 = 1.019), but the intercept exhibits mild negative bias (b_0 = −0.17). The uniform prediction bounds underestimate the observed bounds of the data for high concentrations and dramatically overestimate the width of the distribution at the lower concentrations. Figure 4.2 clearly illustrates that variability is strongly related to concentration. The fitted line in Figure 4.2 is based on the Rocke and Lorenzato model and does a reasonable job of tracking the observed sample standard deviations. Figure 4.3 displays the WLS regression and corresponding nonconstant prediction bands. Figure 4.3 reveals that the WLS estimated regression coefficients are b_0 = 0 and b_1 = 1.003. The WLS regression model places increased weight on the lower concentrations and therefore eliminates the bias observed for the OLS estimator and the prediction bands now fit the observed deviations from the fitted calibration line extremely well. From Figures 4.2 and 4.3 and equation (4.20) we can now compute the parameters of the

44 CALIBRATION-BASED REGRESSION MODELS

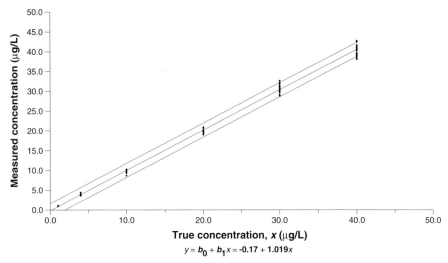

$y = b_0 + b_1 x = -0.17 + 1.019x$

Figure 4.1 OLS regression with 99% prediction interval for benzene.

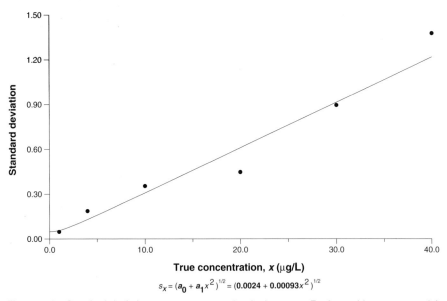

$s_x = (a_0 + a_1 x^2)^{1/2} = (0.0024 + 0.00093 x^2)^{1/2}$

Figure 4.2 Standard deviation versus concentration for benzene: Rocke and Lorenzato model.

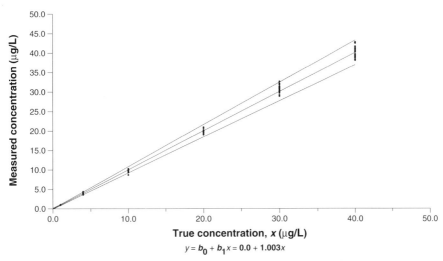

Figure 4.3 WLS regression with 99% prediction interval for benzene: Rocke and Lorenzato model.

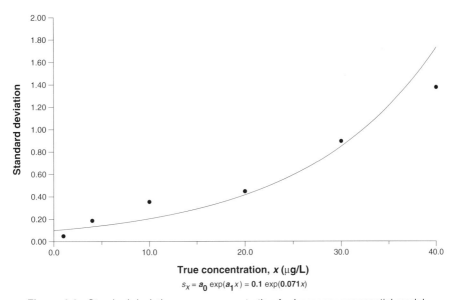

Figure 4.4 Standard deviation versus concentration for benzene: exponential model.

46 CALIBRATION-BASED REGRESSION MODELS

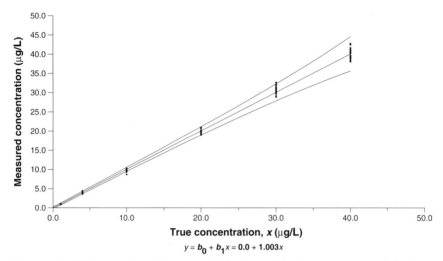

Figure 4.5 WLS regression with 99% prediction interval for benzene: exponential model.

Rocke and Lorenzato model as

$$\hat{\sigma}_e^2 = 0.0024$$

and

$$\hat{\sigma}_\eta^2 = \log_e \frac{1 + \sqrt{1 + 4(0.00093)/1.003^2}}{2} = 0.0009$$

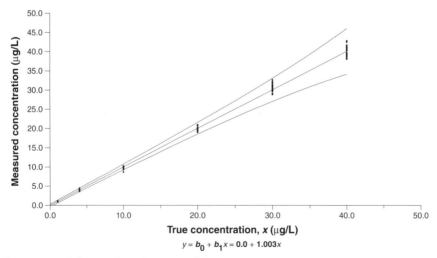

Figure 4.6 WLS regression with 99% confidence 99% coverage tolerance interval for benzene: exponential model.

4.10 ILLUSTRATION

As such, the expected standard deviation at a concentration of 20 μg/L is

$$s_{x=20} = \sqrt{\hat{\sigma}_e^2 + b_1^2 x^2 e^{\hat{\sigma}_\eta^2}(e^{\hat{\sigma}_\eta^2} - 1)}$$
$$= \sqrt{0.0024 + 1.003^2(20^2)e^{0.0009}(e^{0.0009} - 1)} = 0.604$$

which agrees closely with Figure 4.2.

Figure 4.4 illustrates the fit of the exponential model to the standard deviations observed at each concentration. For this example, the fit of the model to the data is slightly better than the Rocke and Lorenzato model, yet Figure 4.5 reveals that the WLS estimates of β_0 and β_1 are identical and the prediction bands are quite similar. Finally, Figure 4.6 displays the WLS 99% coverage 99% confidence tolerance intervals fitted to the same data (using the exponential model). Interestingly, the intervals are only slightly wider, yet the tolerance intervals will include 99% of the entire population of deviations from the fitted calibration line with 99% confidence, whereas the prediction interval will produce only 99% coverage on average.

5

SINGLE-CONCENTRATION-BASED DETECTION LIMIT METHODS

5.1 INTRODUCTION

There are a myriad of terms for limits of detection [e.g., method detection limit (MDL), detection limit (DL), limit of detection (LOD), instrument detection limit (IDL), practical method detection limit (PMDL)]; however, their conceptual foundations as well as implementation algorithms are often quite diverse, leading to tremendous inconsistency in practice. To provide a framework to compare and contrast these various methodologies, the pioneering definitions of Currie (1968) are used, since they form the basis of most approaches. Currie defined two levels: the decision limit and the detection limit. The *decision limit* is the measured concentration "at which one may decide whether or not the result of an analysis indicates detection." The *detection limit* is the true concentration "at which a given analytical procedure may be relied upon to lead to a detection." From a statistical perspective, the decision limit is the critical value for a test of the null hypothesis "analyte absent" versus the alternative hypothesis "analyte present." Note that the detection limit is a direct function of the selection of the critical value.

Operationally, the decision limit or critical level L_C is the measurement above which the response signal significantly differs from zero. When a measurement exceeds L_C, we can make the binary decision "detected." Note that when the true concentration is zero, the probability of the correct decision, "not detected," is $1 - \alpha$, where α is the *Type I error rate* or *false positive rate* of the statistical test (e.g.,

$\alpha = 0.01$). Assuming a symmetric distribution of measurement errors, when the true concentration is equal to L_C the probability of reporting it below L_C is 50%. This is termed the *Type II error rate* or *false negative rate* (β). To control both Type I and II errors, Currie then developed the detection limit L_D. When the true concentration is L_D, the Type II error rate (β) using L_C as the decision limit is small. For example, if we assume that $\alpha = \beta = 0.01$, another way of stating this is that 99% of the measurements for samples not containing the analyte will be less than the decision limit L_C and 99% of the measurements with true concentration at the detection limit L_D will exceed the decision limit L_C.

Investigators have defined the limit of detection to include control of both false positive and false negative rates, varying levels of confidence, uncertainty in the concentration mean and variance and in the calibration function that relates instrument response to concentration, variance components related to multiple instruments and analysts, application to multiple future detection decisions, and nonconstant variance and baseline instrument response (Clayton et al., 1987; Currie, 1968; Gibbons et al., 1988; Hubaux and Vos, 1970). However, these operational definitions may cloud the distinction between decision limits and detection limits (see, e.g., Glaser et al., 1981).

Single-concentration designs involve spiking a series of n samples with a fixed concentration of a compound and using variability in instrument response at that concentration to derive a decision limit, detection limit, or both. In the following, the major strategies for estimating decision limits and detection limits from single-concentration designs are described.

5.2 KAISER–CURRIE METHOD

Based on developments due to Kaiser (1956, 1965, 1966), Currie (1968) described a two-stage procedure for calculating the detection limit L_D. At the first level of analysis, Currie computes the decision limit L_C as

$$L_C = z_{1-\alpha}\sigma_0 \tag{5.1}$$

where σ_0 is the population standard deviation of the response signal when the true concentration (x) is zero (i.e., the standard deviation of the net signal found in the population of blank samples) and $z_{1-\alpha}$ is a multiplication factor based on the $(1-\alpha)100$ percentage point of the standardized normal distribution. For example, the one-sided 99% point of the normal distribution is 2.33; therefore, the decision limit for a 1% false positive rate is defined as

$$L_C = z_{1-\alpha}\sigma_0 = 2.33\sigma_0 \tag{5.2}$$

As discussed, when the true concentration is equal to L_C and the measurement error distribution is symmetric, the Type II error rate at the decision limit is 50%. That is, we have a 50% chance of declaring that the analyte is not present when the analyte in fact is present at the decision limit. To provide an acceptable Type II error rate, Currie defined the detection limit (L_D) as

$$L_D = L_C + z_{1-\beta}\sigma_D \tag{5.3}$$

where σ_D is the population standard deviation of the response signal at L_D (or net response signal after subtracting the background signal) and β is the acceptable Type II error rate (i.e., false negative rate).

Currie points out that if we make the simplifying assumption that $\sigma_0 = \sigma_D$ (i.e., the variability of the signal is constant in the range of L_C to L_D) and that the risk of false positive and false negative rates is equivalent (i.e., $z_{1-\alpha} = z_{1-\beta} = z$), the detection limit is simply

$$L_D = L_C + z_\beta\sigma_D = z(\sigma_0 + \sigma_D) = 2L_C \tag{5.4}$$

or twice the decision limit. For $\alpha = \beta = 0.01$, the detection limit is therefore $4.66\sigma_0$. If the true concentration is L_D, the probability of a measured value below L_C is 1%.

In reviewing Currie's method it is critically important to note that the only case considered is the one in which the population values σ_0 and σ_D are known. In practice, however, population values are never known and the methods described in following sections are required to incorporate uncertainty in sample-based estimates of these statistics in decision limit and detection limit estimates.

5.3 U.S. EPA 40CFR, PART 136 (GLASER et al.)

Glaser et al. (1981) defined the method detection limit (MDL) as "the minimum concentration of a substance that can be identified, measured and reported with 99% confidence that the analyte concentration is greater than zero" and that "on average, 99% of the trials measuring the analyte concentration of the MDL must be significantly different from zero." The first definition is in terms of true concentration and could therefore be interpreted as L_D. Similarly, if one interprets "significantly different from zero" as greater than L_C, which is reasonable, the MDL as conceptually defined is also consistent with L_D. Unfortunately, their derivation converges asymptotically to an estimate of the 99th percentile of the distribution of measurements at concentration $x = 0$, which is L_C. These definitions have been adopted by the U.S. EPA (1984b) and the MDL is used in establishing all published detection limits under EPA regulations for environmental monitoring of drinking water, groundwater, and surface water. In addition, all laboratories performing analyses under these regulations must be certified on the basis of meeting published MDLs using this method.

In constructing a model for the MDL, Glaser et al. (1981) initially assume that variability is a linear function of concentration, such that

$$s_x = b_0 + b_1 x \tag{5.5}$$

where s_x is the standard deviation of n replicate analyses at concentration x and b_0 and b_1 are the intercept and slope of the linear regression of standard deviation on concentration. They define the MDL to be that concentration x for which

$$t_x = \frac{x}{s_x/\sqrt{n}} \tag{5.6}$$

holds when t is set equal to the 99th percentile of Student's t-distribution with $n-1$ degrees of freedom. At this point Glaser et al. make two simplifying assumptions. First, they set b_1 to zero; therefore,

$$\text{MDL} = \frac{t_{[0.01, n-1]} b_0}{\sqrt{n}} \tag{5.7}$$

Second, they assume that $s_x = b_0/\sqrt{n}$; therefore,

$$\text{MDL} = t_{[0.01, n-1]} s_x \tag{5.8}$$

where s_x is defined as the standard deviation of n analytical replicates spiked at a single concentration x.

There are numerous problems with this model. First, having started with a conceptual definition that sounds promisingly, like L_D, they derive a mathematical definition that corresponds to L_C. This highlights the major distinction between the MDL and the other methods reviewed here. Second, having started with the useful generalization, that variance is allowed to be a function of concentration, they remove it by assuming that $b_1 = 0$ and using a single concentration to estimate s_x. To estimate b_0 and b_1 would require multiple samples at each of several concentrations (i.e., a calibration design). Third, there are inexplicable inconsistencies about whether $s_x = b_0$ [per equation (5.5) after you set $b_1 = 0$] or $b_0/\sqrt{n}$ as asserted by Glaser et al. (1981). Of course, $b_0/\sqrt{n}$ is the standard error of measurement at $x = 0$, not the standard deviation at concentration x.

It seems odd that the initial assumption of this model is that variability is dependent on concentration, yet in practice the standard deviation observed for seven replicates is used regardless of the concentration at which the sample is spiked. In practice, laboratories demonstrate that they can achieve low state and federally mandated MDLs (used as decision limits, not detection limits) by spiking at progressively lower concentrations. Glaser et al. suggest that the spiking concentration should be close to the true MDL; however, since the true MDL is unknown, the spiking concentration is often set at risk-based standards [e.g., a maximum contaminant level (MCL)] and can be well below the true MDL. Since Glaser et al. assume that the s_x is a linear function of concentration, this practice can lead to a gross underestimate of the true MDL.

It should be noted that despite the obvious limitations of the statistical derivation of the MDL, the underlying conceptual framework foreshadows some of the ideas of calibration designs. Conceptually, the underlying model assumes that the measured value is normally distributed with mean equal to the true concentration x and standard deviation $b_0 + b_1 x$. This conceptual model leads directly to many of the statistical results for calibration designs but is intractable for single concentration designs since b_0 and b_1 are unknown and cannot be estimated from a single concentration. Unfortunately, the EPA continues to use the MDL as the basis for regulatory detection decisions. The MDL also forms the basis for the approach that the EPA has taken to quantification, the minimum level (ML), which is computed as 3.18 (MDL) (U.S. EPA, 1993).

5.4 PREDICTION LIMITS

Prediction limits were apparently not considered in the context of single-concentration designs; nonetheless, they are more relevant to making detection decisions than either the MDL or the Kaiser–Currie method. A prediction limit provides a certain level of confidence (e.g., 99%) of including the next single measurement based on a previous sample of n measurements. If we consider a single future detection decision, the decision limit based on a prediction limit (Hahn and Meeker, 1991) for a normally distributed instrument response is

$$L_C = t_{[n-1,\alpha]} s_0 \sqrt{1 + \frac{1}{n}} \tag{5.9}$$

where s_0 is an estimate of σ_0 (i.e., the standard deviation in blank samples). For the seven replicate samples required by the EPA, the decision limit becomes

$$L_C = 3.14 s_0 \sqrt{1 + \tfrac{1}{7}} = 3.36 s$$

where $\alpha = 0.01$. Assuming that $s_0 = s_D = s$ and $\alpha = \beta$, the detection limit can be approximated by

$$L_D = 2 t_{[n-1,\alpha]} s \sqrt{1 + \frac{1}{n}} = 2 L_C \tag{5.10}$$

however, technically, this is an invalid use of the central t-distribution since the statistic $\sqrt{n}\, x/s$ has a noncentral t-distribution when $x > 0$ (i.e., in computing L_D, $x = L_C$). In practice, this is a useful approximation.

5.5 TOLERANCE LIMITS

In practice, a limit of detection is computed and applied to a large and potentially unknown number of future detection determinations. As such, the prediction limit

Table 5.1 EPA 40CFR MDL at various spiking concentrations of silver in distilled water using Method 1638

Concentration (ppt)	SD	MDL
0	0.47	1.47
10	0.67	2.09[a]
20	0.66	2.06
50	1.08	3.40
100	2.23	7.00
200	15.32	48.10[a]
500	9.05	28.43
1000	21.43	67.29

[a]Meets EPA's 5:1 criterion.

described in Section 5.4 may not be relevant to many situations because the error rate pertains to a single future detection decision. Alternatively, the prediction limit will provide $(1 - \alpha)100\%$ coverage on average. An alternative is to base the decision limit on a statistical tolerance limit that will contain a proportion of all future measurements (e.g., 99%) with a specified level of confidence (e.g., 99%). Tolerance limits are well known in the statistical literature and have been described in detail by Guttman (1970), with corresponding tables relevant to this application. In this context the decision limit becomes

$$L_C = K_{(P,\alpha)} s_0 \tag{5.11}$$

where values of K for varying coverage proportions (P) and confidence levels $(1 - \alpha)$ are available in tabular form (Gibbons, 1994; Guttman, 1970; Hahn and Meeker, 1991).

Following EPA guidance using seven replicate samples, the decision limit would be $4.64 s_0$ and would provide 95% confidence of including 99% of all future measurements when the true concentration is zero (see Table 5.1). Of course, as the number of background samples n increases, the size of the multiplier decreases. Use of tolerance limits to compute the detection limit is not straightforward; however, an approximate method has been developed (Gibbons et al., 1991).

5.6 SIMULTANEOUS CONTROL OF FALSE POSITIVE AND FALSE NEGATIVE RATES

The distinction between decision limits and detection limits is based on simultaneous control of both false positive and false negative rates. When the population parameters σ_0 and σ_D are known, as was assumed by Currie, repeated application of the normal multiplier $z_{1-\alpha}$ followed by $z_{1-\beta}$ is appropriate. In practice, however, σ_0 and σ_D are unknown and are replaced by their sample-based estimates s_0 and s_D

provided by a sample of n spiked measurements. In this case repeated application of Student's t-distribution (i.e., $t_{1-\alpha}$ and $t_{1-\beta}$) is incorrect, although it can be viewed as an approximation. In computing L_C, the true concentration $x = 0$ and the false positive rate of the decision limit is based on Student's t-distribution. In computing L_D, the true concentration $x = L_C$ and the statistic $\sqrt{n}\,x/s_0$ has a noncentral t-distribution on $n-1$ degrees of freedom (Clayton et al., 1987; Gibbons et al., 1991). Relevant tabled values of the noncentrality parameters $\phi(n-1, \alpha, \beta)$ are also available (Clayton et al., 1987; Gibbons et al., 1991). For the case of $n = 7$ measurements and false positive and false negative rates of 1%, the limit of detection is $\phi(6, 0.01, 0.01)s_0 = 6.64s_0$. Although this limit of detection considers both false positive and false negative rates, it pertains only to a single future detection decision. Gibbons et al. (1991) have further generalized the result to include coverage of a specified proportion of all future values and have provided a corresponding table of multipliers (see also Gibbons, 1994).

5.7 LIMITATIONS

The critical assumption underlying single-concentration designs is that variability is homogeneous in the range of possible spiking concentrations and the true limit of detection. This assumption is implicit in use of observed spiking concentration variance as an estimator of the true variance at the limit of detection. This assumption is rarely realized in practice. As a typical example, consider the calibration function presented in Figure 5.1. For the purpose of illustration, actual concentrations are presented on the x-axis and measured concentrations are presented on the y-axis.

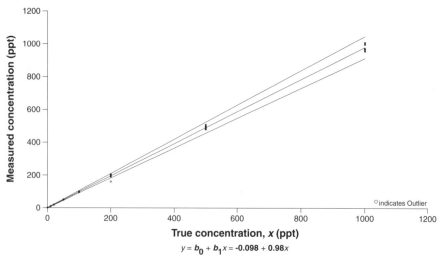

Figure 5.1 WLS regression with 99% prediction interval for silver (Method 1638): Rocke and Lorenzato model.

In practice, the y-axis should represent actual peak area ion counts prior to calibration. Ignoring this step will lead to ignoring uncertainty in the calibration function, never known with certainty.

Figure 5.1 displays calibration data for silver using EPA Method 1638 in distilled water. These data were obtained under controlled nonblind conditions that are not representative of routine practice. Spiking concentrations were 0, 10, 20, 50, 100, 200, 500, and 1000 ppt. The centerline that passes through these data represents the weighted least-squares (WLS) estimate of the linear calibration regression line. The curvilinear functions that provide an interval surrounding the data are WLS prediction bands and provide 99% confidence of including the next single measurement at a given concentration.

Inspection of Figures 5.1 and 5.2 reveals that variability increases with increasing concentration. In fact, this same pattern was observed for each of eight other compounds. The prediction bands (Figure 5.1) and observed standard deviations (Figure 5.2) are nicely fit by the Rocke and Lorenzato model. Of the 56 measurements (i.e., seven replicates of each of the eight spiking concentrations) one measurement is beyond the prediction bands, and therefore not inconsistent with the 99% confidence level. The implication of this finding is that the driving force behind limits of detection from single concentration designs is the concentration at which the samples are spiked. To illustrate this point, Table 5.1 displays the EPA/Glaser et al. MDL estimator applied to the data in Figure 5.1 at various concentrations. Table 5.1 reveals that EPA's MDL estimator will yield values ranging from 67.29 to 1.47 ppt from exactly the same data. If a new analytical method was developed and believed

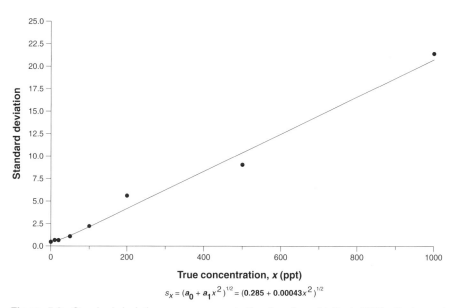

$$s_x = (a_0 + a_1 x^2)^{1/2} = (0.285 + 0.00043 x^2)^{1/2}$$

Figure 5.2 Standard deviation versus concentration for silver (Method 1638): Rocke and Lorenzato model.

to have lower limits of detection, presumably it would be tested by spiking at a correspondingly lower concentration. Thus, regardless of the true analytical properties of the method, a lower MDL will be found.

Anticipating this problem to some extent, Glaser et al. suggest both a single-step and an iterative approach to computing the MDL (also described in EPA's adoption of the procedure). For the single-step approach they recommend spiking samples at a concentration between one and five times the best estimate of the true MDL. They note that it is possible that an incorrect MDL could be obtained even if the spiking concentration were less than five times the computed MDL, illustrated in the higher values in Table 5.1. This logic requires that we know the actual MDL, which never happens in practice. In this case they recommend an iterative approach in which successively lower concentrations are computed until the pooled variance stabilizes. They note that convergence of the iterative procedure depends on the closeness of the estimated MDL to the calculated MDL. They also note that if the calculated MDL is lower than the background level of the analyte, the iterative procedure cannot be used.

Inspection of Table 5.1 reveals that spiking concentrations of 200 ppt and 10 ppt both meet EPA's criterion of a spike to MDL ratio of 5:1 or less. Presumably, if we had started at higher concentrations, and iteratively decreased the spiking concentration, we would obtain an MDL of 48 ppt. Alternatively, if we started at zero and worked up, we would have obtained an MDL of 2 ppt. This arbitrariness is unacceptable in a detection limit estimator and can lead to both false positive and false negative regulatory impact decisions based on samples spiked at higher or lower concentrations. For this reason, single-concentration decision and detection limits are of little practical value, despite their widespread use.

5.8 ILLUSTRATION

Returning to the silver data example of Section 5.7 and using the first (largest) concentration ($x = 200$ ppt) that meets the EPA 5:1 criterion, application of the previously described single-concentration estimators are now illustrated. At $x = 200$ ppt the observed standard deviation of the $n = 7$ replicates is 15.32 ppt. Assuming that the true value of $\sigma = 15.32$ Currie's critical level is $L_C = 2.33\sigma = 35.7$ ppt and the detection limit is $L_D = 4.66\sigma = 71.4$ ppt. Of course, these estimates are too low given that σ_0 and σ_D are unknown. As shown previously (Table 5.1), the EPA MDL is MDL $= 3.14s = 48.1$ ppt. Using the correct form of the MDL (i.e., a prediction limit) we obtain $L_C = 3.36s = 51.5$ ppt for the critical level and $L_D = 6.72s = 103.0$ ppt for the detection limit. Use of the more statistically justified noncentral t-distribution leads to the adjusted detection limit $L_D = 6.64s = 101.7$ ppt, which is slightly more conservative. These results clearly confirm that the MDL is an estimate of L_C and not L_D, as is often claimed. In terms of a multiple-use tolerance limit (95% confidence and 99% coverage), the critical level is $L_C = 4.64s = 71.1$ ppt, which is larger than the other critical levels. The reason is that the tolerance limit provides at least 99% coverage of the distribution 95% of the time, whereas the prediction limit only provides average coverage of at least 99% (i.e., 50% of the time).

6

SINGLE-CONCENTRATION-BASED QUANTIFICATION LIMIT METHODS

6.1 INTRODUCTION

Quantitative determination in analytical chemistry is a two-stage process. First, we must make the binary decision of whether or not the compound is present in the sample. Second, in the event that the compound is present, we must determine if the estimated concentration supports quantitative determination. By comparison to the literature on analytical detection, far less literature exists on the second stage of the decision process [i.e., quantification (also called quantitation)]. Currie (1968) originally described the determination limit (L_Q) as the concentration at which the relative standard deviation is 10% (i.e., a signal-to-noise ratio of 10:1). Gibbons et al. (1992) have provided a more statistically rigorous estimator of L_Q that accommodates effects of nonconstant variance, uncertainty in the parameters of the recovery curve, and uncertainty in the sample-based estimator of the true population variance. More recently, the U.S. EPA (1993) has defined the *minimum level* (ML) as the method detection limit (defined in 40 CFR, Part 136, App. B) multiplied by a factor of 3.18. EPA is using the ML as a quantification limit above which measured concentrations can be compared directly to regulatory standards and compliance decisions can be made. Gibbons et al. (1997a,b) have developed a calibration-based alternative to the ML for quantitative determination which they call the *alternative minimum level* (AML). In this chapter, however, we focus on quantification limit estimators for single samples.

6.2 CURRIE'S DETERMINATION LIMIT

Currie (1968) originally described the *determination limit* (L_Q) as the concentration (*x*) at which the relative standard deviation (RSD) is 10% (i.e., a signal-to-noise ratio of 10:1):

$$L_Q = 10\sigma_{L_Q} \tag{6.1}$$

In many ways, this definition remains the most general and useful in that it requires the signal-to-noise ratio at or above the quantification level to be small. Whether small corresponds to 10% or 15% or 20% is somewhat arbitrary; however, the basic idea can be extended to any % RSD, and the choice of RSD can be based on data quality objectives (DQOs). A major limitation of the method is that in Currie's original formulation, he provided no method for accommodating nonconstant variance and assumed that the true population standard deviation σ at L_Q was known, which is never the case. To actually solve for L_Q in equation (6.1) would require (1) a model for the relationship between σ and x, (2) a model for the relationship between true and measured concentration, and (3) iterative solution of equation (6.1). Note, however, that in practice it is not possible to solve equation (6.1) if the rate of change in σ as a function of x is greater than 0.1.

6.3 ACS LIMIT OF QUANTITATION

The American Chemical Society (ACS; Keith et al., 1983) defined the *limit of quantitation* (LOQ) as the concentration that is 10 standard deviation units above the average blank response:

$$\text{LOQ} = \bar{x}_0 + 10s_0 \tag{6.2}$$

where $\bar{x}_0$ is the average blank measurement and s_0 is the standard deviation of the blank measurements. As written, the model makes no assumptions regarding the relationship between variability and concentration, since both $\bar{x}$ and s are based on blank samples. In many cases, however, instruments censor measured concentrations of less than zero (and even instrument responses such as peak areas below a peak area rejection level); therefore, it may not even be possible to obtain a valid estimate of s_0. Unlike Currie's L_Q, the RSD at the LOQ will typically be much higher than 10%, making quantitative determination doubtful at best. Also note that if $\bar{x}_0 < 0$, the LOQ can be negative, which is meaningless.

6.4 EPA PRACTICAL QUANTIFICATION LIMIT

The EPA (1987) defined the *practical quantitation limit* (PQL) as "the lowest level achievable by good laboratories within specified limits during routine laboratory operating conditions." This rather vague definition has been operationally defined by the EPA as 5 or 10 times the MDL, or the concentration at which 75% of the laboratories in a between-laboratory study report concentrations within ±20% of the true value, or the concentration at which 75% of the laboratories report concentrations within ±40% of the true value. The first operational definition is arbitrary and depends completely on the validity of the corresponding MDL. The second and third operational definitions are somewhat better; however, the between-laboratory studies are often done at a single concentration [e.g., maximum contaminant level (MCL)] in experienced government laboratories that "knew they were being tested with standard samples in distilled water without matrix interferences" (U.S. EPA, 1987). Furthermore, it is unclear whether all measurements made by a single laboratory must be within ±20% or if this criterion can be satisfied by just one or two measurements. In practice, the PQLs published by the EPA have taken on fixed values that were determined by consensus (e.g., the PQLs for most volatile organic priority pollutant compounds are either 5 or 10 $\mu g/L$).

6.5 EPA MINIMUM LEVEL

More recently, USEPA (1993) has defined the *minimum level* (ML) as

$$\text{ML} = 3.18(\text{MDL}) = 3.18(3.14)s_x = 10s_x \tag{6.3}$$

All of the limitations noted previously for the MDL apply equally to the ML. The ML is actually more of a detection level than a quantification level since the MDL is an estimate of the critical level. An alternative view is that the ML represents the concentration for which the RSD is 10%; however, this will rarely be realized in practice since variability at 10 times s_x may be quite different (typically larger) than variability at the values selected for the spiking concentration x.

6.6 ILLUSTRATION

To illustrate computation of the various methods, we return to the Silver data presented in Section 5.7 (see Figure 5.1). Table 6.1 displays the EPA/Glaser et al. MDL and ML estimators applied to the data in Figure 5.1 at various concentrations. Table 6.1 reveals that the EPA's ML estimator will yield values ranging from 4.7 to 214.3 ppt from exactly the same data. Inspection of Table 6.1 reveals that spiking concentrations of 200 ppt and 10 ppt both meet the EPA's criterion of a spike to MDL ratio of 5:1 or less. Presumably, if we had started at higher concentrations and iteratively decreased the spiking concentration, we would obtain an ML of 153 ppt. Alternatively, if we started at zero and worked up, we would have obtained an ML of 7 ppt.

Table 6.1 EPA 40CFR MDL at various spiking concentrations of silver in distilled water using Method 1638

Concentration	SD	MDL	ML
0	0.47	1.47	4.7
10	0.67	2.09	6.7[a]
20	0.66	2.06	6.6
50	1.08	3.40	10.8
100	2.23	7.00	22.3
200	15.32	48.10	153.2[a]
500	9.05	28.43	90.5
1000	21.43	67.29	214.3

[a]Meets EPA's 5:1 criterion.

6.7 DISCUSSION

Similar to problems encountered in the estimation of detection limits, use of single-concentration designs for estimating quantification limits can also yield biased and misleading results. Although they are easy to compute, the results are so heavily dependent on the concentration that the analyst has selected to perform the experiment that the resulting estimates are of little practical value. The more skilled analyst will follow EPA guidance and spike at successively lower and lower concentrations; however, as illustrated in the example, this too can produce a wide range of possible estimates. Note, however, that this iterative approach of selecting spiking concentrations leads naturally to a calibration-based approach where the appropriate spiking concentration (including zero) can be selected in advance based on prior knowledge of the range at which detection and quantification decisions are hypothesized to be possible. Application of the calibration-based estimators using WLS even allows the analyst to choose too wide a range of concentrations without serious consequence.

7

CALIBRATION-BASED DETECTION LIMIT METHODS

7.1 INTRODUCTION

As described in Chapter 4, an alternative method for estimating detection limits is based on a calibration design. In this case, a series of samples are spiked at known concentrations in the range of the hypothesized detection limit, and variability is determined by examining deviations of the actual response signals from the fitted regression line of response signal for a known concentration. In these designs it is generally assumed that deviations from the fitted regression line are normally distributed.

There are two major advantages of calibration designs over single-concentration designs. First, single-concentration designs ignore uncertainty in the calibration function that relates measured to true concentration. Second, when variability in measured concentration is a function of true concentration, the estimated detection limit based on a single-concentration design will depend in large part on the spiking concentration selected by the analyst. If different laboratories select different spiking concentrations, it will appear that they have different analytical capabilities when they may, in fact, be identical. This is not a problem for calibration-based detection limits since the relationship between variability and concentration can explicitly be modeled and incorporated into the detection limit estimator.

Proponents of single-concentration designs suggest that nonconstant variance can only lead to overestimates of the true detection limit since variability at a concentration greater than zero will be larger than variability when the analyte is absent. In many cases this is true, and artificially high detection limits can result. However, if a very low spiking concentration is selected, gross underestimation of the detection limit is also possible. Modern analytical instruments are often designed to censor measured concentrations less than zero. If the spiking concentration is close to zero, we would expect up to 50% of the measured concentrations to be negative, however, if the instrument assigns a concentration of zero to these samples, the estimated standard deviation will be a gross underestimate of the true population standard deviation. To make matters worse, many instruments use peak area rejection criteria so that even positive concentrations of low magnitude will also be assigned a measured concentration of zero. Again, this practice will lead to even further underestimates of variability and corresponding detection limits. Calibration designs are immune to such problems since concentrations both above and below the true detection limit are included and non-constant variance accommodated. The primary disadvantage of calibration-based detection limit estimators is that they are more complicated to compute and typically will require a computer program capable of unweighted least-squares (ULS) regression and/or weighted least-squares (WLS) regression. Computer programs for these computations are now becoming readily available (see Gibbons et al., 1997a,b).

7.2 MODELS FOR CONSTANT VARIANCE (HUBAUX AND VOS)

Hubaux and Vos (1970) were the first to apply the theory of statistical prediction to the problem of detection limit estimation. Beginning from a calibration design in which response signals are determined for analyte-containing samples with concentrations throughout the range L_C to L_D, they constructed a 99% prediction interval for the calibration regression line (see Table 7.1).

The prediction limit is exactly of the form given in equation (4.29), and the decision limit is defined as the value of the prediction limit for zero concentration (i.e., $x = 0$) given in equation (4.30) (see Table 7.1). The detection limit is defined as the point at which we can have 99% confidence that the response signal is greater than L_C; therefore, Hubaux and Vos suggest that the response signal be obtained graphically by locating the abscissa corresponding to L_C on the lower prediction limit (see Table 7.1). A more direct solution for L_D is obtained by solving a quadratic equation in x for given y, in our case L_C. A reasonably compact solution is

$$L_D = \frac{\hat{x} + (t_{[1-\alpha, n-2]} s_{y \cdot x}/b_1) \sqrt{[(n+1)/n](1-c^2) + \hat{x}^2/\sum_{i=1}^{n} x^2}}{1 - c^2} \quad (7.1)$$

where

$$c^2 = \frac{t^2 s_b^2}{b_1^2}$$

$$= \frac{1}{\sum_{i=1}^{n}(x_i - \bar{x})^2} \left(\frac{t s_{y \cdot x}}{b_1}\right)^2$$

$$\hat{x} = \frac{y_C - \bar{y}}{b_1}$$

7.2 MODELS FOR CONSTANT VARIANCE (HUBAUX AND VOS)

Table 7.1 Example calculations for Hubaux–Vos (OLS) L_C and L_D for benzene (μg/L)

Step	Equation	Description
1	$s_x = s$ $= 0.753$	Assumes constant variance
2	$\mathrm{Var}(y_0) = s^2 \dfrac{1 + 1/n + (0 - \bar{x})^2}{\Sigma[(x - \bar{x})^2]}$ $= 0.567 \left[\dfrac{1 + 1/130 + (0 - 17.754)^2}{25{,}396.123} \right]$ $= 0.578$	Variance of a predicted measurement at $x = 0$
3	$L_C = \dfrac{t}{b_1} [\mathrm{Var}(y_0)]^{1/2}$ $= \dfrac{2.356}{1.019}(0.578)^{1/2}$ $= 1.759$	Hubaux–Vos critical level
4	$\mathrm{Var}(y_C) = s^2 \dfrac{1 + 1/n + (L_C - \bar{x})}{\Sigma[(x - \bar{x})^2]}$ $= 0.567 \left[\dfrac{1 + 1/130 + (1.759 - 17.754)}{25{,}396.123} \right]$ $= 0.577$	Variance of a predicted measurement at L_C
5	$L_D = L_C + \dfrac{t}{b_1}[\mathrm{Var}(y_C)]^{1/2}$ $= 1.759 + \dfrac{2.356}{1.019}(0.577)^{1/2}$ $= 3.515$	Hubaux–Vos detection limit

The quantity $c = ts_b/b_1$ is related to the significance test for b. In the present context, b_1 will be highly significant. As such, c will be small, c^2 will be negligible, and the prediction limit becomes (approximately)

$$L_D = \hat{x} + \frac{ts_{y \cdot x}}{b_1}\sqrt{1 + \frac{1}{n} + \frac{\hat{x}^2}{\sum_{i=1}^{n}(x_i - \bar{x})^2}} \tag{7.2}$$

This early method assumes that variability is constant throughout the range of concentrations used in the calibration design. If this assumption is violated, a variance-stabilizing transformation such as the square-root transformation might be applied and the assumption of constant variance may be reevaluated. The choice of the square-root transformation is not at all arbitrary. Many response signals are essentially sums of ion counts. In such cases, the Poisson distribution applies and is consistent with the common observation that concentration and variability are proportional (i.e., the Poisson mean and variance are identical). The square-root transformation is used to bring about normality for data arising from a Poisson process (Snedecor and Cochran, 1980). The more modern approach involves a WLS solution which is described in a following section.

7.3 PROCEDURE DUE TO CLAYTON AND CO-WORKERS

Clayton et al. (1987) point out that the method due to Hubaux and Vos is appropriate to establish the decision limit (L_C) but not to establish the detection limit (L_D). Their argument is that under the null hypothesis (i.e., $x = 0$) the false positive rate of the detection rule is achieved by specifying an observed measurement threshold corresponding to L_C,

$$\hat{y} = b_0 + s_0 t_{[n-2,\alpha]} \sqrt{\frac{1 + 1/n + \bar{x}^2}{\sum_{i=1}^{n}(x_i - \bar{x})^2}} \qquad (7.3)$$

where $t_{[n-2,\alpha]}$ is the upper 100α percentage point of Student's t-distribution on $n-2$ degrees of freedom. Under the alternative hypothesis (i.e., $x > 0$) the detection rate or power of the test must be obtained from the noncentral t-distribution. Thus they derive the detection limit as a function of the noncentrality parameter (ϕ) of the noncentral t-distribution with $n-2$ degrees of freedom and specified values of α and β (e.g., $\alpha = \beta = 0.01$):

$$\phi = \frac{(x - \bar{x}) b_1}{s_{y \cdot x}} \sqrt{\frac{1 + 1/n + \bar{x}^2}{\sum_{i=1}^{n}(x_i - \bar{x})^2}} \qquad (7.4)$$

The estimate of L_D may be found directly as

$$L_D = \frac{\phi s_{y \cdot x}}{b_1} \sqrt{\frac{1 + 1/n + \bar{x}^2}{\sum_{i=1}^{n}(x_i - \bar{x})^2}} \qquad (7.5)$$

As in the case of the Hubaux and Vos method, this approach also assumes constant variance throughout the calibration range.

Coleman (1993) has pointed out that this derivation ignores uncertainty in the calibration line, since only deviations from the regression line at the mean spiking concentration $\bar{x}$ are considered. In general, $\bar{x}$ is not equivalent to the concentration that corresponds to the $(1-\alpha)100\%$ upper prediction limit for the response signal corresponding to a concentration of $x = 0$, the quantity considered in the derivation by Hubaux and Vos (1970). In addition, Coleman points out that as n increases, the difference between central t and noncentral t becomes negligible, rendering detection limit estimates within a few percentage points of each other.

7.4 GENERALIZATION TO MULTIPLE FUTURE DETECTION DECISIONS

The methods of sections 7.2 and 7.3 solve the problem of predicting an interval that will contain a single future measurement with specified Type I and Type II error rates. Ideally, a calibration of this type would be performed and a corresponding detection limit estimated for each evaluation of a new test sample. This is rarely the case. For example, in the context of groundwater monitoring, the U.S. EPA has determined the detection limits experimentally for a variety of classes of compounds (see U.S. EPA, 1987), and these regulatory limits [i.e., decision limits computed using the method of Glaser et al. (1981)] are used regardless of the true limit for a laboratory. As such, detection decisions for numbers of test samples are being made on the basis of results obtained from a single analytical study in a single laboratory using a single analyst and a single instrument as well as involving a single, pure analyte in a pristine (noninterfering) matrix: typically, reagent water. Furthermore, the detection limits were computed assuming that the task was to make a detection decision in a single future sample.

As shown in Chapter 4, a simultaneous tolerance interval with $P\%$ coverage and $(1 - \alpha)100\%$ confidence level can be constructed for the entire calibration line using the method described by Lieberman and Miller (1963). As illustrated, a special case of computing the upper tolerance limit when $x = 0$ (i.e., the test sample does not contain the analyte in question) can provide an analogous result to the Hubaux and Vos method of estimating L_C, with the added benefit of including a specified proportion P of all future measurements with confidence $(1 - \alpha)100\%$. The solution of this equation corresponds to Currie's notion of the decision limit L_C. Solving the quadratic equation in x for $y = L_C$ yields the decision limit:

$$L_C = \frac{c + \dfrac{s_{y \cdot x}(2F_{2,n-2}^{1-\alpha/2})^{1/2}}{b_1 \sqrt{\sum_{i=1}^{n}(x_i - \bar{x})^2}} \sqrt{c^2 + \dfrac{b_1^2 \sum_{i=1}^{n}(x_i - \bar{x})^2 - s_{y \cdot x}^2 \, 2F_{2,n-2}^{1-\alpha/2}}{nb_1^2}}}{1 - \dfrac{s_{y \cdot x}^2 \, 2F_{2,n-2}^{1-\alpha/2}}{b_1^2 \sum_{i=1}^{n}(x_i - \bar{x})^2}} \quad (7.6)$$

where

$$c = \hat{x} + \frac{s_{y \cdot x}}{b_1} \Phi(P) \left(\frac{n - 2}{\alpha/2 \chi_{n-2}^2} \right)^{1/2}$$

$$\hat{x} = \frac{y_C - \bar{y}}{b_1}$$

66 CALIBRATION-BASED DETECTION LIMIT METHODS

For example, if $P = 0.95$ and $\alpha = 0.01$, we will have 99% confidence that 95% of the population of future measurements that do not contain the analyte in question will be below the decision limit.

7.5 NONCONSTANT VARIANCE

The previous detection limit calculations assume that the variance is homogeneous throughout the range of the calibration function. This assumption is rarely justifiable (see Clayton et al., 1987; Gibbons et al., 1991). In practice, the variation in response signal is dependent on concentration, aßs will be illustrated in a following section. If violations of this assumption are ignored, the result will be an overestimate of the variability at low levels and therefore an overestimate of the detection limit. In the following sections, a variety of approaches to the nonconstant variance problem are examined.

7.6 VARIANCE PROPORTIONAL TO CONCENTRATION

When variance is proportional to concentration, L_C and L_D can be computed noniteratively via WLS. Gibbons et al. (1991) suggest a noniterative approximation based on the assumption that variance is proportional to concentration. This assumption leads to the following WLS estimates:

$$b_{0w} = \frac{\sum_{i=1}^{n}(y_i/x_i)\sum_{i=1}^{n}x_i - n\sum_{i=1}^{n}y_i}{\sum_{i=1}^{n}(1/x_i)\sum_{i=1}^{n}x_i - n^2} \tag{7.7}$$

$$b_{1w} = \frac{\sum_{i=1}^{n}(1/x_i)\sum_{i=1}^{n}y_i - n\sum_{i=1}^{n}(y_i/x_i)}{\sum_{i=1}^{n}(1/x_i)\sum_{i=1}^{n}x_i - n^2} \tag{7.8}$$

$$s_w^2 = \frac{\sum_{i=1}^{n}(y_i^2/x_i) - b_{0w}\sum_{i=1}^{n}(y_i/x_i) - b_{1w}\sum_{i=1}^{n}y_i}{n - 2} \tag{7.9}$$

In this case, s_w^2 represents the weighted mean square of the residuals from the fitted calibration line. At a given point x on the calibration function, the variance is $s_w^2 x$. To compute L_C, we can set the weight equal to the lowest nonzero concentration x_{min}:

$$L_C = \frac{ts_w}{b_{1w}}\sqrt{x_{min} + \frac{1}{n} + \frac{(x_{min} - \bar{x})^2}{\sum_{i=1}^{n}(x_i - \bar{x})^2}} \tag{7.10}$$

where t is the upper $(1 - \alpha)100$ percentage point of Student's t-distribution on $n - 2$ degrees of freedom. To compute L_D, we can set the weight equal to L_C:

$$L_D = L_C + \frac{ts_w}{b_{1w}}\sqrt{L_C + \frac{1}{n} + \frac{(L_C - \bar{x})^2}{\sum_{i=1}^{n}(x_i - \bar{x})^2}} \tag{7.11}$$

7.7 MODELING OBSERVED VARIANCE

When the proportionality assumption is inadequate, the next level of sophistication involves modeling the observed variance as a function of concentration and using the inverse of the estimated variances as weights in the WLS regression. Two good choices for the variance model are the Rocke and Lorenzato model and the exponential model as described in Chapter 4. In this case the WLS estimates are

$$b_{1w} = \frac{\sum_{i=1}^{n}[(x_i - \bar{x}_w)y_i/k_i]}{\sum_{i=1}^{n}[(x_i - \bar{x}_w)^2/k_i]} \tag{7.12}$$

$$b_{0w} = \bar{y}_w - b_{1w}\bar{x}_w \tag{7.13}$$

$$\bar{y}_w = \frac{\sum_{i=1}^{n}(y_i/k_i)}{\sum_{i=1}^{n}(1/k_i)} \tag{7.14}$$

$$\bar{x}_w = \frac{\sum_{i=1}^{n}(x_i/k_i)}{\sum_{i=1}^{n}(1/k_i)} \tag{7.15}$$

and the weight $k_i = s_{x_i}^2$ is the estimated variance for sample i (Gibbons et al., 1997a; Oppenheimer et al., 1983; Zorn et al., 1997). The weighted residual variance is

$$s_w^2 = \frac{\sum_{i=1}^{n}[(y_i - \hat{y}_{wi})^2/k_i]}{n - 2} \tag{7.16}$$

and the estimated variance for a predicted value $\hat{y}_{wj}$ is

$$V(\hat{y}_{wj}) = s_w^2\left[k_j + \frac{1}{\sum_{i=1}^{n}(1/k_i)} + \frac{(x_j - \bar{x}_w)^2}{\sum_{i=1}^{n}(x_i - \bar{x}_w)^2/k_i}\right] \tag{7.17}$$

where k_j is the estimated variance at concentration x_j. An upper $(1 - \alpha)100\%$ confidence interval for $\hat{y}_{wj}$ (i.e., a prediction interval for a new measured concentration or instrument response at true concentration x_j) is

$$\hat{y}_{wj} + t\sqrt{\text{Var}(\hat{y}_{wj})} \tag{7.18}$$

where t is the upper $(1 - \alpha)100$ percentage point of Student's t-distribution on $n - 2$ degrees of freedom. The WLS estimate of L_C is therefore

$$L_C = \frac{ts_w}{b_{1w}}\sqrt{s_{L_C}^2 + \frac{1}{\sum_{i=1}^{n}(1/k_i)} + \frac{(L_C - \bar{x}_w)^2}{\sum_{i=1}^{n}(x_i - \bar{x}_w)^2/k_i}} \tag{7.19}$$

68   CALIBRATION-BASED DETECTION LIMIT METHODS

and the WLS estimate of L_D is

$$L_D = L_C + \frac{ts_w}{b_{1w}}\sqrt{s_{L_D}^2 + \frac{1}{\sum_{i=1}^{n}(1/k_i)} + \frac{(L_D - \bar{x}_w)^2}{\sum_{i=1}^{n}(x_i - \bar{x}_w)^2/k_i}} \qquad (7.20)$$

Note that in order to compute L_C and L_D, we must have estimates of $s_{L_C}^2$ and $s_{L_D}^2$ which are often unavailable and must be estimated using a model of standard deviation versus concentration: for example, an exponential model of the form

$$s_x = a_0 e^{a_1 x} \qquad (7.21)$$

The simplest approach to estimating the parameters of the exponential model is to compute the OLS or WLS regression of concentration (x) on $\log_e(s_x)$, which yields the coefficients a_0 and a_1 in equation (7.21) (see Snedecor and Cochran, 1980). Alternatively, the model suggested by Rocke and Lorenzato (1995) can be used, which reduces to

$$s_x = \sqrt{a_0 + a_1 x^2} \qquad (7.22)$$

To compute a_0 and a_1, compute the OLS or WLS regression of concentration on the sample variance estimates s_x^2.

The final estimates of L_C and L_D are obtained from simple repeated substitution beginning from $L_C = 0$ and $L_D = L_C$ until convergence (i.e., change of less than 10^{-4} in estimates of L_C and L_D on successive iterations).

7.8   ITERATIVELY REWEIGHTED LEAST SQUARES

As noted in Chapter 4, a further improvement over use of estimated variances as weights is to model the sum of squared residuals as a function of concentration,

$$s^2(x) = \frac{\sum_{i=1}^{m(x)}(\hat{y}_w - y_i)^2}{m(x)} \qquad (7.23)$$

where

$$\hat{y} = \hat{\beta}_{0w} + \hat{\beta}_{1w} x \qquad (7.24)$$

and $m(x)$ is the number of replicates for concentration x.

Since the sum of squared residuals depends on b_{0w} and b_{1w} and the current estimates for b_{0w} and b_{1w} depend on the estimated variance function which produces the weights, parameters of the variance function and the calibration function can be estimated jointly using an iterative solution. For example, beginning with the WLS estimates of Section 7.7, compute the residual variance at each concentration. The regression of the residual variances on concentration yields the new parameters of the Rocke and Lorenzato model, and the inverses of these residual variances are the corresponding weights that are in turn used to obtain improved estimates of β_{0w} and β_{1w}. This process can be continued until convergence.

7.9 ILLUSTRATION

To illustrate the various estimators described in this chapter, we return to the example benzene data presented in Chapter 4 (see Table 4.1). Table 7.1 and Figure 7.1 display computational details and graphical depiction of the Hubaux–Vos model, which assumes constant variance and provides corresponding Hubaux–Vos estimates of L_C and L_D. Figure 7.1 reveals that if we assume constant variance, in order to incorporate larger variability at higher concentrations, variability is overestimated at the lower concentrations ($L_C = 1.759$ and $L_D = 3.515$ μg/L). In contrast, the approximate noniterative WLS prediction limit approach described by Gibbons et al. (1991) and Gibbons (1991, 1996) provides a much better fit to the data observed for both high and low true concentrations ($L_C = 0.344$ and $L_D = 0.546$ μg/L; see Table 7.2 and Figure 7.2).

Table 7.3 and Figure 7.3 present results of the more complete iterative WLS solution based on modeling the observed variances using the Rocke and Lorenzato model. Results are, approximately one-half the size of the approximate solution ($L_C = 0.129$ and $L_D = 0.259$ μg/L). Tables 7.4 and 7.5 and Figures 7.4 and 7.5

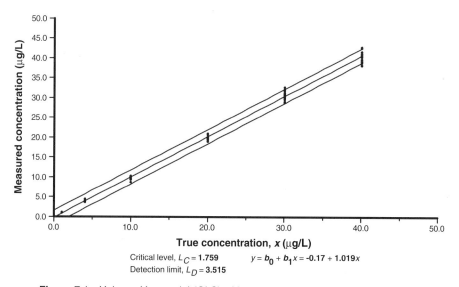

Figure 7.1 Hubaux–Vos model (OLS) with 99% prediction interval for benzene.

Table 7.2 Example calculations for WLS proportional variance model (prediction interval) L_C and L_D for benzene (μg/L)

Step	Equation	Description
1	$s_x = (s_w^2 x)^{1/2}$ $= (0.022x)^{1/2}$	Variance proportional to concentration
2	$\mathrm{Var}(y_0) = s_w^2 \dfrac{x_{\min} + 1/n + (x_{\min} - \bar{x})^2}{\sum[(x - \bar{x})^2]}$ $= 0.022 \left[\dfrac{1.0 + 1/130 + (1.0 - 1.72)^2}{95{,}513.492} \right]$ $= 0.022$	Variance of a predicted measurement at $x_{\min}$
3	$L_C = \dfrac{t}{b_1} [\mathrm{Var}(y_0)]^{1/2}$ $= \dfrac{2.356}{1.01} (0.022)^{1/2}$ $= 0.344$	Critical level
4	$\mathrm{Var}(y_C) = s_w^2 \dfrac{L_C + 1/n + (L_C - \bar{x})^2}{\sum[(x - \bar{x})^2]}$ $= 0.022 \left[\dfrac{0.344 + 1/130 + (0.344 - 1.72)^2}{95{,}513.492} \right]$ $= 0.007$	Variance of a predicted measurement at L_C
5	$L_D = L_C + \dfrac{t}{b_1} [\mathrm{Var}(y_C)]^{1/2}$ $= 0.344 + \dfrac{2.356}{1.01} (0.007)^{1/2}$ $= 0.546$	Detection limit

present corresponding approximate and iterative WLS estimates based on 99% coverage and 99% confidence tolerance intervals. For the approximate noniterative solution $L_C = 0.463$ and $L_D = 0.777$ μg/L, and for the iterative solution $L_C = 0.191$ and $L_D = 0.385$ μg/L. The difference in the iterative and approximate solutions is most likely due to the fact that the estimates of L_C and L_D are below the lowest spiking concentration. It is interesting to note that we pay a very small price for the added flexibility of the tolerance intervals.

7.10 BETWEEN-LABORATORY DETECTION ESTIMATE

The *interlaboratory detection estimate* (IDE) was developed to represent the detection capability of an analytical method measuring a particular analyte in a particular matrix for a population of *qualified laboratories*, taking into account all of the

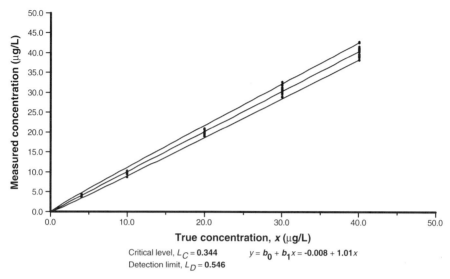

Figure 7.2 WLS proportional variance model with 99% prediction interval for benzene.

measurement variation associated with routine measurement using that method. To be *qualified,* laboratories must meet analytical method requirements and must survive laboratory screening based on results from an interlaboratory study (see ASTM D2777).

The IDE incorporates the two desirable properties: a targeted (minimum acceptable) statistical power of detection and a targeted (maximum acceptable) rate of false positives. Specifically, the IDE is computed to be the lowest concentration at which there is 90% confidence that a single measurement from a laboratory selected from the population of qualified laboratories represented in an interlaboratory study will have a true detection probability of at least 95% and a true nondetection probability of at least 99% (when measuring a blank sample). There are several reasons to conduct an interlaboratory study. A typical practical objective for conducting an interlaboratory study to compute the IDE is to be able to make statements such as the following with respect to the computed value, IDE:

- With 90% confidence, one should expected qualified laboratories to be 95% certain of detecting this analyte in this matrix using this method under routine measurement conditions (i.e., they can achieve 95% power of detection at the IDE).
- With 90% confidence, one should expect qualified laboratories to be 99% certain that there will be no detection of the analyte when a blank sample is measured, simultaneous with achieving 95% power.

Another objective for such a study could be method validation (perhaps under the auspices of a voluntary consensus standards organization, or to achieve broadly

Table 7.3 Example calculations for WLS Rocke and Lorenzato model (prediction interval) L_C and L_D for benzene (μg/L)

Step	Equation	Description
1	$s_x = (a_0 + a_1 x^2)^{1/2}$ $= (0.0024 + 0.00093 x^2)^{1/2}$	R-L regression model for standard deviation versus concentration (x);
2	$\mathrm{Var}(y_C) = s_w^2 s_{L_C}^2 + 1/\Sigma(1/k) + \dfrac{(L_C - \bar{x})^2}{\Sigma[(x - \bar{x})^2/k]}$ $= 1.209\left[0.002 + 1/9464.969 + \dfrac{(0.129 - 1.72)^2}{95{,}513.492} \right]$ $= 0.003$	Variance of a predicted measurement at L_C
3	$L_C = \dfrac{t}{b_1}[\mathrm{Var}(y_C)]^{1/2}$ $= \dfrac{2.356}{1.003}(0.003)^{1/2}$ $= 0.129$	Critical level
4	$\mathrm{Var}(y_D) = s_w^2 s_{L_D}^2 + 1/\Sigma(1/k) + \dfrac{(L_D - \bar{x})^2}{\Sigma[(x - \bar{x})^2/k]}$ $= 1.209\left[0.002 + 1/9464.969 + \dfrac{(0.259 - 1.72)^2}{95{,}513.492} \right]$ $= 0.003$	Variance of a predicted measurement at L_D
5	$L_D = L_C + \dfrac{t}{b_1}[\mathrm{Var}(y_D)]^{1/2}$ $= 0.129 + \dfrac{2.356}{1.003}(0.003)^{1/2}$ $= 0.259$	Detection limit

accepted regulatory status for a government agency). Additionally, one might want to determine the lowest level of quantitation (IQE).

Note that the IDE procedure is not restricted to a 90%, nor to 95% and 99% rates. Since the IDE is a random variable, as is any detection limit, these rates (and the confidence level) are "targeted," it cannot be claimed that they are attained *exactly* for any particular IDE.

7.10.1 Interlaboratory Detection Estimate and the ASTM

The IDE is a unique example of a calibration-based detection limit that has been developed within the consensus process of a respected voluntary standards body (ASTM). It was authored principally by David Coleman with considerable input. It

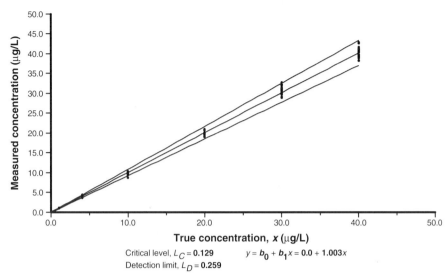

Figure 7.3 WLS Rocke and Lorenzato model with 99% prediction interval for benzene.

was approved in 1997 as Standard Practice D6091. The American Society for Testing and Materials (ASTM) is a nonprofit body founded in 1898. It is one of the world's largest and most respected voluntary consensus standards organizations, with about 9100 standards. ASTM has about 35,000 diverse members, including many representatives of federal government agencies (e.g., Geological Survey, Army Corps of Engineers, Nuclear Regulatory Commission, Coast Guard, and EPA), along with representatives of industry, consulting firms, and (to a lesser extent) universities. ASTM involves 132 committees, one of which is D19 (Water), founded in 1932. Within each committee are several subcommittees, such as the one from which D6091 originated, D19.02, "General Specifications, Technical Resources, and Statistical Methods," chaired in the 1990s by Paul Britton of the EPA. To develop standards, a subcommittee will typically form a Task Group, which any member of the committee can join and which anyone (ASTM member or not) can attend. The IDE was the product of many hours of deliberation by the Task Group on Detection and Quantitation, chaired by Nancy Grams.

7.10.2 IDE Approach

The IDE is based loosely on Currie's approach to L_C and L_D, sharing these assumptions and concepts:

- Measurement errors for multiple laboratories, at multiple concentrations, are independent and normally distributed.
- L_C can be computed as a detection threshold (critical value) to have desired (maximum allowable) level of expected false positives.

Table 7.4 Example calculations for WLS proportional variance model (tolerance interval) L_C and L_D for benzene (μg/L)

Step	Equation	Description
1	$s_x = (s_w^2 x)^{1/2}$ $= (0.022x)^{1/2}$	Variance proportional to concentration
2	$L_C = \dfrac{(s_w^2 x_{min})^{1/2}}{b_1}\left[(2F_{2,n-2}^{1-\alpha})^{1/2}\left\{\dfrac{1/n + (x_{min} - \bar{x})^2}{\Sigma[(x-\bar{x})^2]}\right\}^{1/2} + Z\left(\dfrac{n-2}{{}^{\alpha}\chi_{n-2}^2}\right)^{1/2}\right]$ $= \dfrac{(0.022 \times 1.0)^{1/2}}{1.01}\left\{(2 \times 3.77)^{1/2}\left[\dfrac{1/130 + (1.0 - 1.72)^2}{95,513.492}\right]^{1/2} + 2.57\left(\dfrac{130-2}{98.576}\right)^{1/2}\right\}$ $= 0.463$	Critical level
3	$L_D = L_C + \dfrac{(s_w^2 L_C)^{1/2}}{b_1}\left[(2F_{2,n-2}^{1-\alpha})^{1/2}\left\{\dfrac{1/n + (L_C - \bar{x})^2}{\Sigma[(x-\bar{x})^2]}\right\}^{1/2} + Z\left(\dfrac{n-2}{{}^{\alpha}\chi_{n-2}^2}\right)^{1/2}\right]$ $= 0.463 + \dfrac{(0.022 \times 0.463)^{1/2}}{1.01}\left[(2 \times 3.77)^{1/2}\left[\dfrac{1/130 + (0.463 - 1.72)^2}{95,513.492}\right]^{1/2} + 2.57\left(\dfrac{130-2}{98.576}\right)^{1/2}\right]$ $= 0.777$	Detection limit

Table 7.5 Example calculations for WLS Rocke and Lorenzato model (tolerance interval) L_C and L_D for benzene (µg/L)

Step	Equation	Description
1	$s_x = (a_0 + a_1 x^2)^{1/2}$ $= (0.0024 + 0.00093 x^2)^{1/2}$	R-L regression model for standard deviation versus concentration (x)
2	$L_C = \dfrac{s_w}{b_1}\left[(2F^{1-\alpha}_{2,n-2})^{1/2} \left\{ \dfrac{1/\Sigma(1/k) + (L_C - \bar{x})^2}{\Sigma[(x-\bar{x})^2]} \right\}^{1/2} \right] + s_{L_C} Z\left(\dfrac{n-2-p}{{}^\alpha\chi^2_{n-2}}\right)^{1/2}$ $= \dfrac{1.099}{1.003}\left\{ (2 \times 3.77)^{1/2}\left[\dfrac{1/9464.969 + (0.191 - 1.72)^2}{95{,}513.492}\right]^{1/2} \right\} + 0.049 \times 2.57\left(\dfrac{130 - 2 - 2}{96.822}\right)^{1/2}$ $= 0.191$	Critical level
3	$L_D = L_C + \dfrac{s_w}{b_1}\left[(2F^{1-\alpha}_{2,n-2})^{1/2}\left\{\dfrac{1/\Sigma(1/k) + (L_D - \bar{x})^2}{\Sigma[(x-\bar{x})^2]}\right\}^{1/2}\right] + s_{L_D} Z\left(\dfrac{(n-2-p)}{{}^\alpha\chi^2_{n-2}}\right)^{1/2}$ $= 0.191 + \dfrac{1.099}{1.003}\left\{(2 \times 3.77)^{1/2}\left[\dfrac{1/9464.969 + (0.385 - 1.72)^2}{95{,}513.492}\right]^{1/2}\right\} + 0.05 \times 2.57\left(\dfrac{130 - 2 - 2}{96.822}\right)^{1/2}$ $= 0.385$	Detection limit

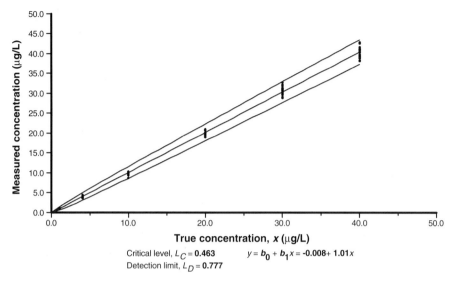

Figure 7.4 WLS proportional variance model with 99% confidence 99% coverage tolerance interval for benzene.

- The IDE is the same as L_D, a detection limit equal to the concentration at which the expected (minimum acceptable) power of detection can be achieved.

Additionally, the IDE makes the following assumptions with respect to a population of qualified labs and a trace-level range of concentrations, including blanks (zero-concentration samples):

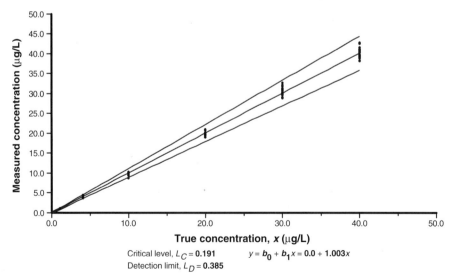

Figure 7.5 WLS Rocke and Lorenzato model with 99% confidence 99% coverage tolerance interval for benzene.

- ASTM's "Standard Practice for Method Validation," D2777, is adequate for describing requirements for an interlaboratory study and defining qualified laboratories.
- Collectively, the relationship between measurements and true concentrations can be modeled by a straight line. This relationship is called the *mean recovery curve* (or *mean response curve*). Ideally, the mean recovery curve would have slope = 1 and would pass through the origin. In practice, recovery is rarely equal to 100%, and there can be biases due to matrix effects and other effects.
- The standard deviation of aggregated measurements is either constant over a finite concentration range or its relationship with true concentration can be modeled by a straight line, exponential, or other suitable monotonic function. This relationship is called the *precision function* or *interlaboratory standard deviation* (ILSD) *model*.

Note that there are three direct consequences to allowing a nonconstant precision function:

1. The procedure is made more complex by the need to select and fit a nonconstant precision function.
2. To fit the mean recovery model, weighted least squares should be (and is) used in place of ordinary least squares. This also adds complexity but has the beneficial effect of discounting the influence of noisier data.
3. In general, explicit recognition that measurement variation decreases with decreasing concentration can result in a lower detection limit than does the constant variation assumption. Conversely, measurement variation that increases rapidly with increasing concentration can result in a higher detection limit.

Overview of the Steps of the IDE Standard Practice

1. Determine the IDE study plan, design, and protocol.
   (a) Choose the analyte, matrix, and method.
   (b) Choose the IDE study design and protocol based on the anticipated interlaboratory standard deviation (ILSD) model (*precision function*).
   (c) Choose a protocol for interlaboratory study.
   (d) Choose allowable sources of variation.
2. Conduct the IDE study, screen the data, and choose models.
   (a) Conduct the study (following ASTM D2777, "Method Validation").
   (b) Screen the data (following ASTM D2777).
   (c) Identify and fit the interlaboratory standard deviation model. Evaluate and fit ILSD models, in the order (1) constant, (2) straight-line, (3) exponential, or other.
   (d) Fit the mean recovery model.

3. Compute the IDE.
   (a) Compute the recovery critical value, Y_C.
   (b) Compute L_C, the true concentration critical value.
   (c) Compute the IDE, using the ILSD and mean recovery models.

For a detailed case study, the reader is referred to the IDE Standard Practice (D6091) itself.

8

CALIBRATION-BASED QUANTIFICATION LIMIT METHODS

8.1 INTRODUCTION

Just as detection limit estimators can be generalized to calibration designs, quantification limit estimators are available in the calibration setting as well. The primary advantages of the calibration approach are: (1) the variance can be modeled explicitly as a function of concentration and incorporated into the quantification limit estimator (eliminating dependence on the specific choice of a single spiking concentration), (2) uncertainty in the calibration function parameters can be incorporated, and (3) the relative standard deviation (i.e., signal-to-noise ratio) can be estimated directly at the quantification limit or the quantification limit can be defined in terms of a particular % RSD value. In the following sections, several different approaches to estimating L_Q from calibration-based data are described.

8.2 ALTERNATIVE MINIMUM LEVEL

Gibbons et al. (1997a, b) derived an estimate of L_Q based on calibration data that paralleled EPA's single concentration-based ML (see Chapter 6). The fundamental problem with the ML is not its definition (i.e., $10s_x$) but the fact that the sample standard deviation s_x is a moving target and depends in large part on the choice of spiking concentration x.

80 CALIBRATION-BASED QUANTIFICATION LIMIT METHODS

As a conceptual foundation, Gibbons et al. (1997a) followed Currie's definition of the determination limit,

$$L_Q = 10\sigma_Q \qquad (8.1)$$

where σ_Q represents the true population standard deviation at the determination limit L_Q. The L_Q value is based on the idea that quantitative determinations can be confidently made when the signal-to-noise ratio is 10:1 [i.e., a relative standard deviation (RSD) of 10%] (Currie, 1968). Unfortunately, in the real world, we never know the population standard deviation (i.e., it must be estimated from a sample of n measurements), and in many cases, the proportionality constant between the standard deviation and concentration is larger than 10% in the practical calibration range, and therefore a 10% RSD may not be achievable or achievable only at unacceptably high levels.

A reasonable alternative to both approaches is to compute the standard deviation at the lowest concentration that is differentiable from zero and use that standard deviation in computing the ML. In this way, the spiking concentration is no longer arbitrary and the effect of nonconstant variance is incorporated up to the lowest nonzero concentration. Of course, this alternative ML (AML) will not guarantee a RSD = 10% at the AML; however, we can compute the actual % RSD at the ML by modeling the association between s and x. The algorithm is as follows:

1. Using calibration data, model the relationship between s and x, for example, an exponential function

$$s_x = a_0 e^{a_1(x)} \qquad (8.2)$$

The exponential model $s_x = a_0 e^{a_1 x}$ generally provides excellent fit to the data observed and it used in the following discussion; however, the Rocke and Lorenzato model is also an excellent choice. Discussion of parameter estimation for these two variance function models is presented in Chapters 4 and 7, and in Chapter 11 for the between-laboratory case.

2. Compute the weighted least-squares regression of measured concentration or instrument response (y) on true concentration (x) for the linear model

$$\hat{y} = b_{0w} + b_{1w} x_i \qquad (8.3)$$

as described in Chapter 4.

3. Iteratively compute a multiple-use estimate of L_C using as an approximation

$$y_C = K_{0.95, 0.99} s_{L_C} + b_{0w} = K_{0.95, 0.99} a_0 e^{a_1(y_c - b_{0w})/b_{1w}} + b_{0w} \qquad (8.4)$$

8.2 ALTERNATIVE MINIMUM LEVEL

Table 8.1 Factors (K) for constructing one-sided normal tolerance limits ($\bar{x}+Ks$) for 95% confidence and 95% and 99% coverage

n	95% Coverage	99% Coverage	n	95% Coverage	99% Coverage
4	5.144	7.042	20	2.396	3.295
5	4.210	5.749	21	2.371	3.262
6	3.711	5.065	22	2.350	3.233
7	3.401	4.643	23	2.329	3.206
8	3.188	4.355	24	2.309	3.181
9	3.032	4.144	25	2.292	3.158
10	2.911	3.981	30	2.220	3.064
11	2.815	3.852	35	2.166	2.994
12	2.736	3.747	40	2.126	2.941
13	2.670	3.659	50	2.065	2.863
14	2.614	3.585	60	2.022	2.807
15	2.566	3.520	80	1.965	2.733
16	2.523	3.463	100	1.927	2.684
17	2.486	3.414	500	1.763	2.475
18	2.453	3.370	∞	1.645	2.326
19	2.423	3.331			

where $K_{0.95, 0.99}$ is the 95% confidence 99% coverage one-sided normal tolerance limit factor for n observations, where in this case, n is the total number of measurements (see Table 8.1). When the true concentration is zero, we can have 95% confidence that 99% of the measurements will be less than L_C. L_C is the corresponding true concentration:

$$L_C = \frac{y_C - b_{0w}}{b_{1w}} \quad (8.5)$$

Note that although it is not required for computing the AML, a Currie type of estimate of the detection limit (L_D) can also be obtained by iteratively computing

$$y_D = y_C + K_{0.95, 0.99} s_{L_D} = y_C + K_{0.95, 0.99} a_0 e^{a_1(y_D - b_{0w})/b_{1w}} \quad (8.6)$$

and L_D is the corresponding true concentration:

$$L_D = \frac{y_D - b_{0w}}{b_{1w}} \quad (8.7)$$

4. Compute the standard deviation at the L_C from step 1 (i.e., s_{L_C}): for example,

$$s_{L_C} = a_0 e^{a_1(L_C)}$$

5. Compute the measured concentration at 10 times the standard deviation at the L_C:

$$y_Q = 10 s_{L_C} + b_{0w} \qquad (8.8)$$

6. The AML is then computed as

$$x_Q + \frac{t}{b_{1w}} \sqrt{V(y_Q)} \qquad (8.9)$$

where $x_Q = (y_Q - b_{0w})/b_{1w}$ and

$$\operatorname{Var}(y_Q) = s_w^2 \left[s_{y_Q}^2 + \frac{1}{\sum_{i=1}^n (1/k_i)} + \frac{(x_Q - \bar{x}_w)^2}{\sum_{i=1}^n (x_i - \bar{x}_w)^2/k_i} \right] \qquad (8.10)$$

The value t is the upper 99th percentile of Student's t-distribution on $n - 2 - p$ degrees of freedom, where p is the number of unknown parameters in the standard deviation model. Alternatively, when n is large (e.g., $n > 25$) the AML can be approximated by

$$\text{AML} \sim \frac{x_Q + t s_{y_Q}}{b_{1w}} \qquad (8.11)$$

Illustration To illustrate the computation of the AML, return to the silver data presented in Section 5.7 (see Figures 5.1 and 5.2). Inspection of Figure 5.2 reveals that the Rocke and Lorenzato model fits the observed data quite well. The various steps in the computation are displayed in Table 8.2. Inspection of the table reveals that the estimated AML is 7.276 ppt, and this provides a relative standard deviation of 7.78%. Note that a wide range of MLs are possible for these data. For example, spiking at 200 ppt the ML is 153 ppt, whereas spiking at 10 ppt yields 6.66 ppt. Both spiking concentrations meet EPA's 5:1 criterion and would therefore represent valid estimates of the ML.

8.3 ALTERNATIVE FORMULATIONS

An alternative that departs somewhat from the AML is to estimate the true concentration that leads to a measured relative standard deviation of 10% or 20% or whatever is achievable. Requiring 10% will not always lead to a solution since the rate of change in standard deviation relative to concentration may be greater than 0.1. The solution for the case of nonconstant variance involves iterative solution of

$$y_Q = \frac{1}{r} \sqrt{\operatorname{Var}(\hat{y}_Q)} \qquad (8.12)$$

8.3 ALTERNATIVE FORMULATIONS

Table 8.2 Example calculations for the alternative minimum level of silver (ppt, Method 1638: Rocke and Lorenzato model

Step	Equation	Description
1	$s_x = (a_0 + a_1 x^2)^{1/2}$ $= (0.285 + 0.00043 x^2)^{1/2}$	R-L regression model for standard deviation versus concentration (x)
2	$L_C = K_{.95,.99}\, s_x + b_0$ $= 2.833 s_x - 0.098 = 1.546$	Result of iterative solution for Currie (1968)-type critical level or EPRI CMDL
3	$s_{L_C} = (a_0 + a_1 L_C^2)^{1/2}$ $= (0.285 + 0.00043 \times 2.389)^{1/2} = 0.535$	Estimated standard deviation at the critical level L_C
4	$L_D = L_C + K_{.95,.99}\, s_x$ $= 1.546 + 2.833 s_x = 3.1$	Result of iterative solution for Currie (1968)-type detection limit
5	$s_{L_D} = (a_0 + a_1 L_D^2)^{1/2}$ $= (0.285 + 0.00043 \times 9.609)^{1/2} = 0.538$	Estimated standard deviation at the detection limit L_D
6	$y_Q = 10 s_{L_C} + b_0$ $= 10 \times 0.535 - 0.098 = 5.25$	Averaged measured concentration (or instrument response) at 10 times S_{L_C}
7	$x_Q = \dfrac{y_Q - b_0}{b_1}$ $= \dfrac{5.25 + 0.098}{0.98} = 5.456$	True concentration corresponding to y_Q where b_0 and b_1 are the intercept and slope of the recovery equation; see the recovery curve
8	$s_{x_Q} = (a_0 + a_1 x_Q^2)^{1/2}$ $= (0.285 + 0.00043 \times 29.768)^{1/2} = 0.546$	Estimated standard deviation at x_Q
9	$\text{AML} = x_Q + \dfrac{t\, s_w}{b_1} \left\{ \dfrac{s_{x_Q}^2 + 1/\Sigma(1/k) + (x_Q - \bar{x})^2}{\Sigma[(x - \bar{x})^2/k]} \right\}^{1/2}$ $= 5.456 + \left(2.402 \times \dfrac{1.328}{0.98}\right)$ $\left[\dfrac{0.298 + 1/71.562 + (5.456 - 14.258)^2}{66812.09}\right]^{1/2}$ $= 7.276$ ppt	The AML

(Continued)

Table 8.2 (*Continued*)

Step	Equation	Description
10	$s_{AML} = (a_0 + a_1 \cdot AML^2)^{1/2}$ $= (0.285 + 0.00043 \times 52.946)^{1/2} = 0.555$	Estimated standard deviation at the AML
11	$\% \text{ RSD} = 100 \left(\dfrac{s_{AML}/AML}{b_1} \right)$ $= 100 \left(\dfrac{0.555/7.276}{0.98} \right) = 7.78\%$	Relative standard deviation at the AML

where r represents the relative standard deviation (e.g., $r = 0.1$ for a 10% RSD). The estimate of L_Q is then

$$L_Q = \frac{y_Q - b_{0w}}{b_{1w}} \tag{8.13}$$

When applied to the silver data application of equations (8.12) and (8.13) yield $L_Q = 7.836$ ppt for an RSD of 10%, which is quite similar to the AML.

8.4 ESTIMATOR DUE TO GIBBONS AND CO-WORKERS

Gibbons et al. (1992) have provided a direct estimator of L_Q under the assumption that square-root transformation of both x and y homogenizes variance. This leads to the predicted concentration x_α^*, for which

$$\hat{y} = \frac{100}{\alpha} s(\hat{y}) \tag{8.14}$$

a predicted response $(\hat{y})$ that is $100/\alpha$ times its estimated standard deviation $[s(\hat{y})]$. Let y_α^* denote the value of $\hat{y}$ that satisfies equation (8.14). Gibbons et al. (1992) solved equation (8.14) for y_α^* and showed that a good approximation can be obtained as

$$x_\alpha^* = \bar{x} + \frac{(100/\alpha) s_{y \cdot x} - \bar{y}}{b} \tag{8.15}$$

$$L_Q = (x_\alpha^*)^2 + 0.632456 x_\alpha^* \tag{8.16}$$

(see also Gibbons, 1994). Gibbons and co-workers make several observations regarding this estimator. First, they suggest that with small numbers of measurements, $s_{y \cdot x}$ may be a poor estimator of the true population standard deviation $\sigma_{y \cdot x}$ and we may wish to replace $s_{y \cdot x}$ with its upper confidence limit.

Second, the previous definition of $\alpha/100$ times the standard deviation applies to the metric in which the PQL is estimated. In the present context, a square-root transformation is used to stabilize variance across the calibration line while preserving linearity of the function. An RSD of 10% in this transformed response metric actually corresponds to a 20% RSD in the raw concentration metric. Therefore, obtaining a 10% RSD in the metric of measured concentrations would require a 5% RSD in the transformed response scale.

Application of this estimator to the silver data yields an L_Q of 22.159 ppt for a 10% RSD in the transformed metric. This value is somewhat higher than the previous estimators, indicating that in this example, the square-root transformation failed to bring about constant variance at the lowest concentrations.

8.5 INTERLABORATORY QUANTITATION ESTIMATE

The interlaboratory quantitation estimate (IQE) was developed to represent the quantitation (also called *quantification*) capability of an analytical method used to measure a particular analyte in a particular matrix, for a population of qualified laboratories, taking into account all of the measurement variation associated with routine measurement using that method. To be qualified, laboratories must meet analytical method requirements and must survive laboratory screening based on results from an interlaboratory study (see ASTM D2777).

Similar to the interlaboratory detection estimate (IDE), the IQE is a unique example of a calibration-based quantitation limit that has been developed within the consensus process of a respected voluntary standards body (ASTM). As with the IDE, the IQE was authored principally by David Coleman with considerable help from members of the D19 Water Committee's Task Group on Detection and Quantitation, chaired by Nancy Grams. The IQE was approved in 1999, as a standard practice D6512.

The IQE can be considered to be a more rigorous version of the alternative minimum level described in Section 8.2, and is similar to the method described in section 8.3. The added rigor comes from the IQE's explicit precision function modeling at the quantitation limit concentration. Recall that the AML itself is a more rigorous version of the EPA's minimum level (ML), correcting some technical deficiencies. All three of these limits are based on the arbitrary but reasonable definition of quantitation as being the concentration at which the relative standard deviation (RSD) = 10% (i.e., the QL = 10σ), as discussed earlier in the chapter. This definition has been promulgated by Currie, by Larry Keith in shortcourses for the American Chemical Society, and by others. The key differentiators among 10σ QLs are (1) the sources of variation allowed in σ, and (2) the rigor in estimating σ. Additionally, since it is considered common knowledge in analytical chemistry that not all methods can achieve an RSD as low as 10% (especially older methods), the IQE is defined for any of these three RSD levels: 10%, 20%, or 30%. It is recommended that the lowest attainable of these three be published, and the IQE practice explicitly recognizes that an RSD value of 10%, or 20%, or even 30% may not be achievable. Alternatively, the user can select any achievable RSD level that is less than 10%. A RSD > 30% is not permitted, since as the RSD increases, the limit becomes more appropriately interpreted as a detection limit than a quantitaion limit. In chapter 9 we develop a stricter definition of quantitation based on ensuring at least one significant digit in the measurement reported. Note that for a measurement to be above ML, AML, IQE, or any 10σ-type quantitation limit *does not* ensure that a reported measurement has even a single significant digit in the measurement.

IQE can also be regarded as a companion to the IDE described in Chapter 7, sharing these characteristics with the IDE.

- Based on data from an interlaboratory study, as prescribed by ASTM Standard Practice, D2777 (Method Validation). Also uses data screening from the same practice.
- May require extension of the interlaboratory study design to low concentration levels, including blanks.
- Involves systematic selection and fitting of an interlaboratory standard deviation (ILSD) model (also known as a precision function) that characterizes the approximate mathematical relationship between the measurement standard deviation and the concentration.
- Calculations are based on both the coefficient estimates for the ILSD model and the mean response function (assumed to be a straight line, but with slope and intercept not necessarily equal to 1 and 0, respectively).

8.5.1 IQE Approach

The IQE is loosely based on Currie's approach to L_Q sharing these assumptions and concepts:

- Measurement errors for multiple laboratories, at multiple concentrations, are independent and normally distributed.
- L_Q can be computed so that the estimated RSD = 10% (or 20% or 30% in the case of IQE). However, Currie assumed that σ was constant within the low ranges of measurement. In the IQE practice, σ is merely hypothesized to be constant so that the hypothesis can be tested and (often) rejected.

Additionally, similar to the IDE, the IQE makes the following assumptions with respect to a population of qualified labs, and a trace-level range of concentrations, including blanks (zero concentration samples):

- ASTM's "Standard Practice for Method Validation," D2777, is adequate for describing requirements for an interlaboratory study and for defining qualified laboratories.
- Collectively, the relationship between measurements and true concentrations can be modeled by a straight line. As with the IDE, this relationship is called the *mean recovery curve*.

- The standard deviation of aggregated measurements is either constant, or its relationship with true concentrations can be modeled by one of the following: a straight line, the hybrid function [$\sigma^2 = g^2 + (hT)^2$ for nonnegative constants g and h and true concentration T], or some other suitable monotonic function. As with the IDE, this relationship is called the *precision function* or the *interlaboratory standard deviation* (ILSD) *model*.

Hybrid refers to the synthesis of two behaviors that hold for many combinations of method, analyte, and matrix: At the lowest concentrations, measurement errors are predominantly additive, and the standard deviation is nearly constant with respect to concentration, whereas at higher concentrations, measurement errors may be predominantly multiplicative, and the standard deviation increases linearly with concentration. As described earlier in this chapter, the hybrid model is identical to the model developed by Jim Kanzelmeyer (through years of analysis of interlaboratory measurement data) and incorporated in ASTM standard E1763. (Standard Guide for Interpretation and Use of Results from Interlaboratory Testing of Chemical Analysis Methods). It is also closely related to the model derived independently by Rocke and Lorenzato (1995).

Overview of the Steps of the IQE Standard Practice Consider the interlaboratory standard deviation (ILSD) models (dropping the error term):

Constant: $s = g$
Straight-line: $s = g + hT$
Hybrid: $s = [g^2 + (hT)^2]^{1/2}$
Recovery model is $y = a + bT$.

1. Conduct outlier rejection and lab ranking according to ASTM D2777.
2. Plot reported measurements (y) versus true concentration (T) to ensure that there are no obvious data issues to resolve.
3. For each T, compute $s =$ (adjusted) sample standard deviation of y values.
4. Plot s versus T. Fit a straight line by ordinary least squares.
5. If the slope is not statistically significant, use the constant model; fit the straight-line recovery model by OLS. Go to step 10.
6. Compute residuals, $r = [s - (\text{predicted } s)]$, and plot versus T.
7. If there is no evidence of curvature (formal test available), use the straight-line model; compute the weights. Go to step 9.
8. If curvature is present, use nonlinear least squares to fit the hybrid model (or other suitable model); compute the weights.
9. Fit the straight-line recovery model by weighted least squares.
10. Compute $IQE_{z\%}$, using the formula appropriate for the ILSD model, where $Z = \%$ RSD; can compute for any $Z < 10$, or $Z = 10, 20,$ or 30, provided that a solution exists and is within the range of the data.

CALIBRATION-BASED QUANTIFICATION LIMIT METHODS

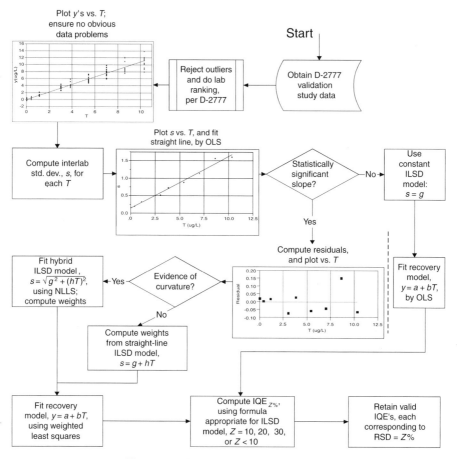

Figure 8.1 Flowchart for IQE computation.

Constant model: $\text{IQE}_{z\%} = (100/Z)(g/b)$
Straight-line model: $\text{IQE}_{z\%} = g/[b(Z/100) - h]$
Hybrid model: $\text{IQE}_{z\%} = g/[(bZ/100)^2 - h^2]^{1/2}$

These steps are summarized in Figure 8.1. Some differences between IDE and IQE follow (some of which will probably lead to future revisions in the IDE):

- IQE has more detailed and specific suggestions for the interlaboratory study design.
- IQE includes the bias-correction factor for the sample standard deviation. IDE does not.
- IQE is more formal than IDE regarding the process by which the precision model is selected.

- If the data indicate curvature in the precision function, the IQE suggests using the hybrid model, whereas the IDE suggests using the exponential model, although each practice allows models other than those suggested explicitly. While both the hybrid and the exponential models are probably adequate for many method/analyte/matrix combinations, the hybrid has stronger technical justification (as established by Rocke and Lorenzato, 1995) and a record of applications documented in ASTM E1763.
- The IDE is based on statistical tolerance limits, thus accounting explicitly for estimation uncertainty as well as measurement variation in computing at what concentration one can confidently separate absence ("black") from presence ("white"). The IQE is essentially an arbitrary (but sensible) point estimate in a portion of the real number line for measurements that have "shades of gray." In such a region it is harder to justify accounting for estimation uncertainty in addition to measurement variation.

9

SIGNIFICANT DIGITS

9.1 INTRODUCTION

In this chapter we tie together several seemingly disparate technical topics: (1) measurement uncertainty, (2) calibration, (3) significant digits, (4) detection, and (5) quantitation. These topics are often treated independent of one another in textbooks and the literature, but they should not be so treated. The *statistical characterization of measurement* uncertainty is a common thread running through all these topics. The first section establishes the link between the first two of these topics, the next two sections lay some groundwork, and the remaining sections complete the links among the remainder of the topics. Finally, we make recommendations regarding statistically sound formats for reporting measurements with uncertainty. It can be argued that these formats are technically superior to the use of significant digits, detection limits, and quantitation limits, but they are not as simple.

9.2 MEASUREMENT UNCERTAINTY INTERVALS FROM CALIBRATION

What is the uncertainty associated with a measurement? In previous chapters we have provided the concepts and computations for statistical intervals about the calibration line. Most useful in the calibration context are the various forms of prediction intervals and statistical tolerance intervals, both of which take the general form

$$x = \frac{y-a}{b} \pm \frac{s}{b} F(x)$$

where $F(x)$ is a function that achieves its minimum at $x = \bar{x}$, and for well-balanced calibration designs with $n \gg \phi F(x)$ is nearly constant with respect to x over the range of concentration in the study. Recall from Chapter 4 for prediction intervals, that

$$F(x) = t_{[n-2, 1-\alpha/2]} \sqrt{\frac{1 + (1/n) + (x - \bar{x})^2}{Sxx}}$$

where $Sxx = \Sigma(x - \bar{x})^2$.

Recall from Chapter 3 that for statistical tolerance intervals, with a little rewriting,

$$F(x) = 2 F_{2, n-2}^{1-\alpha/2} \sqrt{\frac{(1/n) + (x - \bar{x})^2}{Sxx}} + \Phi(P) \sqrt{\frac{n - 2}{\alpha/2 \chi_{n-2}^2}}$$

where $F_{2, n-2}^{1-\alpha/2}$ and $^{\alpha/2}\chi_{n-2}^2$ are critical values from the F and chi-square distributions, respectively, and $\Phi(P)$ is the two-sided critical value from the normal distribution, sometimes written, $\Phi(1 - P/2)$.

Examination of these formulas shows that, indeed, $F(x)$ achieves its minimum at $x = \bar{x}$, due to the term in the numerator, $x - \bar{x}$. Additionally, for well-balanced calibration designs (i.e., $\bar{x}$ at or near the center of the calibration range), $F(x)$ is nearly constant with respect to x (to a degree that increases with n, through the factor $1/Sxx$; see Figure 9.1).

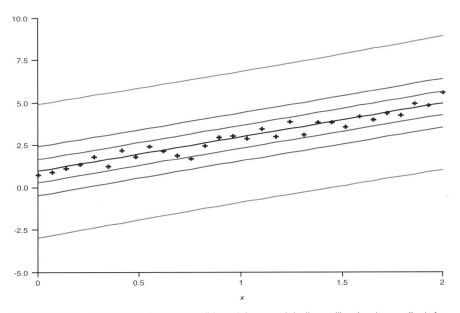

Figure 9.1 Prediction intervals (95% confidence) from straight-line calibration (center line), for $n = 4$ (outermost), $n = 9$, and $n = 30$ (innermost; all 30 points shown).

Thus the measurement uncertainty associated with x is

$$\text{measurement uncertainty} = \pm \frac{s}{b} F(x)$$

which depends on n and α (and P, for a statistical tolerance interval), but depends very little on x. The *relative* measurement uncertainty is

$$\text{RMU} = \pm \frac{s}{b} \frac{F(x)}{x}$$

Given the uncertainty formula, it is appropriate to report a measurement with its uncertainty (or relative uncertainty) and associated confidence level $1 - \alpha$ (and P, if appropriate). Recommended reporting formats are developed further in Section 9.7.

9.3 COMMON USE OF SIGNIFICANT DIGITS

"A number should be displayed with *all* of its significant digits and *only* its significant digits." This is a statement that is commonly accepted (although not necessarily observed in practice). Some form of this statement is the scientist's or technician's oath before taking the witness stand (figurative or literal) to report a measurement, as in: "The truth, the whole truth, and nothing but the truth."

For example, one might represent a measurement, M, in scientific notation as $M = v.xyz \times 10^k$ (where v, x, y, and z each represent a digit, not necessarily unique, and k is an integer). Note that the value of k is immaterial to any discussion of significant digits. Given this representation, M would be assumed to have four significant digits, each displayed. One would also assume that there is information content in the fourth digit, z, but the true thousandths-place digit may be uncertain. For example, the uncertainty could be $\pm 0.002 \times 10^k$, or $\pm 0.004 \times 10^k$. Truly, how many significant digits are such numbers? The common answer is *four,* since there is information content in four digits. However, the correct answer is *between three and four* if we use a literal definition of *significant,* to mean *not* attributable to chance.

Specifically, it is sensible to assert that the measurement $M = v.xyz \times 10^k$ has *greater* than *three* significant digits if and only if the least significant digit, z, has some information content. More specifically, M would have greater than three significant digits if the measurement uncertainty interval, at some acceptable confidence level, did *not* contain all 10 decimal digits, zero through nine. Otherwise, z is not telling us anything; it should not be presented as a significant digit.

9.4 PAIR OF PROBLEMS DUE TO HAVING TEN FINGERS AND THEIR SOLUTIONS

Consider how much easier life would be if we each had only two fingers, one on each hand! Naturally, it would be difficult to play the piano, grip a pen, or do

9.4 PAIR OF PROBLEMS DUE TO HAVING TEN FINGERS AND THEIR SOLUTIONS   93

"cat's cradle," but significant digits would be much simpler! We would probably be using binary arithmetic—unless a forward-thinking person were to win the public discourse to persuade everyone to use octal or hexadecimal (a term whose etymology would prevent it from ever being coined). Using binary arithmetic simplifies the issues of significant digits and rounding conventions because there are only two choices: 0 and 1. However, regardless of the arithmetic number base, there are thorny issues to be resolved. [For example, one issue is whether to assume that there are additional nonzero digits to the right of the last displayed value, and whether to follow a convention or assume worst case (or typical case) regarding *computations* involving values with known numbers of significant digits, etc.]. Because it is somewhat arcane, further discussion of significant digits with binary arithmetic is not developed.

In this section we turn our attention constructively to two problems with the conventional whole-number approach to significant digits. The proposed solutions to these problems accomplishes two things, in order. The first solution strengthens the notion of significant digits by introducing the concept and explicit calculation of *fractions* of significant digits. The second solution provides a sanity-check reminder of the importance of thinking of precision in terms of relative uncertainty, not just absolute uncertainty. The result is a bounding relationship relating the relative uncertainty to the minimum and maximum number of significant digits.

9.4.1 Order-of-Magnitude Uncertainty Range: Fractions of Significant Digits

We first consider:

Problem 1   For any set of numbers that all have the same number of significant digits (under the conventional definition), the uncertainty range in those numbers can nearly span an order of magnitude (i.e., a 10:1 ratio between largest uncertainty and smallest uncertainty). Consider the two extremes: a wide uncertainty interval and a narrow uncertainty interval. Using the example $M = v.xyz \times 10^k$, the information content in z can be low: M may have an uncertainty range up to but not including $\pm 0.005 \times 10^k$ (i.e., its total uncertainty interval width could be nearly 0.01×10^k). Any wider band would have three or fewer significant digits, because with a wider band, z could plausibly be any digit, zero through nine. Alternatively, the information content of z may be high: M may have uncertainty as small as $\pm 0.0005 \times 10^k$ (i.e., its uncertainty interval could be as narrow as 0.001×10^k). Any narrower band would mean that M would have more than four significant digits, so the next digit should be displayed. Thus the total uncertainty interval width can be as wide as nearly 0.01×10^k and as narrow as 0.001×10^k. This result holds for any value of k, and it can be generalized for any nonnegative number of significant digits.

Solution   Use *fractions* of significant digits. The solution we propose is to introduce the concept and computation of *fractions* of significant digits:

94 SIGNIFICANT DIGITS

Definition of Significant Digits, Including Fractions A number expressed in scientific notation (i.e., $v.xyz \times 10^k$, with mantissa $v.xyz$ and exponent k) is said to have w *significant digits* if the mantissa has an uncertainty interval of width 5×10^{-w}. For example, if w is a whole number, the mantissa has $w - 1$ digits to the right of the decimal point. Note that there is nothing that restricts w to be a whole number.

This definition implies that a number has w significant digits when the mantissa has a *total* uncertainty interval width of $(5 + 5) \times 10^{-w} = 10 \times 10^{-w}$. We shall find that the inverse-function relationship between uncertainty range and significant digits is the easiest way to compute fractional significant digits:

Corollary Definition of Significant Digits, Including Fractions When the mantissa of a number has a total uncertainty interval width of q, the number has $w = -\log_{10}(q) + 1$ significant digits. Note that q is required to be positive, but w is not.

Some conventional examples of applying the inverse relationship examples (with digit $v = 4$) are:

$(4.xyz \pm 0.0005) \times 10^k$ has $-\log_{10}(2 \times 0.0005) + 1$ = 4 significant digits

$(4.xyz \pm 0.005) \times 10^k$ has $-\log_{10}(2 \times 0.005) + 1$ = 3 significant digits

Since there is no reason to confine w to whole numbers,

$(4.xyz \pm 0.0008) \times 10^k$ has $-\log_{10}(2 \times 0.0008) + 1$ = 3.8 significant digits

$(4.xyz \pm 0.002) \times 10^k$ has $-\log_{10}(2 \times 0.002) + 1$ = 3.4 significant digits

$(4.xyz \pm 0.004) \times 10^k$ has $-\log_{10}(2 \times 0.004) + 1$ = 3.1 significant digits

Thus a commonsense result is achieved for any given measurement value with an uncertainty interval: As the width of the uncertainty interval decreases (perhaps due to more precise measurement systems or calibrations), the number of significant digits increases. The conventional definition of a whole number significant digits results in a step function for the relationship between significant digits and uncertainty. Under this new, proposed definition, we have instead a smooth, monotonic function that exactly passes through the well-defined (whole-number) points of the conventional definition, but interpolates in-between those points (see Figure 9.2).

9.4.2 Order-of-Magnitude Relative Uncertainty: Bounding Relationship

We next consider a limitation (rather than a problem) of significant digits: the fact that significant digits are based on, and reflect, the absolute magnitude of uncertainty in a measurement, not the relative uncertainty. Thus two different measurements, $M1$

9.4 PAIR OF PROBLEMS DUE TO HAVING TEN FINGERS AND THEIR SOLUTIONS 95

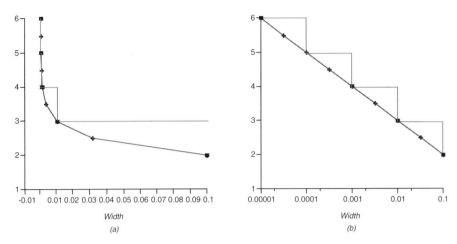

Figure 9.2 Conventional (whole numbers) significant digits and fractional significant digits versus width of uncertainty interval: (a) linear horizontal scale; (b) log horizontal scale.

and $M2$, with the same number of significant digits (possibly fractional), may have very different levels of *relative* measurement uncertainty (RMU). This limitation can be stated succinctly as follows:

Problem 2 The mantissa (e.g., $v.xyz$ for measurement M) can range over an order of magnitude. M could be any number from 1.000×10^k to 9.999×10^k. If the uncertainty in M is fixed, different values of M will have different relative uncertainty, but M will still have four significant digits.

Some examples with 3.4 significant digits (but different levels of RMU) follow:

(a) $(4.xyz \pm 0.002) \times 10^k$ has 3.4 significant digits; RMU = $0.002/4.xyz$ = 0.05% [$3.4 = -\log_{10}(2 \times 0.002) + 1$, as seen above]
(b) $(1.000 \pm 0.002) \times 10^k$ also has 3.4 significant digits; RMU = $0.002/1.000$ = 0.2%
(c) $(9.999 \pm 0.002) \times 10^k$ also has 3.4 significant digits; RMU = $0.002/9.999$ = 0.02%

These examples show that *relative* measurement uncertainty can span nearly an order of magnitude if the measurement values span nearly an order of magnitude, holding the uncertainty interval constant (i.e., the same number of significant digits).

Relationship Between Significant Digits and RMU Generalizing examples (b) and (c), it can be shown that a number with w significant digits satisfies the following inequality:

$$0.5 \times 10^{-w} < \text{RMU} \leq 5 \times 10^{-w} \qquad (9.1)$$

Thus, whereas absolute uncertainty and significant digits have an exact functional relationship [$w = -\log_{10}(q) + 1$], RMU and significant digits have a *bounding relationship* with one another (i.e., they are related through an inequality).

For example, consider a number with one significant digit ($w = 1$). It satisfies the relationship 5% ≤ RMU < 50%. If the RMU exceeds 50%, there cannot be as much as one significant digit, and in fact there is only one value with RMU = 50% that has one significant digit: 1×10^k, which could be written, $(1 \pm 0.5) \times 10^k$. All other values with RMU = 50% have less than one full significant digit [e.g., $(2 \pm 1) \times 10^k, \ldots, (9 \pm 4.5) \times 10^k$ each have a fraction of a significant digit]. There is only one value with (approximately) 5% RMU that has one significant digit: $(9.99 \ldots \pm 0.5) \times 10^k$. All other values with RMU = 5% have *more* than one significant digit [e.g., $(1 \pm 0.05) \times 10^k, \ldots, (9 \pm 0.45) \times 10^k$ have between one and two significant digits].

Another way to express the bounding relationship is to transform inequality (9.1) by multiplying through by 2 and taking $\log_{10}$:

$$-w < \log_{10}(2\text{RMU}) \leq -w + 1$$

Then, defining the signal-to-noise ratio,

$$\text{SNR} = -\log_{10}(2\text{RMU})$$

and multiplying through by -1, we get

$$w \geq \text{SNR} \geq w - 1$$

Adding 1 to each part of the inequality, then combining with the original ineqality, provides

$$w + 1 > \text{SNR} + 1 \geq w \geq \text{SNR} > w - 1 \qquad (9.2)$$

Therefore, for a measurement to have at least one significant digit ($w = 1$), we must have 1 > SNR, therefore, RMU must be no more than 50%. On the other side of the boundary, if relative uncertainty is less than 5%, the number will necessarily have more than one significant digit (possibly two).

9.5 DEFINITION OF DETECTION: AT LEAST ZERO SIGNIFICANT DIGITS

In this section we integrate the results of Sections 9.2 to 9.4 to produce a rather startling, but sensible, interpretation of detection. This is the second step of tying together some of the disparate topics mentioned at the beginning of the chapter: measurement uncertainty (and implicitly, calibration), significant digits, and detection.

In Section 9.2 a simple general expression was presented for characterizing measurement uncertainty:

$$\text{measurement uncertainty} = \pm \frac{s}{b} F(x)$$

9.5 DEFINITION OF DETECTION: AT LEAST ZERO SIGNIFICANT DIGITS

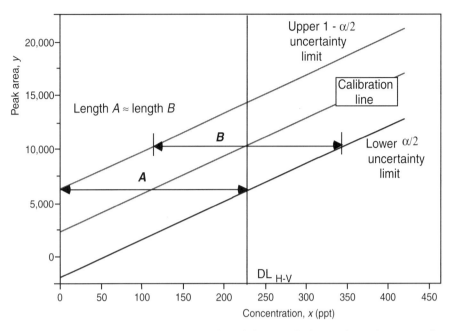

Figure 9.3 Computation of a Hubaux–Vos–based detection limit, ensuring at least zero significant digits.

where for sufficiently large and well-balanced calibration designs, $F(x)$ is a function that is nearly constant with respect to x. Given that $F(x)$ is nearly constant for either prediction or statistical tolerance intervals, consider the behavior of measurement uncertainty and its relation to detection near the low end of the measurement range (see Figure 9.3). We will also assume that we are interested in using symmetric two-sided intervals. The Hubaux–Vos approach to detection (and analogously, the approach based on statistical tolerance intervals) involves the steps below, presented graphically, but with equivalent mathematical development:

1. Compute the calibration line by ordinary least squares.
2. Compute the prediction interval about that line.
3. Find where the upper prediction interval (i.e., upper uncertainty limit) intersects the vertical (response) axis. This is the detection threshold. Draw a horizontal line through that point (line segment A in Figure 9.3).
4. Find where the threshold line intersects the lower prediction interval (i.e., lower uncertainty limit). This concentration is the Hubaux–Vos detection limit, $DL_{H\text{-}V}$.
5. Compute the uncertainty interval, in concentration units, about $DL_{H\text{-}V}$ (line segment B in Figure 9.3).

98 SIGNIFICANT DIGITS

We now make the following observations:

- The assumption that $F(x)$ is nearly constant with respect to x means that line segments A and B are of approximately equal length. This holds for prediction intervals (as in Hubaux–Vos) and for statistical tolerance intervals.
- Since the uncertainty interval is symmetric about the calibration line, the width of the interval at the detection limit is approximately equal to $\pm 50\% \times$ (detection limit).
- Therefore, at the detection limit, RMU $\approx 50\%$.
- Detection limits can be lowered by replacing individual measurements with the mean of replicate measurements, as well as by using alternative analytical techniques, such as different concentrations or dilutions, different solvents, and different retention times.

Applying RMU $\pm 50\%$ to inequality (9.1), we get

$$0.5 \times 10^{-w} < 0.5 \leq 5 \times 10^{-w}$$

and isolating w, we find that $w \geq 0$. In other words, *at the detection limit, there are at least zero significant digits.* This suggests an alternative definition for the detection limit: The detection limit (DL_{0D}) is the minimum concentration at which (at a specified confidence level) there are at least zero significant digits in the measurement. The subscript, 0D stands for "zero significant digits."

This result is not as startling as it may at first seem (although it begs the question, "How much does one pay a laboratory for providing measurements with at least zero significant digits?"). First, if one accepts the concept of a fraction of a significant digit, one should accept the notion of a *dwindling fraction* of a significant digit, approaching zero. Second, the concept of a detection limit can be described (consistent with the development in previous chapters) as that concentration that is so low that one can marginally make the most basic, binary decision (at an acceptable confidence level): yes, the analyte is present ("1"), or no, it is not present ("0"). There is no lower nonzero level of information content. Third, we note that the result does not depend on what type of interval is selected, provided that it is symmetric and two-sided (if a statistical tolerance interval is used, the symmetry should extend to the population as well as the confidence level).

9.6 DEFINITION OF QUANTITATION: AT LEAST ONE SIGNIFICANT DIGIT

As discussed in Chapter 2, there are diverse names, definitions, and procedures to estimate quantitation limits (QLs). In this section we develop a different approach, one similar to that for the detection limit, DL_{0D}, developed in Section 9.5. The bounding relationships (9.1) and (9.2) are used to provide a sensible definition of the quantitation limit (and quantitation), as follows: The quantitation limit (QL_{1D})

9.6 DEFINITION OF QUANTITATION: AT LEAST ONE SIGNIFICANT DIGIT

is the lowest concentration at or above which measurements have at least one significant digit (at specified confidence, $1 - \alpha$), and equivalently, have limited relative measurement uncertainty, RMU = 5%. Equivalently, there is a lower bound on the signal-to-noise-ratio (SNR) ≥ 1 at the QL. Note that in QL_{1D}, the subscript 1D stands for "one significant digit."

9.6.1 Use of QL_{1D}

This definition has the unfortunate quality that it is appealing to *customers* of analytical measurements, but it can be anathema to *providers* of such measurements. From the customers' perspective, who wouldn't want to have assurance of at least one significant digit in a paid-for measurement? Most laboratories provide two, three, or more digits in their lab results reports, all the way down to their stated detection limit. Unfortunately, most such low-level results are no more than random numbers.

In fact, many analytical methods (particularly soil and air methods, and some methods for measuring organics in water) cannot achieve relative standard deviation (RSD) = 10%, so a traditionally computed (10σ) quantitation limit (QL) is unattainable. Most laboratories cannot achieve even one significant digit near their QL, far less near their stated detection limit. The requirement to have RMU $\leq$ 5% at QL_{1D} implies that the quantitation limit must be at approximately $2[1/(5\%)]\sigma = 40\sigma$, or higher (using 2 as the smallest plausible factor for a prediction interval or statistical tolerance interval). This corresponds to RSD = 2.5%, which is four times as strict as the 10σ QL.

This alternative definition of the quantitation limit has a statistical basis but is also presented with the intent to bring some hard realism into reporting practices for analytical measurements. If a customer is paying for low-level measurements, the laboratory is publishing measurements with two, three, or four digits, and the customer is using those measurements to make decisions, perhaps with the legal obligation to report those measurements to a regulatory or enforcement agency, all parties should be fully aware of the quality of the measurements. Publishing a quantitation limit below which one significant digit cannot be assured is a potentially helpful precautionary act.

9.6.2 Development of QL_{1D}

Recall inequality (9.1), involving relative measurement uncertainty and number of significant digits (w):

$$0.5 \times 10^{-w} < \text{RMU} \leq 5 \times 10^{-w}$$

As seen above, this inequality asserts that a measurement with one significant digit ($w = 1$) satisfies the inequality, $0.05 < \text{RMU} \leq 0.5$, and to ensure at least one significant digit, one must have RMU <0.05, so one can define QL_{1D} as the lowest concentration for which this bound holds. Equivalently, recall inequality (9.2), involving the signal-to-noise ratio and the number of significant digits:

$$w + 1 > \text{SNR} + 1 \geq w \qquad \text{SNR} \geq w - 1$$

A measurement with one significant digit satisfies the inequality, $1 \geq \text{SNR} > 0$, and to ensure at least one significant digit, one must have $\text{SNR} \geq 1$.

Therefore, to compute QL_{1D}, the following steps are taken, making the same assumptions as for the detection limit, DL_{0D} (nearly constant, symmetric, two-sided intervals, and acceptable confidence level, $1 - \alpha$, and acceptable P, if a statistical tolerance interval is used):

1. Compute the calibration line by ordinary least squares.
2. Compute the prediction interval or statistical tolerance interval about that line.
3. Compute the relative measurement uncertainty (RMU) or the signal-to-noise ratio (SNR), as a function of concentration (x):

$$\text{RMU}(x) = \pm \frac{s}{b} \frac{F(x)}{x}$$

where $F(x)$ is defined according to the interval used,
or

$$\text{SNR}(x) = \log_{10}[2\text{RMU}(x)]$$

4. Find QL_{1D}, which is the solution to $\text{RMU}(x) = 5\%$, or (equivalently) $\text{SNR}(x) = 1$.

9.6.3 Graphical Example of Computing QL_{1D}

Figure 9.4 is based on real data and uses a 94% confidence prediction interval about the calibration line. The figure illustrates how $\text{RMU}(x)$ and $\text{SNR}(x)$ change,

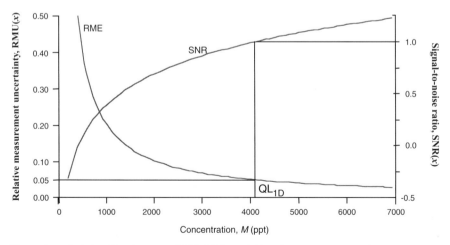

Figure 9.4 Relationships between RMU, SNR, and QL_{1D} for measurement of aroclor 1248 in river water by gas chromatography ($n = 34$, $\alpha = 6\%$).

monotonically, with x, and how in order to ensure measurements with at least one significant digit, they achieve the required equality with their lower bounds, at QL_{1D}. We see that $QL_{1D} \approx 4100$ ppt ≈ 4 ppb. We would use this estimate to say that at or above 4 ppb, the first digit of a reported measurement is correct with 94% confidence (after rounding to one digit). As with any statistical interval, the actual coverage is unknown and will vary from calibration to calibration, but on average, 94% or more of the reported measurements will have the correct first digit, after rounding.

Note that for some measurements, QL_{1D} may not exist. The author (D.C.) was involved in evaluating one analytical measurement system that had maximum SNR $= \frac{1}{3}$ (at some selected confidence level). The chemist supervising the use of the system tacitly approved reporting measurements to three places, but unfortunately it turned out that *not even a single significant digit could be obtained*. Having a third of a significant digit *is* superior to mere detection; SNR $= \frac{1}{3}$ implies that RMU $\approx 23\%$, so one could report, for example, 4 ± 1. Note, however, that just as is the case with detection limits, quantitation limits can be reduced by replacing individual measurements with the mean of replicate measurements, as well as by using alternative analytical techniques, such as different concentrations or dilutions, different solvents, and different retention times.

9.7 REPORTING MEASUREMENT AND UNCERTAINTY

In previous sections in this chapter, we have tied together significant digits, detection limits, and quantitation limits. The link between these has been measurement uncertainty, based on intervals that can be calculated from ordinary least-squares fitting for straight-line calibration. In this section we recommend formats for reporting measurements with uncertainty. These formats are technically superior to the use of significant digits, detection limits, and quantitation limits, but they are not as simple.

9.7.1 Adorning a Measurement with Uncertainty

Measurements should be reported in such a way that people use them to make correct decisions. An unadorned measurement (i.e., without any indication of uncertainty, e.g., 7 ppt cyanide) fails this principle. Better is a measurement accompanied by a statistically sound detection limit and quantitation limit. For example, a lab could report: 7 ppt cyanide; $DL_{0D} = 0.1$ ppt; $QL_{1D} = 1$ ppt. In this case it would be clear that the concentration was at quantifiable level, and with high confidence, the "7" digit is correct. It would be a little more difficult to make decisions if the report was: 7 ppt cyanide; $DL_{0D} = 1$ ppt; $QL_{1D} = 10$ ppt. In this case there is very high confidence that cyanide is present in the example, but the measurement uncertainty is too high to be confident that the "7" digit is correct.

An even better reporting format is a measurement with an explained uncertainty interval. For example: 7 ppt cyanide, with ± 1 ppt prediction interval at 95%

confidence. The latter reporting format can help the customer limit his or her risk comparing the measurement to a standard, such as a regulatory limit. As seen in Chapter 3, an interval is equivalent to a hypothesis, provided that the two are matched with respect to distribution assumption, type of test interval, selection of α (and β, if relevant), and one-sided versus two-sided. Therefore, assuming the right matching, we can correctly judge 7 ppt to be statistically significantly higher than 5 ppt.

9.7.2 Measurement, Standard Deviation, Radical, and Degrees of Freedom: MSRD Format

There is a way, beyond the formats presented above, to maximize the usefulness of the measurement to the customer, at the cost of greater complexity. One such approach we propose is the *MSRD format,* with the following definitions:

- S = standard deviation is the unit uncertainty that applies to M.
- R = radical factor that scales S according to the design

$$R = \sqrt{\frac{1 + (1/n) + (x - \bar{x})^2}{Sxx}}$$

- D = degrees of freedom is the number of degrees of freedom associated with S; usually, $D = n - 2$, where n is the number of points in the calibration study.

When measurements are reported in the MSRD format, the customer can construct whatever uncertainty interval is relevant for him or her. For example, a lab could report:

$$M = 7 \text{ ppt}; S = 1 \text{ ppt}; R = 1.01; D = 18.$$

Recall from Section 9.1 that the general form for uncertainty intervals is

$$x = \frac{y - a}{b} \pm \frac{S}{b} F(x)$$

where for prediction intervals,

$$F(x) = t_{[1-\alpha/2, n-2]} \sqrt{\frac{1 + (1/n) + (x - \bar{x})^2}{Sxx}}$$

and for statistical tolerance intervals,

$$F(x) = 2F_{2, n-2}^{1-\alpha/2} \sqrt{\frac{(1/n) + (x - \bar{x})^2}{Sxx}} + \Phi(P) \sqrt{\frac{n - 2}{\alpha/2 \chi_{n-2}^2}}$$

9.7 REPORTING MEASUREMENT AND UNCERTAINTY

From these formulas and the values of M, S, R, and D, the customer can create any desired interval. For example, a 95% confidence uncertainty interval, based on a prediction interval, is as follows: Predicted value is likely to fall within

$$M \pm SR\, t_{[D, 0.975,]} = 7 \pm 1 \times 1.01 \times 2.1 = 7 \pm 2.1 \text{ ppt at 95\% confidence}$$

Alternatively, the customer can create a 95% confidence uncertainty interval for the central 90% of the population; for example, based on a statistical tolerance interval, it is as follows: The central 90% is likely to fall within

$$M \pm S \left[2F_{2,D}^{0.975} \sqrt{R^2 - 1} + \Phi(0.9) \sqrt{\frac{D}{{}^{0.025}\chi_D^2}} \right]$$

$$= 7 \pm 1 \times [2 \times 4.56 \times 0.1 + 1.28 \times 1.48] = 7 \pm 2.8 \text{ ppt}$$

for 95% confidence of enclosing 90% of the population.

10

EXPERIMENTAL DESIGN OF DETECTION AND QUANTIFICATION LIMIT STUDIES AND RELATED STUDIES

10.1 INTRODUCTION

Detection and quantitation studies are among the simplest kinds of studies to design in that they are single-factor (the only independent variable is spiking concentration), the appropriate response is known [e.g., peak area, log(*PA*), or voltage], and there is usually some prestudy knowledge of the relationship between the response and analyte concentration (e.g., straight line, quadratic, or rational function). One of the most basic decisions to be made when contemplating a measurement capability study for an analytical method is whether to make it a within-laboratory study or a between-laboratory study. (We prefer the terms *within-laboratory* and *between-laboratory* in place of *intralaboratory* and *interlaboratory* because of the similarity of the latter two terms, easily leading to confusion.) In some cases, the choice of type of study is clearly dictated by the use of results. For example, a between-laboratory study is required to produce a detection limit or quantitation limit for a *population* of laboratories represented by those in the study. More generally, a between-laboratory study is required for demonstration of the practicality, precision, and bias of a method for possible inclusion as an accepted standard in a voluntary consensus standards organization such as ASTM, because such standards are used in laboratories across the country. On the other hand, a within-laboratory study is sufficient to produce a detection limit or quantitation limit for the particular laboratory studied. A within-laboratory study is called for if laboratory management wants to determine whether there is consistency across analysts and pieces of comparable equipment and if target values are needed for routine QA/QC. If a university or analytical services company has a few different laboratories of unknown capability, an intermediate study might be appropriate: All laboratories might be included, but they do not represent any population other than themselves.

10.2 BASIC CONSIDERATIONS AND RECOMMENDATIONS

The key issue in study design (whether for a within-laboratory study or for a between-laboratory study) is the selection of concentrations. For detection and quantitation, the guidelines given below should be followed (see also ASTM, 1997, 1998, 2000).

1. Replicate blanks should be included.
2. There should be at least one spiked concentration below the lowest limit to be computed. Since that limit is generally not known prior to the study, a conservative estimate should be used. Call the lowest spike concentration, C_1.
3. The maximum spike concentration (C_{max}) should be selected based on the objectives of the study. In particular, it should be decided whether the study should attempt to span the working range of the method or should just span a trace-concentration range. Generally, there should be replicates of the highest-concentration sample.
4. Intermediate spike concentrations should be selected. The number and distribution of concentrations depend on objectives, but for a detection/quantitation study, there should be more samples at low concentrations than at high. Without loss of generality, suppose that the estimated quantitation limit (QL) is 1 ppz (part per zillion). Recommended designs are given below, but the number of spike concentrations is somewhat arbitrary. More or fewer spike concentrations could be used, provided that there are at least five:
   (a) The semigeometric design (used for wide-range studies, where $C_{max}/C_1 >> 10$) has five or more replicates at each spike level (e.g., 0, $\frac{1}{2}$, 1, 2, 4, 8, 16, 32 *or* 0, $\frac{1}{3}$, 1, 3, 10, 40, 200, 1000 ppz). Note the geometric progression.
   (b) The equispaced design has five or more replicates at each spike level (e.g., 0, $\frac{1}{2}$, 1, 1.5, 2, 2.5, 3 *or* 0, $\frac{1}{2}$, 1, 2, 3, 4, 5, 6 *or* 0, $\frac{1}{2}$, 1, 5, 10, 15, 20, 25 ppz). Note that all of these designs are referred to as *equispaced* even though the spacing between concentrations is equal only for the first of these designs.
   Note: Minor perturbations can be made to the "nice round numbers" of these designs, especially to make it less obvious to study participants what the results

"should be." For example, for the semigeometric design, one might choose the concentrations 0, 0.6, 1.05, 1.98, 4.07, 7.7, 15.9, 32.1.

    (c) The Youden pair design (discussed in detail in Section 10.4) has one or more blanks plus samples at pairs of spiked concentrations based on the semigeometric or equispaced design. Each design spike concentration is replaced by a pair at a small proportional deviation ($<\pm 10\%$). For example, a small semigeometric design might become: 0, 0.55, 0.6, 1, 1.1, 1.9, 2, 4, 4.3, 7.5, 8.

See Figure 10.1 for a graphical depiction of an example of these designs.

5. Rules for analyte identification, whether or not there will be blank subtraction, and result-reporting procedures should be determined and communicated prior to the study.
6. Rules should be set for how to report nondetects and how to identify and deal with outliers.
7. Samples should be run blind (see Section 10.8). Note that this might be rendered impossible if special instructions must be followed to ensure the integrity of the data.
8. Samples should be run in random order.
9. Samples should be run by several different analysts on different days, using as many systems as is practical.

There are many other chemistry-dictated and logistical aspects to planning a successful within-laboratory or interlaboratory detection/quantitation study (see ASTM, 1998). A few guidelines are:

1. Develop written and tested sample-preparation procedures, taking into account holding times and preservatives. Consider the pros and cons of having each laboratory do its own spiking, dilution, or concentration.
2. Ensure comparability of matrices that will be treated as identical; one laboratory's DI water may not be the same as another's.
3. Ensure homogeneity and purity of the analyte. Will traceable standards be used?
4. If designing a study that involves multiple analytes analyzed by the same method in the same matrix, consider a cocktail-mix strategy: Let every sample contain a mixture of analytes, preferably at unequal and unpredictable levels. For example, concentrations of analytes A, B, and C can be represented as a triplet [e.g., (10, 0, 5) would represent a sample with 10 ppb of analyte A, 0 ppb of analyte B (i.e., a blank sample with respect to B), and 5 ppb of analyte C]. A semigeometric, cocktail-mix design might then be: (4,1,2); (0,2,0); ($\frac{1}{2}$,8,1); (1,$\frac{1}{2}$,4); (2,0,8); (8,0,$\frac{1}{2}$). Naturally, forethought must be given to possible chemical reactions or physical interactions (e.g., flocculation) of analytes and whether their responses can be independently identified (e.g., separable peaks).

Selection of Calibration Standards The first two design recommendations made above are also recommended for calibration standards:

- The semigeometric design [Note that (except for the use of a blank), the semigeometric design is equispaced in log(concentration).]
- The equispaced design.

The semigeometric design is generally preferred, because in order to span orders of magnitude in concentration (generally required for most analytical methods; three or more orders of magnitude are not uncommon), it uses far fewer different levels than does the equispaced design.

Some design considerations for selection of calibration standards are slightly different than for within- or between-laboratory studies. In particular:

- Replicates are not needed unless the calibration function will be evaluated for lack of fit or for the significance of coefficients.
- Usually, the matrix for a calibration standard is different than for a study sample. The former is typically a solvent or highly purified matrix, eliminating the need for sample preparation (e.g., extraction, concentration, or dilution), whereas the latter may be a matrix of practical interest, perhaps causing interference problems with the method and analyte(s) being studied.
- A Youden pair design is *not* recommended for calibration, since the advantages obtained by pairing concentrations in a study are not enjoyed in calibration. Unlike analysis of data from a within- or between-laboratory study, there is no good way to take advantage of the degree of freedom assumed available for precision calculation, by virtue of the concentrations being close. Therefore, pairing standards results in an unnecessary doubling of the number of spike concentrations.

General Guidelines for Selection of Calibration Standards

- The minimum number of levels will be dictated by how the calibration data will be used. The key factor is how inverse prediction will be carried out:

  — If a kth-order polynomial will be fit, there must be calibration standards for at least $k + 1$ different levels of concentration, and more levels are needed if the function is to be evaluated for lack of fit (Caulcutt and Boddy, 1983). A good rule of thumb is to have at least $K + 3$ different levels. Replicate measurements are needed at half or more of the levels if the function is to be evaluated for lack of fit or significance of coefficients. Note that unless there is a good theoretical reason to do so, it is dangerous to fit high-order (e.g., quartic, quintic) polynomials. High-order polynomials tend to overfit data, resulting in high absolute bias in prediction (interpolating between calibration standards). Note that high R^2 is *not* a sufficient indicator of good prediction. Coefficients of high-order polynomials are also vulnerable to outliers.

- If a k-degree spline will be fit, or the user will do piece-wise fitting with k-degree polynomials, there must be at least $k + 1$ different levels of concentration for each segment of the spline. This minimum number of levels should be increased for each spline constraint. For example, there might be forced equality of the slopes of the fit at the boundaries between regions.
- If inverse prediction will be done by table look-up with linear interpolation, calibration standards should be as close together as possible.

• Calibration standards should have an accuracy pedigree (exceptions are discussed below). This is usually accomplished by traceability, ultimately to a national or international standards body such as NIST, ISO, or BSI. Traceability may be indirect; the standards used for routine calibration might be traceable to secondary standards, which are themselves traceable to primary standards. For years, The National Institute of Standards and Technology has promoted a ladder of traceability, with increasingly strict accuracy bounds as one climbs the ladder. For examples refer to the MTS web page, www.mts.com, where NIST measurement traceability diagrams are provided for physical measurements. Note that accuracy assessment or claims should include the error due to bias *and* imprecision.
• Traceable calibration standards may not exist. For example, PCBs are no longer manufactured, and the ordinary historical Monsanto production runs of various PCB mixtures serve today as de facto standards. Another example: A company might develop—and make standards from—a proprietary chemical mixture with a distinctive analytical "signature" (e.g., combination of peak ratios). These standards can be valuable for calibration even though the chemical composition may not be fully known.
• Calibration standards may be consumable (i.e., used up in the process of measurement); this is generally the case for analytical measurements. If the standards are consumable, then at any given time there should be a *vintage-mix* redundancy in standards, to protect against the possibility of an explainable shift in measured values.

10.3 VARIANCE COMPONENTS

Variance components analysis (VCA) is a special case of analysis of variance (ANOVA). VCA is used to study sources of variance in an analytical method such as matrix, analyst, instrument, and laboratory. In VCA, the factors are considered random factors, or equivalently, they are factors with random effects. This contrasts with the more familiar fixed-effect factors found in traditional ANOVA. An ANOVA with both random and fixed factors is called a *mixed-factor* or *"mixed-effects"* ANOVA. The specification of a model that contains both types of effects is called a *mixed model* (Caulcutt and Boddy, 1983).

Random factors are factors that introduce noise or variability in a study but for which there is a potentially large population of factor levels; there is no particular interest in estimating the effects of individual levels, just the amount of variation (i.e., variance or standard deviation) they contribute. By contrast, fixed factors have a small number of levels, and typically an objective of the ANOVA is to judge the statistical significance and magnitude of each effect (i.e., difference in response due to change in level).

For example, a large, between-laboratory study of arsenic by GC/MS might be designed with the matrix as a factor (e.g., local groundwater). Each laboratory may select its own unique groundwater; each one considered a different (unordered) level of the groundwater factor. A purpose of the study may be to estimate the measurement variation attributable to groundwater, in which case groundwater would be treated as a random factor. For this type of study, there may be no interest in estimating the bias or the significance of the bias introduced by any one laboratory's groundwater compared to the overall mean or compared to the groundwater of another laboratory. Nor are we interested (for this analysis) in *what* is causing such bias. Instead, we want to know that the measurement variation due to groundwater has a standard deviation of 15 ppb (i.e., 60% of the total variance).

By contrast, a similar study (or even the same study) may involve just two laboratories; suppose that there are two EPA laboratories. From the EPA's perspective, the EPA-only subset of the data may constitute a small study of between-EPA-laboratory bias. Fixed-effect ANOVA may be used to estimate the magnitude and statistical significance of the bias between the two laboratories, with an eye to eliminating the cause, or at least knowing how to adjust data from different laboratories. Hence, in this case, the factor labeling "fixed" versus "random"—which affects the type of ANOVA carried out—is in the eye of the beholder (i.e., it depends on the context).

A variance components analysis with random-factor ANOVA typically produces an outcome such as shown in Table 10.1. See standard statistics textbooks for more detail of VCA, and Milliken and Johnson (1984) for an especially thorough treatment. We see in this example that groundwater is the factor that contributes more to total variation than any other factor studied, and this is *not* an artifact of the relatively large number of levels. Note that expected values of VCA results do not

Table 10.1 Variance component analysis

Factor	Levels (ppb)	Standard Deviation	Variance (ppb^2)	% Total Variation	Cumulative %
Groundwater	9 locations	3.87	14	56	56
Instrument	2 or 3 per laboratory, 24 total	2.65	7	28	84
Analyst	2 per laboratory, 18 total	1.73	3	12	96
Residual (error)	—	1.00	1	4	100

depend on the number of levels, provided that the levels chosen are representative of the population of interest. In other words, more levels do not result in a higher (or a lower) percent of total variance. Also, unlike naïve analyses of data from studies with random factors, VCA separates the estimate of the variance contribution of different factors, so that one factor cannot "mark" another.

Recommendations and Cautions with VCA Variance component analysis is based on these assumptions:

- The levels chosen for any random factor are representative of a larger population, whether an actual population or merely hypothetical. Thus levels should be chosen at random. For example, analysts could be selected by choosing every other one in an alphabetical listing. It is inappropriate and misleading to select just the best *or* just the worst analysts—*or* the best *and* worst, *or* for that matter just those with an *average* level of experience or training.
- Technically, the factor effects do not have to be distributed according to a normal distribution, but the *interpretation* of VCA results does depend implicitly on the normal distribution, so the normal distribution must be plausible. Note that it is not usually possible to test this assumption, due to the limited number of levels for any factor. Note also that the assumption is *not* that the distribution of *all measurements* is normal.
- Measurement errors are independent and identically distributed (i.i.d.). This requires that apart from the factors included in the study, all other influential factors are held constant or have their effects neutralized. It is important to realize that VCA estimates of standard deviations can be corrupted to an arbitrary degree by outliers, shifts, and trends. Thus outliers should be prescreened, and removed if possible.
- A key realization of the i.i.d. assumption is that the measurements do not trend over time. Failure of this assumption is one of the most common weaknesses of variance components studies. Shifts and trends should be investigated—preferably in a trial run preceding the actual study—and shifting or trending data should be removed, or the offending nuisance factor might have to be constrained and the study rerun. Note that time is *not* a factor, but it can serve as a surrogate for one or more factors that change with use or with elapsed time (e.g., GC column contamination, filter saturation).

VCA also has some limitations:

- The standard deviation estimates from VCA are noisy; it is rare for them to have even one significant digit. Unfortunately, VCA confidence intervals are too complicated to be used in routine analysis, although modern statistical software packages will compute them.
- A noisy estimate of a low variance component can actually be negative when the typical analysis method for VCA is used: the *method of moments* (Milliken

and Johnson, 1984). Any negative estimate should be set to zero. Better still, an alternative analysis method, *restricted maximum likelihood,* should be used if possible; it does not produce negative estimates. Note that it has been observed by Hocking et al. (1989) that a negative variance component is often a symptom of one or more outlier or the violation of some other VCA assumption(s).

10.4 YOUDEN PAIRS

The Youden pair design (or Youden design) is a time-honored but controversial design named after the renowned chemist/statistician, W. J. Youden (Youden and Steiner, 1975). The Youden pair design is typically used for a between- or within-laboratory study, such as prescribed by the ASTM standard D2777, "Determination of Precision and Bias of Applicable Test Methods of Committee D19 on Water" (ASTM, 1998). The design consists of concentration pairs, as described in Section 10.2 and shown in Figure 10.1.

Assumptions of Youden Pair Designs

- It is assumed that measurements at two close concentrations have the same variability. Thus an extra degree of freedom can be "squeezed" from each pair: Not only can each measurement be used to determine the response versus concentration relationship, but the difference between the measurements in each pair can be used as a single-degree-of-freedom estimate of the measurement error standard deviation at the two concentrations. Note that for analytical data, the standard deviation typically *increases* with increasing concentration, so this assumption should be challenged, especially if the paired concentrations are at their widest allowed separation ($\pm 10\%$).
- If a Youden pair study is conducted with multiple laboratories (analogously, multiple instruments, multiple analysts, or multiple matrices), for any concentration it is assumed that all labs have the same precision. One should also be skeptical of the validity of this assumption, and unfortunately the

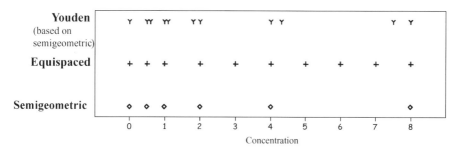

Figure 10.1 Recommended designs (replicates not shown).

Youden pair design with no replicates within a given lab provides no good way to test the assumption. That is because there is only one degree of freedom in the estimate of the standard deviation for each lab at each concentration.
- There must be a way to screen effectively both single-measurement outliers and labs that are biased (high or low). Unfortunately, both types of screening are more difficult with a Youden pair design than with a design containing replicates.

Graphical Presentation and Interpretation of Results from Youden Pair Designs Youden developed a simple but elegant graphical tool for assessing the results of a Youden pair study: the Youden plot. The Youden plot has one point per pair per lab, with a different symbol or color used for each lab. The measurement at the higher spike concentration is plotted versus the measurement of the lower spike concentration. An example is shown in Figure 10.2. In a nutshell, a Youden plot can reveal the following:

- Whether there are extreme outliers (isolated points in the plot, such as the rightmost box in the figure)

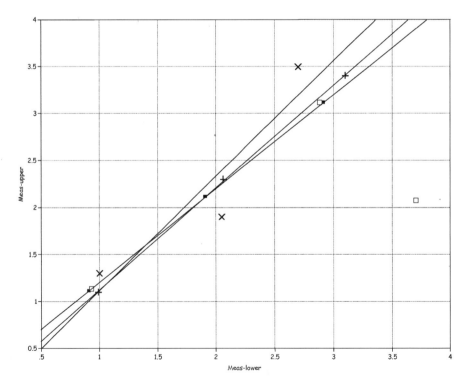

Figure 10.2 Straight-line fit for three of four laboratories: ., +, and × (but not for boxes, with outlying point, A).

- Whether one or more labs stand out as biased from the rest (the points are consistently higher or consistently lower, such as the " + " points compared to the "." points)
- Whether any labs stand out as having better or poorer precision (the points cluster more tightly, or they cluster less tightly, such as noisy X points)
- Whether any labs have measurement "reporting problems," such as an inability to detect or quantitate below a certain concentration, or discretization (only round numbers are reported, e.g., 5, 10, and 15)

Quantitative Analysis of Results from Youden Pair Designs Complementing the graphical presentation of Youden pair design is formal statistical analysis. Youden Pair analysis is a special case of regression. Although it is becoming a historical artifact, one of the appealing aspects of a Youden Pair design is that the basic analysis can be done on a calculator. Youden demonstrated the relatively simple summations and other calculations that could be done to generate estimates of measurement precision and bias (see Youden and Steiner, 1975 and ASTM, 1998).

Because the Youden pair design is efficient (hence, lacks data redundancy), and because regression can hide a multitude of data sins, regression should never be done without graphical assessment. At issue are the usual questions: Are there outliers? If data are from a between-laboratory study, is there evidence that any labs are biased relative to the others. [(ASTM (1998) provides a way to detect such bias, and recommends that the data from suspect laboratories be eliminated from further analysis.]

10.5 MATRIX EFFECTS AND BLANKS

10.5.1 Matrix Effects

An analytical chemist's idea of analytical heaven might involve only deionized or ultrapure water. A hard, practical reality is that our world involves many different types of water: groundwater, drinking water, seawater, lake water, river water, ashpond overflow, noncontact steam plant cooling water, to name a few. What distinguishes these different types of water is different contaminant mixes, whose distinguishing characteristics might be resistivity, pH, or other chemical or physical properties. Similar statements can be made for soil and air, although it is harder to define the ideal matrices for those media.

What are the effects of different matrices on measurements; that is, what does matrix interference look like? One can easily identify four types of direct impact on the ideal-matrix relationship between response [e.g., log(peak area)] and spiking concentration. Let's suppose that for DI-water-based samples, there is a straight-line relationship between log(peak area) and arsenic spiking concentration for a GC/MS system. What might happen to that simple relationship when the matrix is changed, say for ultrapure to another water matrix?

- The slope (i.e., gain factor) could change. Usually, we would expect it to drop (i.e., a loss in sensitivity to the analyte).
- The intercept could change. Usually, we would expect it to increase; that is, the mean of multiple blank measurements would be statistically significantly higher than zero, either because some of the analyte is present as a background, or because the measurement system falsely picks up something in the matrix that contributes to the perceived concentration of analyte.
- The noise of the measurements about the straight line could change. Usually, we would expect the noise to increase because the matrix composition or concentrations of interfering chemicals are not likely to be tightly controlled—if they are controlled at all.
- The nature of the signal versus concentration relationship could change. For example, what was a straight line for DI water might be sigmoidal for ash-pond overflow.

The best way to address the issues of matrix effects practically is to run a two-stage study, as follows:

1. Select the base-case matrix. This might be DI water or another matrix commonly found in samples analyzed by the lab.
2. Run the base-case study. Typically, this will be a within-lab study, with replicate blanks and replicate samples at each of several concentrations.
3. Run samples of an alternative matrix (if any) at selected concentrations of interest. Typically, extreme concentrations will be the most informative. For example, at a minimum, one should run replicate blanks and replicate samples at high concentration.
4. Plot and compare the results for the two matrices. Direct comparisons should reveal whether there are biases (i.e., differences in mean measurements of blanks, or differences in mean measurements of high-concentration samples). Alternatively, the same type of comparison can be made between the estimated slopes and estimated intercepts.
5. Finally, compare the precision for the different matrices. If the measurement error of each of two matrices seems to have approximately constant standard deviations, an F-test can be used to test the statistical significance of the square of the ratio of the RMSEs after a straight-line fit (or whatever common model applies). If there are more than two matrices, a test such as Hartley's F-max test should be used (Milliken and Johnson, 1984).
6. If there are differences in means or precision, particularly if they are practically important and statistically significant, the accuracy and precision results of the base-case study (and/or the standard calibration) should not be quoted for samples with the alternative matrix. Countermeasures should be developed, depending on the nature of the inequivalence. These might be as simple as a system standardization, blank subtraction, or an alternative calibration.

10.5.2 Blanks

If at all possible, blanks should be included in detection, quantitation, or bias and precision studies. Including blanks has several advantages:

- Blanks help expose issues with analyte identification at trace-level concentrations, including possible problems with matrix interference.
- Blanks provide a direct opportunity to observe false positives.
- Blanks help "pin down" the bias and precision models; behavior seen in spiked samples does not always extrapolate down to zero concentration.

The concept of a blank is simple: a sample of pure matrix, with no analyte present. The measurement of such a blank should ideally produce the result: "Concentration = 0%." Unfortunately, analytical chemistry is much more complicated, and care should be taken to select the type of blank(s) most appropriate for the study, those best satisfying study objectives. Although it is true that DI water can have low levels (but not necessarily zero) of any one of several possible contaminants, there is no such guarantee for other types of matrix, and these matrices may have contaminant levels that are clearly *not* zero, and furthermore, may vary over time.

Additionally, because contamination is endemic to trace-level measurement, blanks must be defined with respect to opportunity for exposure vectors. Thus, in addition to the laboratory blank, assumed to be relatively pristine, there are actually different types of laboratory blanks, and there are other types of blanks—useful in some contexts, such as discussed in Keith (1988). A subset is presented here:

Laboratory Blanks

- A *system* or *instrument blank* is not a physical sample at all, but a measurement of the background response to establish a minimum baseline.
- A *solvent* or *calibration blank* is a solvent (possibly containing impurities).
- A *reagent* or *method blank* is a sample containing all analytical reagents, taken through the sample preparation and analysis; used to estimate the effect of contamination.

Field Blanks

- A *trip* or *sampling media blank* is a sample of the sampling media treated as an actual field sample, experiencing all the routine potential sources of contamination: sample bottles, handling, transport funnels, filters, and so on.
- A *matched-matrix blank* is identical to a trip blank except that the sample is made from a matrix similar to the actual (possibly synthetic).

Generally, laboratory blanks should be used for method development and refinement. Laboratory blanks are least noisy and will thus reveal biases introduced by method-related sources of contamination. For studies whose objectives are to determine the capabilities of a method in routine use, a field blank is most appropriate.

10.6 BLIND STUDIES

Blindness is a virtue when it comes to measurement capability studies. The reasons for running a blind study are well known, but the magnitudes of the consequences of not running a study blind are not widely appreciated. A simple blind study is one where the analyst does not know the desired result. Some additional benefits can be enjoyed if further measures are taken:

- Spike concentrations are selected so that the analyst cannot guess the result due to the use of "nice round" concentration levels (e.g., select 47.7 ppb for the spike concentration rather than 50 ppb).
- The analyst is not aware which samples are part of a study and which are not.
- The analyst does not even know that a study is taking place. Although this may seem impractical, it can be accomplished through the aid of an analytical service hired to do proficiency testing by submitting samples to commercial labs "in plain brown-paper packages." The advantage of a blind study is that the analyst, possibly with the best of intentions, has no opportunity:
  — To take special care for these measurements, producing atypical results
  — To tweak the system, especially for the particular concentration (i.e., do an extra rinse before measuring a trace amount)
  — To redo the analysis if the deviation in the measurement is too large
  — To "round favorably" to be closer to the actual concentration

Some limited additional benefit might be achieved through conducting a *double-blind study,* where even the supplier of the samples (perhaps a laboratory manager or laboratory sample dispatcher) does not know the concentrations and does not even know that the samples are part of an internal study.

11

BETWEEN-LABORATORY DETECTION AND QUANTIFICATION LIMIT ESTIMATORS

11.1 INTRODUCTION

The question of whether detection and quantification limits should be laboratory specific or reflect between-laboratory variability has been the focus of considerable debate. On one hand, proponents of within-laboratory estimates contend that there is no reason to inflate the limits of one laboratory to accommodate lack of precision on the part of another laboratory. In contrast, proponents of between-laboratory estimates argue that split samples are commonly sent to multiple laboratories; therefore, the probability of false detection should be 1% regardless of where the sample is sent for analysis. At the heart of this debate is the fact that health-based standards are often set below levels of detection and quantification; therefore, detection and quantification limits often form the basis for regulatory action. Unfortunately, permits are often written with overly optimistic detection and/or quantification limits explicitly set, and samples are routinely split between laboratories of uneven quality.

From a statistical perspective, it is always of interest to understand more fully the total variability and uncertainty in a system and to decompose that variance into its component parts. Having set this goal, the statistical task is far from trivial. Although there is a rich literature on variance components estimation, little work has been done on variance component estimation in the presence of heteroscedasticity, which is quite characteristic of laboratory calibration data. In the following sections, we provide details regarding what is currently known about this interesting problem.

11.2 NAIVE ESTIMATORS

The simplest approach to this problem is to pool data from multiple laboratories and to compute the usual detection and quantification limit estimators for a single laboratory as described in earlier chapters. The ASTM IDE and IQE are based on this approach. Although statistically naive, the resulting estimators are conservative in that the total variance is underestimated by the simple pooling procedure, and corresponding detection and quantification limit estimators will be underestimated. Historically, this has been the only approach taken to this problem.

11.3 APPROXIMATE ESTIMATOR

A somewhat more sophisticated approach to this problem involves fitting a variance component or random-effects model to the data for all labs separately for each concentration. As a result, we can uniquely estimate both the within- and between-laboratory variances at each concentration and obtain the total variance at each concentration as their sum. Gibbons (1994, Chap. 12) provides the necessary equations for both the least-squares estimator, which is appropriate when the same number of samples are available at each concentration from each laboratory, and a maximum likelihood estimator, which is appropriate for the unbalanced case. The maximum likelihood estimator is a special case (a single concentration and a single random effect) of the more general model presented in the following section; therefore, these equations are not presented here as well. Substituting the computed total variances at each concentration for the within-laboratory variances used in the preceding chapters, and using data from all laboratories, will provide a relatively simple approximation to estimating between-laboratory detection and quantification limits.

11.4 MORE SOPHISTICATED APPROACH

Recently, Gibbons and Bhaumik (2001) proposed a direct method for estimating the parameters of calibration curves with nonconstant variance simultaneously in a series of laboratories. The resulting parameter estimates and their associated variance components structure can be used to derive detection and quantification limit estimators analogous to those presented in earlier chapters for single laboratories. Following Gibbons and Bhaumik (2001) we present an overview of this method, beginning with a description of a random-effects model for homogeneous variances and then extending estimation to the case of heterogeneous variances as described by the Rocke and Lorenzato (1995) model.

11.4.1 Random-Effects Model for Homogeneous Errors

Consider the following linear calibration model ($p = 2$ for the intercept and slope of the calibration curve) for the $n_i \times 1$ response vector $\mathbf{y}_i$ for laboratory i, $i = 1, 2, \ldots, N$:

$$\mathbf{y}_i = \mathbf{X}_i \boldsymbol{\beta}_i + \boldsymbol{\varepsilon}_i \quad (11.1)$$

where $\mathbf{y}_i$ is the $n_i \times 1$ vector of responses for laboratory i, $\mathbf{X}_i$ a known $n_i \times p$ design matrix, $\boldsymbol{\beta}_i$ a $p \times 1$ vector of calibration parameters for laboratory i, and $\boldsymbol{\varepsilon}_i$ a $n_i \times 1$ vector of random residuals distributed independently as $\mathcal{N}(0, \sigma^2 \mathbf{I}_{n_i})$. Notice that since the laboratory subscript i is present for the $\mathbf{y}$ vector and the $\mathbf{X}$ matrix, each laboratory can have a varying number of measurements nested within. Estimates $\boldsymbol{\beta}_i$ of the laboratory-specific calibration parameters $\boldsymbol{\beta}_i' = [\beta_{0i}, \beta_{1i}]$ are assumed to be normally distributed with mean vector $\boldsymbol{\mu}_\beta' = [\mu_{\beta_0}, \mu_{\beta_1}]$ and variance–covariance matrix $\boldsymbol{\Sigma}_\beta$ given by

$$\boldsymbol{\Sigma}_\beta = \begin{bmatrix} \sigma^2_{\beta_0} & \sigma_{\beta_0 \beta_1} \\ \sigma_{\beta_0 \beta_1} & \sigma^2_{\beta_1} \end{bmatrix} \quad (11.2)$$

Given these assumptions, the $\mathbf{y}_i$ are marginally distributed as independent normals with mean $\mathbf{X}_i \boldsymbol{\mu}_\beta$ and variance–covariance matrix $\mathbf{X}_i \boldsymbol{\Sigma}_\beta \mathbf{X}_i' + \sigma^2 \mathbf{I}_{n_i}$. Given the normality assumption for the "prior" distribution, $\boldsymbol{\beta}_i \sim N[\boldsymbol{\mu}_\beta, \boldsymbol{\Sigma}_\beta]$, we can use the available data to estimate its parameters ($\boldsymbol{\mu}_\beta$ and $\boldsymbol{\Sigma}_\beta$) as well as the residual variance σ^2. Given provisional estimates of $\boldsymbol{\mu}_\beta$ and $\boldsymbol{\Sigma}_\beta$, the laboratory-specific calibration parameters can be obtained as a function of the laboratory's data and the empirical prior distribution. Following Laird and Ware (1982), the mean and variance–covariance matrix of the conditional distribution of $\boldsymbol{\beta}_i$ given $\mathbf{y}_i$ (i.e., the empirical Bayes estimates of the posterior distribution) are

$$\overline{\boldsymbol{\beta}}_i = (\mathbf{X}_i' \mathbf{X}_i + \sigma^2 \boldsymbol{\Sigma}_\beta^{-1})^{-1} \mathbf{X}_i'(\mathbf{y}_i - \mathbf{X}_i \boldsymbol{\mu}_\beta) + \boldsymbol{\mu}_\beta \quad (11.3)$$

$$\boldsymbol{\Sigma}_{\beta|y_i} = \sigma^2 (\mathbf{X}_i' \mathbf{X}_i + \sigma^2 \boldsymbol{\Sigma}_\beta^{-1})^{-1} \quad (11.4)$$

In the present context, the empirical Bayes estimates represent the expected intercept and slope of the calibration curve for a given laboratory and the corresponding estimates of the posterior covariance matrix represent the degree of uncertainty and covariation in these estimates. To compute these empirical Bayes estimates, we also need estimates of the residual variance σ^2 and the parameters of the prior distribution $\boldsymbol{\mu}_\beta$ and $\boldsymbol{\Sigma}_\beta$. These parameters can be estimated by maximum likelihood in the marginal distribution [maximum marginal likelihood (MML)]. The marginal density of the data $\mathbf{y}_i$ is expressed in the population as the integral,

$$h(\mathbf{y}_i | \sigma^2, \boldsymbol{\mu}_\beta, \boldsymbol{\Sigma}_\beta) = \int_\beta f(\mathbf{y}_i | \boldsymbol{\beta}, \sigma^2) g(\boldsymbol{\beta} | \boldsymbol{\mu}_\beta, \boldsymbol{\Sigma}_\beta) \, d\boldsymbol{\beta} \quad (11.5)$$

where

$$f(\mathbf{y}_i|\alpha;\boldsymbol{\beta},\sigma^2) = (2\pi\sigma^2)^{-n/2}\exp\left[\frac{-(\mathbf{y}_i - \mathbf{X}_i\boldsymbol{\beta})'(\mathbf{y}_i - \mathbf{X}_i\boldsymbol{\beta})}{2\sigma^2}\right] \quad (11.6)$$

and the prior distribution is in our case the bivariate normal density with mean $\boldsymbol{\mu}_\beta$ and variance–covariance matrix $\boldsymbol{\Sigma}_\beta$.

$$g(\boldsymbol{\beta}|\boldsymbol{\mu}_\beta,\boldsymbol{\Sigma}_\beta) = (2\pi)^{-1}|\boldsymbol{\Sigma}_\beta|^{-1/2}\exp[-\tfrac{1}{2}(\boldsymbol{\beta} - \boldsymbol{\mu}_\beta)'\boldsymbol{\Sigma}_\beta^{-1}(\boldsymbol{\beta} - \boldsymbol{\mu}_\beta)] \quad (11.7)$$

The marginal log-likelihood of a sample of N independent laboratories is

$$\log L = \sum_i^N \log h(\mathbf{y}_i|\sigma^2,\boldsymbol{\mu}_\beta,\boldsymbol{\Sigma}_\beta) \quad (11.8)$$

Differentiating the marginal log-likelihood with respect to each of the parameters and setting the first derivatives to zero and solving provides the initial MML solutions as:

$$\hat{\boldsymbol{\mu}}_\beta = \frac{1}{N}\sum_i^N \overline{\boldsymbol{\beta}}_i \quad (11.9)$$

$$\hat{\boldsymbol{\Sigma}}_\beta = \frac{1}{N}\sum_i^N (\overline{\boldsymbol{\beta}}_i\overline{\boldsymbol{\beta}}_i' + \boldsymbol{\Sigma}_{\beta|y_i}) - \hat{\boldsymbol{\mu}}_\beta\hat{\boldsymbol{\mu}}_\beta' \quad (11.10)$$

$$\hat{\sigma}^2 = \left(\sum_i^N n_i\right)^{-1}\sum_i^N \text{tr}[(\mathbf{y}_i - \mathbf{X}_i\overline{\boldsymbol{\beta}}_i)(\mathbf{y}_i - \mathbf{X}_i\overline{\boldsymbol{\beta}}_i)' + \mathbf{X}_i\boldsymbol{\Sigma}_{\beta|y_i}\mathbf{X}_i'] \quad (11.11)$$

The solution is obtained by iterating between the empirical Bayes solutions and the MML solutions until convergence. Notice that as the estimates of the laboratory effects $\boldsymbol{\beta}_i$ and variances $\boldsymbol{\Sigma}_{\beta|y_i}$ approach zero, the equations for the regression coefficients $\boldsymbol{\beta}$ and the error variance σ^2 approach the maximum likelihood solution of these parameters in the usual fixed-effects regression model, namely, $\hat{\boldsymbol{\beta}} = (\sum_{i=1}^N \mathbf{X}_i'\mathbf{X}_i)^{-1}\sum_{i=1}^N \mathbf{X}_i'\mathbf{y}_i$ and $\hat{\sigma}^2 = (\sum_i^N n_i)^{-1}\sum_i^N(\mathbf{y}_i - \mathbf{X}_i\hat{\boldsymbol{\beta}})'(\mathbf{y}_i - \mathbf{X}_i\hat{\boldsymbol{\beta}})$.

11.4.2 Random-Effects Model for Heterogenous Errors

When the variance is not constant, as is typically the case for calibration data, the previous random-effects model for homoscedastic errors no longer applies. There are several approaches to this problem, but a widely accepted approach is to model the squared residuals as a function of a subset of $\mathbf{X}$ (in the calibration example, true concentration) and then to use the estimated variance function to obtain weights which are then used in reestimating the regression coefficients. This process is iterated until convergence, hence the term *iteratively reweighted least squares* (see Carroll and Ruppert, 1988). As noted by Neter et al (1990), the methods of maximum likelihood and weighted least squares lead to the same estimators for generalized multiple linear regression models of the form considered here.

11.4 MORE SOPHISTICATED APPROACH

The approach for weighting the linear fixed-effects model described in Chapter 4 can be extended to the random-effects model described previously to provide a Rocke and Lorenzato (R-L) model suitable for interlaboratory data. To this end we consider the model

$$y_{ijk} = \beta_{0i} + \beta_{1i} x_j e^\eta + e_{ijk} \tag{11.12}$$

where y_{ijk} is the kth measurement at the jth concentration level in the ith laboratory, $i = 1, 2, \ldots, N; j = 1, 2, \ldots, r; k = 1, 2, \ldots, n_{ij}$, and x_{ijk} is the true concentration in the ith laboratory at the jth level for the kth replication. We assume that the ith laboratory has random effects β_{0i} and β_{1i} which follow a bivariate normal distribution with mean $\boldsymbol{\mu}_\beta$ and variance–covariance matrix $\boldsymbol{\Sigma}_\beta$. The likelihood function is now

$$\prod_{i=1}^{N} \prod_{j=1}^{r} \prod_{k=1}^{n_{ij}} \frac{1}{2\pi \sigma_\eta \sigma_e} \int_{-\infty}^{\infty} \exp\left(\frac{-\eta^2}{2\sigma_\eta^2}\right) \exp\left[-\frac{(y_{ijk} - \beta_{0i} - \beta_{1i} x_j e^\eta)^2}{2(\sigma_e^2)}\right] d\eta \tag{11.13}$$

Let $N_i = \sum_j^r n_{ij}$ denote the total number of measurements in laboratory i. Then $\mathbf{W}_i$ is the $N_i \times N_i$ diagonal matrix of weights for laboratory i with elements $w_{ijk} = 1/\hat{s}_{ijk}^2$, where $\hat{s}_{ijk}^2$ is a fitted value from the R-L variance function described in Chapter 4. Premultiplying both sides of the regression model in equation (11.1) by $\mathbf{W}_i^{1/2}$,

$$\mathbf{W}_i^{1/2} \mathbf{y}_i = \mathbf{W}_i^{1/2} \mathbf{X}_i \boldsymbol{\beta}_i + \mathbf{W}_i^{1/2} \boldsymbol{\varepsilon}_i \tag{11.14}$$

leads to the revised MML solutions

$$\hat{\boldsymbol{\mu}}_\beta = \frac{1}{N} \sum_i^N \overline{\boldsymbol{\beta}}_i \tag{11.15}$$

$$\hat{\boldsymbol{\Sigma}}_\beta = \frac{1}{N} \sum_i^N (\overline{\boldsymbol{\beta}}_i \overline{\boldsymbol{\beta}}_i' + \boldsymbol{\Sigma}_{\beta|y_i}) - \hat{\boldsymbol{\mu}}_\beta \hat{\boldsymbol{\mu}}_\beta' \tag{11.16}$$

$$\hat{\sigma}^2 = \left(\sum_i^N n_i\right)^{-1} \sum_i^N \mathrm{tr}[(\mathbf{W}_i^{1/2} \mathbf{y}_i - \mathbf{W}_i^{1/2} \mathbf{X}_i \overline{\boldsymbol{\beta}}_i)(\mathbf{W}_i^{1/2} \mathbf{y}_i - \mathbf{W}_i^{1/2} \mathbf{X}_i \overline{\boldsymbol{\beta}}_i)'$$
$$+ \mathbf{W}_i \mathbf{X}_i \boldsymbol{\Sigma}_{\beta|y_i} \mathbf{X}_i'] \tag{11.17}$$

The estimated variance–covariance matrix of the random-effect parameters $\hat{\boldsymbol{\Sigma}}_\beta$ will always be positive definite and the estimated residual variance $\hat{\sigma}^2$ will always be positive. The weighted mean and variance of the posterior distribution are

$$\overline{\boldsymbol{\beta}}_i = (\mathbf{X}_i' \mathbf{W}_i \mathbf{X}_i + \sigma^2 \boldsymbol{\Sigma}_\beta^{-1})^{-1} \mathbf{X}_i' \mathbf{W}_i (\mathbf{y}_i - \mathbf{X}_i \boldsymbol{\mu}_\beta) + \boldsymbol{\mu}_\beta \tag{11.18}$$

$$\boldsymbol{\Sigma}_{\beta|y_i} = \sigma^2 (\mathbf{X}_i' \mathbf{W}_i \mathbf{X}_i + \sigma^2 \boldsymbol{\Sigma}_\beta^{-1})^{-1} \tag{11.19}$$

On each iteration, improved estimates of β_{0i} and β_{1i} are used to estimate the standard error of the calibration curve at each concentration [see equation (4.15)]. These estimated standard errors are then used to obtain improved estimates of σ_e, σ_η, and so on, until convergence.

Note that upon convergence the estimated variance at concentration j is now given by

$$s^2(x) = \hat{\sigma}^2_{\beta_0} + 2x\hat{\sigma}_{\beta_0\beta_1}e^{\hat{\sigma}^2_\eta/2} + x^2\hat{\sigma}^2_{\beta_1}e^{2\hat{\sigma}^2_\eta} + \hat{\sigma}^2_\varepsilon + \hat{\mu}^2_{\beta_1}x^2e^{\hat{\sigma}^2_\eta}(e^{\hat{\sigma}^2_\eta} - 1) \quad (11.20)$$

See Gibbons and Bhaumik (2001).

11.5 APPLICATIONS

There are several potential applications of these results. First, the empirical Bayes estimates of the weighted mean (11.18) and variance (11.19) of the posterior distribution for a given laboratory can serve as improved estimates of the parameters of that laboratory's calibration function, which "borrow strength" from the data from all laboratories. We would expect increased precision of these empirical Bayes calibration parameter estimates relative to estimates obtained from individual laboratories. This expectation is confirmed in the following illustration (section 11.6). Second, although the estimated mean calibration function is not directly applicable to use in any one laboratory, confidence bounds on the mean calibration curve are useful for determining if a new laboratory is functioning in accordance with those laboratories used in the initial between-laboratory calibration study. For example, an asymptotic normal $(1 - \alpha)100\%$ confidence interval for the overall mean slope of the calibration curve is

$$\hat{\mu}_{\beta_1} \pm z_{1-\alpha/2}\hat{s}_{\beta_1} \quad (11.21)$$

where z is the $100(1 - \alpha/2)$ percentage point of the standard normal distribution. This can be quite useful for quality control purposes.

Third, given a set of between-laboratory calibration data and resulting parameter estimates a simple test of the null hypothesis that $x = 0$ can be constructed as

$$y_C = \hat{\mu}_{\beta_0} + z_{1-\alpha}\hat{s}(0) \quad (11.22)$$

where z is the $100(1 - \alpha)$ percentage point of the standard normal distribution. y_C represents an upper bound on measured concentrations or instrument responses for which the true concentration is $x = 0$. To convert to concentration units, $L_C = (y_C - \mu_{\beta_0})/\mu_{\beta_1}$. L_C represents a between-laboratory version of the critical level (Currie, 1968) that can be used to make the binary decision of whether or not an analyte is present in a sample. The corresponding between-laboratory detection limit becomes

$$L_D = L_C + z_{1-\alpha}\hat{s}(L_D)/\mu_{\beta_1} \quad (11.23)$$

The detection limit L_D can be relied upon to lead to a correct detection decision with $100(1 - \alpha)\%$ confidence. For this application, a normal multiplier is used because at $x = 0$, the variance of the measured concentrations is a function of $\sigma_{\beta_0}^2$ and σ_ε^2, and the distribution of y is approximately normal. Note that to compute L_D an estimate of the variance at L_D is required. As such, the solution is iterative; however, the problem is well conditioned and five repeated substitutions will generally be sufficient for convergence. Use of this between-laboratory detection limit applies only to those cases in which detection decisions are to be made simultaneously in a series of laboratories.

Fourth, as x increases, the distribution of y becomes lognormal and approximate confidence bounds for the true concentration x become

$$\exp[\ln(x) - z_{1-\alpha/2}\hat{s}_\eta] - z_{1-\alpha/2}(\hat{\sigma}_{\beta_0}^2 + 2x\hat{\sigma}_{\beta_0\beta_1}e^{\hat{\sigma}_\eta^2/2} + x^2\hat{\sigma}_{\beta_1}^2 e^{2\hat{\sigma}_\eta^2} + \hat{\sigma}_\varepsilon^2)^{1/2}$$
$$\exp[\ln(x) + z_{1-\alpha/2}\hat{s}_\eta] + z_{1-\alpha/2}(\hat{\sigma}_{\beta_0}^2 + 2x\hat{\sigma}_{\beta_0\beta_1}e^{\hat{\sigma}_\eta^2/2} + x^2\hat{\sigma}_{\beta_1}^2 e^{2\hat{\sigma}_\eta^2} + \hat{\sigma}_\varepsilon^2)^{1/2} \quad (11.24)$$

where $\hat{x} = (y - \mu_{\beta_0})/\mu_{\beta_1}$. These approximate confidence intervals can be used for many purposes, including comparing the underlying true concentration interval to a regulatory standard. Such an approach can eliminate often fruitless debates in which "split samples" are sent to multiple laboratories and some laboratories report a measured concentration in exceedance of a regulatory standard and others do not.

Fifth, a plot of the estimated standard deviation s_x versus concentration can be used to identify the true concentration at which the errors of measurement are sufficiently small to permit routine quantitative determination. Currie (1968) defines the limit of determination L_Q as the concentration at which the signal-to-noise ratio is 10:1. A similar statistic can be obtained by identifying the true concentration at which the overall estimated standard deviation is one-tenth of its magnitude. The estimator described here allows the determination limit to incorporate correctly both within- and between-laboratory variance components.

11.6 ILLUSTRATION

Gibbons and Bhaumik (2001) analyzed experimental data for copper from a between-laboratory study. Samples were prepared by an independent source, randomized and submitted on a weekly basis over a 5-week period. Copper was analyzed by inductively coupled plasma atomic emissions spectroscopy (ICP/AES) using EPA Method 200.7. The data set consisted of five replicates at each of five concentrations (0, 2, 10, 50, and 200 μg/L) in each of seven laboratories (see Table 11.1). The resulting parameter estimates and the means and standard deviations observed and estimated at each true calibration point are displayed in Table 11.2. Table 11.2 reveals that (1) there is positive bias of $\mu_{\beta_0} = 1.0996$, (2) the mean slope is close to unity ($\mu_{\beta_1} = 0.9565$), and (3) variability within

Table 11.1 Between-laboratory data for copper

Laboratory	Replicate	Concentration (μg/L)				
		0	2	10	50	200
1	1	3.000	3.000	14.000	54.000	205.000
	2	2.000	3.000	10.000	51.000	206.000
	3	−1.000	5.000	11.000	52.000	208.000
	4	1.000	2.000	12.000	54.000	211.000
	5	−1.000	2.000	13.000	38.000	195.000
2	1	2.100	8.000	10.000	53.000	188.600
	2	0.300	1.800	12.400	54.600	210.000
	3	2.000	0.700	10.600	50.000	210.000
	4	1.300	4.000	12.000	50.100	214.000
	5	2.000	3.000	11.000	50.000	200.000
3	1	0.800	2.495	10.500	47.660	181.330
	2	−0.185	2.695	10.335	45.390	173.205
	3	0.990	2.410	9.735	44.270	180.560
	4	0.905	1.840	10.245	46.910	183.650
	5	0.365	2.840	10.325	47.240	181.585
4	1	1.661	3.243	12.250	48.140	205.400
	2	1.996	3.432	13.510	54.450	200.400
	3	0.000	9.246	11.160	51.010	199.700
	4	2.993	3.390	13.440	52.860	189.600
	5	2.042	4.109	10.470	48.720	187.700
5	1	0.090	0.860	10.030	50.060	193.400
	2	−2.510	2.680	12.940	50.350	193.470
	3	7.270	−0.400	8.970	49.320	203.160
	4	7.140	4.730	9.610	49.930	190.020
	5	0.280	5.200	9.120	48.080	191.050
6	1	7.226	4.964	4.713	48.242	191.020
	2	−1.000	2.000	10.000	65.000	205.000
	3	0.000	3.000	8.000	45.000	183.000
	4	10.244	6.716	11.101	43.000	185.000
	5	−2.177	8.844	8.249	47.000	182.000
7	1	0.018	1.323	6.000	45.500	162.000
	2	−3.000	4.900	9.088	44.000	181.000
	3	0.000	0.000	14.100	40.000	187.000
	4	−2.000	0.000	6.000	43.000	178.300
	5	−2.000	0.000	7.000	45.986	188.932

laboratories at low concentrations ($\sigma_e = 2.3928$) is roughly four times the variability between labs at a concentration of zero ($\sigma_{\beta_0} = 0.6794$). However, both between-laboratory variability in the slopes $\sigma_{\beta_1} = 0.0399$ and within-laboratory variability at higher concentrations $\sigma_\eta = 0.0365$ yield much larger variability in measurements at higher concentrations (see Table 11.2).

Table 11.2 Parameter estimates and estimated standard deviations: Between-laboratory data for copper (μg/L)

Parameter	MMLE	Mean		Standard Deviation	
		Obs.	Est.	Obs.	Est.
μ_{β_0}	1.0996	0	1.100	2.922	2.699
μ_{β_1}	0.9565	2	3.013	2.363	2.722
σ_{β_0}	0.6794	10	10.664	2.257	2.860
σ_{β_1}	0.0399	50	48.924	4.875	4.287
σ_e	2.3928	200	192.398	12.208	12.286
σ_η	0.0365	—	—	—	—

Table 11.3 displays the empirical Bayes estimates for the parameters of the calibration function in each laboratory, along with the corresponding estimates of precision (i.e., posterior standard deviations). Inspection of equation (11.19) reveals that for a balanced design such as this, the posterior standard deviations are constant across laboratories. Table 11.3 also presents the corresponding parameter estimates and standard errors for iteratively reweighted least-squares analysis applied to the data from each laboratory separately. In general, the calibration curves are estimated with increased precision when the data from all laboratories are used. The

Table 11.3 Empirical Bayes estimates and simple WLS estimates of individual laboratory calibration curve parameters (standard errors): Between-laboratory data for copper (μg/L)

Laboratory	Empirical Bayes		Simple WLS	
	$\hat{\beta}_0$	$\hat{\beta}_1$	$\hat{\beta}_0$	$\hat{\beta}_1$
1	1.568 (0.298)	0.994 (0.012)	1.007 (0.618)	1.007 (0.019)
2	1.705 (0.298)	1.000 (0.012)	1.409 (0.297)	1.006 (0.020)
3	0.485 (0.298)	0.912 (0.012)	0.784 (0.094)	0.906 (0.005)
4	1.626 (0.298)	0.982 (0.012)	2.304 (0.414)	0.973 (0.014)
5	1.215 (0.298)	0.963 (0.012)	1.256 (0.437)	0.963 (0.013)
6	1.161 (0.298)	0.951 (0.012)	1.935 (1.655)	0.937 (0.032)
7	−0.063 (0.298)	0.893 (0.012)	−0.884 (0.564)	0.899 (0.021)

95% confidence interval for the slope of the calibration curve is (0.878, 1.035), which includes all of the individual laboratory estimates.

Using these estimates we can also compute several other useful statistics. For example, the between-laboratory estimate of the critical level L_C (for $\alpha = 0.01$) is given by

$$y_C = \hat{\mu}_{\beta_0} + z_{1-\alpha}\hat{s}(0) = 1.0996 + 2.33\sqrt{0.4616 + 5.7255} = 6.8952 \ \mu g/L$$

in measured concentration units or

$$L_C = \frac{y_C - \hat{\mu}_{\beta_0}}{\hat{\mu}_{\beta_1}} = \frac{6.8952 - 1.0996}{0.9565} = 6.0592 \ \mu g/L$$

in true concentration units. Beginning from a starting value of $L_D = 2L_C$ the detection limit L_D converges to $L_D = 13.2051 \ \mu g/L$. The meaning of this statistic is that at or above L_D, we can have 99% confidence that an analyte containing a sample sent to any one of these seven laboratories will result in a correct detection decision. Note, however, that the detection limit for an individual laboratory may be considerably lower than this between-laboratory estimate.

Figure 11.1 displays the overall standard deviation $s(x)$ and its three components, $Se^2 = \hat{\sigma}_\varepsilon^2$, $Sb^2 = \hat{\sigma}_{\beta_0}^2 + 2x\hat{\sigma}_{\beta_0\beta_1}e^{\hat{\sigma}_\eta^2/2} + x^2\hat{\sigma}_{\beta_1}^2 e^{2\hat{\sigma}_\eta^2}$, and $Sn^2 = \hat{\mu}_{\beta_1}^2 x^2 e^{\hat{\sigma}_\eta^2}(e^{\hat{\sigma}_\eta^2} - 1)$ as a function of concentration. As postulated in the model, the total variance function

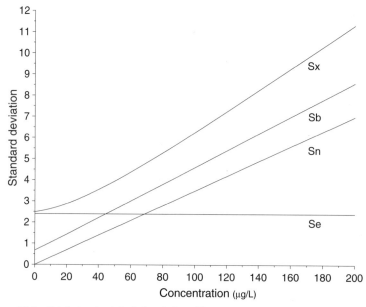

Figure 11.1 Total standard deviation (s_x) and its components (s_b, s_n, and s_e): copper.

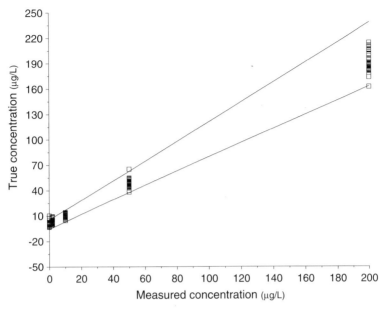

Figure 11.2 Between-laboratory copper calibration data: 99% confidence limits.

has a flat region for near-zero concentrations and is proportional to concentration throughout the remainder of its range. The individual variance components nicely illustrate the effects and magnitude of both within- and between-laboratory uncertainty. Figure 11.2 presents 99% (one-sided) confidence bounds for the entire calibration function. These intervals can be used to determine if a regulatory standard has been exceeded. For example, assume that there is a regulatory standard for copper of 40 µg/L and we have obtained a measured concentration of 50 µg/L from one of the laboratories. For $y = 50$ the lower 99% confidence limit is 37 µg/L, indicating that we cannot reject the null hypothesis that the true concentration is less than or equal to the regulatory standard.

II

DETECTION AND QUANTIFICATION IN THE FIELD

12

COMPARISON OF A SINGLE MEASUREMENT TO A REGULATORY STANDARD

12.1 INTRODUCTION

There is considerable confusion and debate regarding the type of limit (i.e., critical level, detection limit, or quantification limit) that should be used for various environmental monitoring tasks. As a general principle, compliance with environmental criteria should never be based on the results of a single sample. To this end, methods for the comparison of statistical bounds for true concentration based on n measured concentrations are presented in detail in Chapters 18 to 21. Nevertheless, there are many applications in which some type of environmental inference is based routinely on the results of a single measured concentration. To this end, in this chapter we describe the statistical foundations for such applications and which type of limit is appropriate for which type of application. Again, whenever possible, such applications should be based on a series of n such samples so that the effects of analytical, sampling, temporal, and spatial variability can play appropriate roles in the environmental decision-making process.

In this chapter, the three previously described decision levels (critical level L_C, detection limit L_D, and determination limit L_Q) described by Currie (1968) are used to illustrate how various regulatory compliance decisions can be made. In the simplest case of zero discharge, the appropriate limit is the critical level L_C. When the true concentration is zero, values above L_C should be found at a rate of 1% or less. For regulatory standards set at a specific value above zero but below the critical level, the detection limit L_D can be used to determine if the true concentration has exceeded the regulatory standard (STD). The detection limit provides a 99% confidence upper bound on measured concentrations when the true concentration is equal to L_C. For regulatory standards above the critical level but below the determination limit L_Q, the modified detection limit L_D^* can be used to determine if the true concentration has exceeded the standard. L_D^* provides a 99% confidence upper bound on measured concentrations when the true concentration is equal to the regulatory standard. Finally, for regulatory standards that are at quantifiable levels (i.e., regulatory standard $\geq L_Q$) direct comparison of measured concentrations to standards is possible, generally with a low rate of false decisions of either type.

12.2 DETERMINING REGULATORY COMPLIANCE

Before providing operational definitions, the conceptual definitions will be used to show how the three analytical decision levels—the critical level, the detection limit, and the determination limit—relate to making regulatory compliance decisions. In doing so, three general cases are considered: (1) the regulatory standard is zero discharge, (2) the regulatory standard is a specific numerical value greater than zero but less than the determination limit L_Q (i.e., it is set at a nonquantifiable concentration), and (3) the regulatory standard is at a quantifiable level (i.e., the standard is greater than L_Q). From the previous definitions, the following relationships emerge.

12.2.1 Case I: STD = 0

When the regulatory standard is zero discharge, measurements above the critical level will provide 99% confidence that the true concentration is greater than zero. Note that if the true concentration is actually at the critical level, assuming a symmetric distribution of measurement error, in 50% of the cases the measured concentration will actually be below the critical level and the incorrect decision will be made (i.e., a false negative decision in that the true concentration is greater than zero but the critical level is not exceeded). Nevertheless, from a regulatory perspective, if the goal is zero tolerance, the false positive rate associated with measured concentrations above the critical level L_C will be 1% or less. In light of this, the appropriate decision limit for this case is the critical level L_C.

12.2.2 Case II: $0 <$ STD $< L_Q$

When the regulatory standard is greater than zero but less than the determination limit L_Q (i.e., the concentration above which measurements are truly quantifiable), uncertainty in concentrations that are below L_Q is large, and they cannot validly be compared to the standard. Of course, a measured concentration above L_Q is quantifiable and therefore would exceed the standard. In this case, there are two possible alternatives short of simply requiring that only quantifiable measurements (i.e., above L_Q) be compared to the standard.

12.2.3 Case IIa: $0 < \text{STD} \leq L_C$

If the regulatory standard is greater than zero but less than or equal to L_C, we will have at least 99% confidence that the true concentration is greater than the standard if the measured concentration exceeds the detection limit L_D. This follows because if the true concentration is L_C there is only a 1% chance that a measured value will exceed L_D. Since the regulatory standard is less than or equal to L_C (as defined for this case), the probability that the standard is exceeded given a measurement above L_D is 99% or more. Since L_D is always less than L_Q, we will have added environmental protection, since we can provide a valid comparison to a standard even at nonquantifiable levels. In this case we have 99% confidence that the true concentration is greater than the standard, but we have no idea what the true concentration really is. This is quite a different problem than quantification; however, it still represents a valid test of the hypothesis that the true concentration has exceeded the standard.

12.2.4 Case IIb: $0 < L_C < \text{STD} < L_Q$

For this case, let us assume that the standard is greater than L_C but less than L_Q. In this case, measurements at the detection limit L_D could have considerable probability of being observed if the true concentration was at or below the standard; therefore, use of the L_D for determining regulatory compliance is no longer valid as it was in case IIa. If the measured concentration exceeds the determination limit L_Q, then, of course, the standard has been exceeded. However, even in this case, a more environmentally protective solution is possible. For this case, let $L_C = \text{STD}$ (i.e., replace the critical level with the standard). If we compute L_D^* such that the probability of exceeding L_D^* is 1% when the true concentration is equal to the standard, we can have 99% confidence that measurements above L_D^* have true concentration greater than the standard. As in case IIa, above L_D^* we have 99% confidence that the true concentration is greater than the standard, but we have no idea what the true concentration really is.

Note that if the standard is close to L_Q, then L_D^* may in fact be larger than L_Q. For compliance purposes we therefore select either L_D^* or L_Q, depending on which is smaller.

12.2.5 Case III: $L_Q \leq \text{STD}$

For this case the standard is at a quantifiable level. In this case any measurement that exceeds the standard also exceeds L_Q; therefore, the measurement is quantifiable and its concentration can be compared directly to the standard. A measurement above the standard indicates noncompliance. There is, however, an important exception to this rule. If compliance is based on a weekly or monthly average of n measurements and those measurements are a mixture of values above and below the L_Q, the distribution is termed "left censored" (see Cohen, 1961). In this case, concentration information can be used for those measurements above L_Q, but all we know about measurements below L_Q is that their true concentration is less than L_Q.

The EPA has traditionally substituted zero or one-half the detection limit or the detection limit for these measurements and computed the average concentration in the usual way. Two points are important to clarify. First, the censoring point is L_Q and not L_D, since below L_Q we do not have reasonable confidence in the concentration of that particular measurement. Second, depending on the number of measurements and the regulatory context, there may be much better methods for obtaining the average concentration from a left censored distribution than simply imputing an arbitrary value. For a reasonably up to date review of these methods, see Chapter 13.

12.3 CONFIDENCE INTERVAL FOR TRUE CONCENTRATION

Perhaps the best approach of incorporating uncertainty in true concentration, based on the results of a single measured concentration, is to provide a $(1 - \alpha)100\%$ confidence interval for the true concentration.

Recall from Chapter 11 (between-laboratory case) that based on the Rocke and Lorenzato (1995) model, as true concentration x increases, the distribution of measured concentration and/or instrument response y becomes lognormal and approximate confidence bounds for the true concentration x become

$$\exp\left[\ln(x) - z_{1-\alpha/2}\hat{s}_\eta\right] - z_{1-\alpha/2}(\hat{\sigma}_{\beta_0}^2 + 2x\hat{\sigma}_{\beta_0\beta_1}e^{\hat{\sigma}_\eta^2/2} + x^2\hat{\sigma}_{\beta_1}^2 e^{2\hat{\sigma}_\eta^2} + \hat{\sigma}_\varepsilon^2)^{1/2}$$

$$\exp\left[\ln(x) + z_{1-\alpha/2}\hat{s}_\eta\right] + z_{1-\alpha/2}(\hat{\sigma}_{\beta_0}^2 + 2x\hat{\sigma}_{\beta_0\beta_1}e^{\hat{\sigma}_\eta^2/2} + x^2\hat{\sigma}_{\beta_1}^2 e^{2\hat{\sigma}_\eta^2} + \hat{\sigma}_\varepsilon^2)^{1/2} \quad (12.1)$$

where $\hat{x} = (y - \mu_{\beta_0})/\mu_{\beta_1}$. In the single-laboratory case, $\hat{\sigma}_{\beta_0}^2$, $\hat{\sigma}_{\beta_1}^2$ and $\hat{\sigma}_{\beta_0\beta_1}$ are all set to zero, and $\hat{x} = (y - \beta_0)/\beta_1$. These approximate confidence intervals can be used to determine if the underlying true concentration for a given measured concentration or instrument response exceeds a regulatory standard. If the lower confidence limit exceeds the standard, we can have 95% confidence that the true concentration is above the standard. If the upper confidence limit is less than the standard, we can have 95% confidence that the true concentration is below the standard. If the interval contains the standard, the results are equivocal and further monitoring may be required to determine if the standard has been exceeded.

12.4 ILLUSTRATION

Consider the benzene data described in Sections 4.10 and 7.9. Results of fitting the Rocke and Lorenzato model to these data revealed WLS estimates of $\sigma_\varepsilon = 0.05$, $\sigma_\eta = 0.03, \beta_0 = 0.00, \beta_1 = 1.00, L_C = 0.13\ \mu g/L$, and $L_D = 0.26\ \mu g/L$ (Section 7.9). The estimated AML was 0.62 $\mu g/L$ with associated RSD of 8.38%. Assume that we have a regulatory standard for benzene of 0.4 $\mu g/L$ and that we have obtained a newly measured concentration of 0.5 $\mu g/L$. Has the standard been exceeded?

First, we know that the measured concentration exceeds the critical level (L_C), so we have 99% confidence that the true concentration is greater than zero. Second, the measured concentration is also greater than the detection limit (L_D), so we have 99% confidence that the true concentration is greater than $L_C = 0.13$ µg/L. Third, the measured concentration did not exceed the quantification limit (AML = .62 µg/L), so our knowledge of the true concentration is equivocal. Use of these three limit estimates leaves us somewhat in the dark. The 95% confidence interval for true concentration is

$$\exp[\ln(0.5) - 1.96(0.03) - 1.96(0.05)]$$
$$\exp[\ln(0.5) + 1.96(0.03) + 1.96(0.05)] = (0.37, 0.63)$$

which includes the standard of 0.4 µg/L. This result indicates that with this single sample we cannot reject the null hypothesis that the true concentration is less than or equal to the standard.

13

CENSORED DATA

13.1 INTRODUCTION

One of the most difficult problems in analysis of environmental monitoring data involves the incorporation of nondetects into estimates of summary statistics (e.g., mean and standard deviation) and corresponding tests of hypotheses and interval estimates. More often than not, environmental monitoring data consist of a mixture of results that can and cannot be quantified accurately. In practice, the censoring mechanism is the detection limit; values below are reported as ND or $<L_D$ to signify that they were not found in the sample. All other values are reported as a concentration. Based on earlier chapters, one should immediately note that this is the wrong procedure. The L_Q and not the L_D should be the censoring mechanism since values above the L_D and below the L_Q are detected but not quantifiable. Using the L_D as the censoring point produces data with widely varying levels of uncertainty violating the assumption of homoscedasticity (i.e., constant measurement variation) which is assumed by all of the previous statistical theory and methods. Even with an agreed-upon censoring point, there is considerable controversy regarding the appropriate method or methods for incorporating the censored data in computing summary statistics, testing hypotheses, and computing interval estimates. This is not at all surprising since the correct choice of method depends on both the degree of censoring (e.g., 20% versus 80% nondetects) and the type of application (e.g., computing the mean versus computing a prediction limit from data that are a mixture of quantifiable and nonquantifiable measurements), as well as ease of use. Additionally, the controversy can be fueled by an inclination toward a particular favorable outcome.

13.2 CONCEPTUAL FOUNDATION

Assume that there is a population of true concentrations from which we have drawn a sample of size n. For convenience, also assume that variation in the sampled population can be represented by a continuous probability distribution for which a fraction of the true concentrations are essentially zero. This partial loss of information occurs because of censoring imposed by limits of detection and/or quantification.

For example, Davis (1994) points out that we may assume an underlying distribution as in Figure 13.1, but what we observe is the distribution in Figure 13.2, where the vertical line represents a point mass at $L_D/2$ containing the probability content of the region $<L_D$. In practice, the measurements are often coarsely rounded so that the observed frequency distribution looks like Figure 13.3. Davis (1993) points out that in real-world application the true underlying model in Figure 13.3 is unknown; therefore, different approaches will yield widely different results, depending on the degree to which they rely on the assumed distribution. This is even more critical in environmental monitoring applications (e.g., groundwater monitoring) in which repeated application of tail probabilities are used to control the overall site-wide false positive rate (i.e., prediction limits). How well a censored data estimator works in the center of the distribution (e.g., to estimate mean concentration) is often a poor index of how well that method will work in the tails of the distribution (e.g., to estimate a 99% confidence prediction limit for a new single measurement). In the following sections, several methods are described, and some general recommendations are provided.

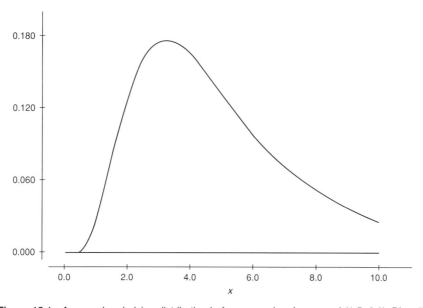

Figure 13.1 Assumed underlying distribution before censoring. Lognormal (1.5, 0.6), DL = 5.

138 CENSORED DATA

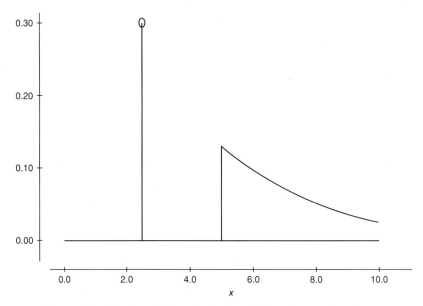

Figure 13.2 Actual underlying distribution with Type I censoring. DL = 5.

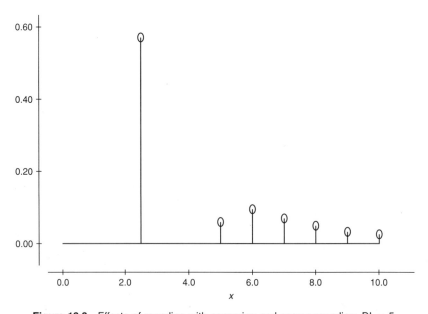

Figure 13.3 Effects of rounding with censoring and coarse rounding. DL = 5.

13.3 SIMPLE SUBSTITUTION METHODS

Historically, the EPA has advocated that nondetected measurements be replaced by one-half the detection limit (e.g., $L_D/2$) and data be analyzed as if all measurements were observable with equal precision. Other substitution values for nondetects required or advocated by the EPA include zero (for drinking water contaminants in the public water supplies) and one-half the *Superfund quantitation limit* (SQL) (for contaminants being evaluated in Superfund risk assessments). Of course, the assumption of equal precision is demonstrably false, but the method is computationally simple and is often adequate for most practical purposes if the detection frequency is 80% or more. For example, Haas and Scheff (1990) found that the simple substitution method was negatively biased (i.e., too low) for low degrees of censoring but positively biased thereafter (i.e., too high). Bias became large for samples of size 10 when detection frequencies fall below 80%. Kushner (1976) found similar results for bias in estimated mean concentrations when 20% or more of the samples were nondetects. Several other studies have also shown that simple substitution methods perform poorly in recovering true population statistics as compared to other statistical procedures (Gilliom and Helsel, 1986; Gleit, 1985; Helsel, 1990; Helsel and Cohn, 1988; Helsel and Gilliom, 1986).

Surprisingly, however, McNichols and Davis (1988) and Davis (1993) suggest that the simple substitution technique (e.g., $L_D/2$) maintains reasonable control over site-wide false positive rates evaluated through repeated application of prediction limits, for nondetect proportions up to 80%. They report that adding verification resamples not only aids in controlling the site-wide false positive rate but also in increasing robustness of the resulting tests. As they point out, this result is vastly different from the conclusions of others and more recent guidance, which has focused on estimation rather than hypothesis testing.

The simple substitution method may be enhanced in a couple of ways. Gleit (1985) suggests substituting nondetects with expected order statistics from a fitted distribution. The approach involves initially fitting the distribution to the quantifiable measurements, and substituting $L_Q/2$ for the nondetects. Gleit suggests iteratively refitting and resubstituting until the parameter estimates converge. An alternative approach suggested by Gilliom and Helsel (1986) involves substitution of random variates drawn from a right-truncated normal distribution with the working parameters. Both of these estimates should lead to improved results over the simple substitution method to the extent that the assumed distribution is correct.

13.4 MAXIMUM LIKELIHOOD ESTIMATORS

For a left singly censored normal distribution, Cohen (1959, 1961) derived the maximum likelihood estimator for the mean and variance of the overall distribution as

$$\hat{\mu} = \bar{x}' - (\bar{x}' - x_0)\lambda(g, h) \tag{13.1}$$

$$\sigma^2 = s^{2\prime} + (\bar{x}' - x_0)^2 \lambda(g, h) \tag{13.2}$$

where $\bar{x}'$ is the mean of the measured values, $s^{2'}$ the variance of the measured values, h the proportion of nondetects, x_0 the censoring point (e.g., $L_D/2$, and

$$g = \frac{s^{2'}}{(\bar{x}' - x_0)^2}$$

Cohen (1961) provided values of $\lambda(g, h)$ for values of g up to 1, Schneider (1986) up to $g = 1.48$, and Schnee et al. (1985) up to $g = 10$, which is reproduced in Table 13.1.

Haas and Scheff (1990) have developed the following power series expansion that fits these tabulated values to within 6% relative error:

$$\log_e \lambda(g, y) = \frac{0.182344 - 0.3756}{g + 1} + 0.10017g + 0.78079y - 0.00581g^2$$
$$- 0.06642y^2 - 0.0234gy + 0.000174g^3$$
$$+ 0.001663g^2y - 0.00086gy^2 - 0.00653y^3 \quad (13.3)$$

Table 13.1 Values of $\lambda(g, h)$ for values of g up to 10

$\hat{\gamma}$	h (Proportion Censored)								
	0.1	0.2	0.3	0.4	0.5	0.6	0.7	0.8	0.9
.0	.1102	.2426	.4021	.5961	.8368	1.145	1.561	2.176	3.283
.1	.1180	.2574	.4233	.6234	.8703	1.185	1.608	2.229	3.345
.2	.1247	.2703	.4422	.6483	.9012	1.222	1.651	2.280	3.405
.3	.1306	.2819	.4595	.6713	.9300	1.257	1.693	2.329	3.464
.4	.1360	.2926	.4755	.6927	.9570	1.290	1.732	2.376	3.520
.5	.1409	.3025	.4904	.7129	.9826	1.321	1.770	2.421	3.575
.6	.1455	.3118	.5045	.7320	1.0070	1.351	1.806	2.465	3.628
.7	.1499	.3207	.5180	.7502	1.0300	1.380	1.841	2.507	3.679
.8	.1540	.3290	.5308	.7676	1.0530	1.408	1.875	2.548	3.730
.9	.1579	.3370	.5430	.7844	1.0740	1.435	1.908	2.588	3.779
1.0	.1617	.3447	.5548	.8005	1.0950	1.461	1.940	2.626	3.827
2.0	.1932	.4093	.6547	.9382	1.2740	1.686	2.217	2.968	4.258
3.0	.2182	.4609	.7349	1.0490	1.4200	1.870	2.447	3.255	4.625
4.0	.2395	.5052	.8038	1.1460	1.5460	2.031	2.649	3.508	4.952
5.0	.2585	.5450	.8653	1.2310	1.6590	2.175	2.829	3.736	5.249
6.0	.2757	.5803	.8912	1.3090	1.7620	2.307	2.995	3.945	5.522
7.0	.2916	.6134	.9729	1.3820	1.8570	2.428	3.149	4.140	5.778
8.0	.3065	.6442	1.0210	1.4490	1.9460	2.543	3.293	4.322	6.018
9.0	.3205	.6733	1.0670	1.5130	2.0310	2.650	3.430	4.495	6.245
10.0	.3337	.7009	1.1100	1.5730	2.1100	2.753	3.559	4.660	6.462

Source: Reproduced with permission from J. Schmee, D. Gladstein, and W. Nelson, *Technometrics* **27**, 2(1985):119–128. Copyright 1985 by the American Statistical Association. All rights reserved.

13.4 MAXIMUM LIKELIHOOD ESTIMATORS

where

$$y = \log_e \frac{h}{1 - h}$$

This approximation is most useful for routine computer applications of this method.

Saw (1961) noted that the maximum likelihood estimators proposed by Cohen (1959) were biased, and derived the first-order bias correction terms for the case in which a constant proportion of observations are censored rather than all observations below the censoring point (e.g., L_Q). Schneider (1986) provided simple computational formulas

$$B_\mu = -\exp\left[2.692 - \frac{5.439(n - n_0)}{n + 1}\right] \quad (13.4)$$

$$B_\sigma = -\left[0.312 + \frac{0.859(n - n_0)}{n + 1}\right]^2 \quad (13.5)$$

where n is the total number of measurements, n_0 the number of censored observations, and the unbiased mean and standard deviation are given by

$$\hat{\mu}_u = \hat{\mu} - \frac{\hat{\sigma} B_\mu}{n + 1} \quad (13.6)$$

$$\hat{\sigma}_u = \hat{\sigma} - \frac{\hat{\sigma} B_\sigma}{n + 1} \quad (13.7)$$

Haas and Scheff (1990) suggest that this correction applies to the fixed censoring point case as well.

Illustration As an example, consider the data in Table 13.2 for TOC from a single background monitoring well over 10 quarterly monitoring events. Inspection

Table 13.2 Historical TOC example

Year	Quarter	TOC (mg/L)
1990	1	5
	2	7
	3	<1
	4	3
1991	1	<1
	2	4
	3	6
	4	5
1992	1	<1
	2	6

of Table 13.2 reveals that 30% of the data are censored at the quantification limit of 1 mg/L. Simple substitution of $L_Q/2$ for the three censored values yields $\bar{x} = 3.75$ and $s = 2.50$. To compute the maximum likelihood estimates note that $h = 0.3$ (i.e., the proportion censored), the mean and standard deviation of the seven uncensored values are $\bar{x}' = 5.14$ and $s' = 1.35$.
Therefore,

$$g = \frac{1.35^2}{(5.14 - 1)^2} = 0.1063$$

Inspection of Table 13.1 reveals that $\lambda(g, h) = 0.425$ (i.e., interpolating between $g = 0.1$ and $g = 0.2$). Using the approximation suggested by Haas and Scheff (1990) yields

$$y = \log_e \frac{0.3}{1 - 0.3} = -0.8473$$

$$\log_e \lambda(g, y) = 0.182344 - \frac{0.3756}{0.1063 + 1}$$
$$+ 0.10017(0.1063) + 0.78079(-0.8473) - 0.00581(0.1063)^2$$
$$- 0.06642(-0.8473)^2 - 0.0234(0.1063)(-0.8473)$$
$$+ 0.000174(0.1063)^3 + 0.001663(0.1063)^2(-0.8473)$$
$$- 0.00086(0.1063)(-0.8473)^2 - 0.00653(-0.8473)^3$$
$$= 0.427$$

which is in close agreement with the tabulated value. The maximum likelihood estimators are

$$\hat{\mu} = 5.14 - (5.14 - 1.0)\,0.427 = 3.37$$
$$\hat{\sigma} = 1.35^2 + (5.14 - 1.0)^2\,0.427 = 3.02$$

Adjusting for bias in the estimators yields

$$B_\mu = -\exp\left[2.692 - \frac{5.439(10 - 3)}{10 + 1}\right] = -0.463$$
$$B_\sigma = -\left[0.312 + \frac{0.859(10 - 3)}{10 + 1}\right]^2 = -0.737$$
$$\hat{\mu}_\mu = 3.37 - \frac{3.02(-0.463)}{10 + 1} = 3.50$$
$$\hat{\sigma}_\mu = 3.02 - \frac{3.02(-0.737)}{10 + 1} = 3.22$$

Interestingly, the simple substitution method yielded an overestimate of the mean (i.e., 3.75 versus 3.50) and an underestimate of the standard deviation (i.e., 2.50

versus 3.22) relative to the bias adjusted maximum likelihood estimates. The biased estimates were approximately 5% lower than the bias adjusted estimates.

13.5 RESTRICTED MAXIMUM LIKELIHOOD ESTIMATORS

To produce a computationally simple estimator for singly censored samples, Persson and Rootzen (1977) combined the method of maximum likelihood and the method of moments. Their estimator is quite close to the maximum likelihood estimator of Cohen (1959) but somewhat simpler to compute. Use of this method in analysis of environmental data was first considered by Haas and Scheff (1990). The restricted maximum likelihood estimators (RMLEs) are given by

$$\hat{\sigma} = \frac{1}{2}\left(\frac{\lambda a}{k} + \left[\left(\frac{\lambda a}{k}\right)^2 + 4\frac{b}{k}\right]^{1/2}\right)$$

$$\hat{\mu} = x_0 - \lambda \hat{\sigma}$$

where

$$\lambda = \Phi^{-1}\left(\frac{n_0}{n}\right)$$

$$k = n - n_0$$

$$a = \sum_{i=n_0+1}^{n} (x_i - x_0)$$

$$b = \sum_{i=n_0+1}^{n} (x_i - x_0)^2$$

Note that and $\Phi^{-1}(n_0/n)$ is the inverse normal (i.e., value of the standardized normal deviate for cumulative probability n_0/n). These estimators are biased at low levels of censoring; therefore, the correction described previously (Saw, 1961) can also be applied to the restricted maximum likelihood estimators.

All that is required to compute these estimators is a method of evaluating the inverse normal cumulative density function Φ^{-1}, which can be approximated as

$$\Phi^{-1}(p) \sim \begin{cases} 1.238T(1 + 0.0262T) & \text{if } p \geq 0.5 \\ -1.238T(1 + 0.0262T) & \text{if } p < 0.5 \end{cases}$$

where

$$T = (-\log_e[4p(1-p)])^{1/2}$$

(see Maindonald, 1984). For example, with $p = 0.95$, the approximation yields $\Phi^{-1}(0.95) = 1.648$, where the correct value is 1.645. In general, two decimal places of accuracy are guaranteed, which should be sufficient for this purpose.

Illustration Using the TOC data from Table 13.2, we obtain

$$\sum_{i=n_0+1}^{n} x_i = 36$$

$$\sum_{i=n_0+1}^{n} x_i^2 = 196$$

$$\sum_{i=n_0+1}^{n} x_i - x_0 = 29$$

$$\sum_{i=n_0+1}^{n} (x_i - x_0)^2 = 131$$

$$k = 10 - 3 = 7$$
$$\lambda = -0.523$$

leading to the estimates

$$\hat{\sigma} = \frac{1}{2}\left(\frac{-0.523(29)}{7} + \left[\left(\frac{-0.523(29)}{7}\right)^2 + 4\left(\frac{131}{7}\right)\right]^{1/2}\right) = 3.38$$

$$\hat{\mu} = 1 - (-0.523)(3.38) = 2.77$$

The nearly unbiased estimators are

$$\hat{\mu}_\mu = 2.91 \quad \text{and} \quad \hat{\sigma}_\mu = 3.60$$

which are reasonably close to the maximum likelihood estimates of $\hat{\mu}_\mu = 3.50$ and $\hat{\sigma}_\mu = 3.22$.

13.6 LINEAR ESTIMATORS

Although maximum likelihood estimators are optimal with respect to estimating variances, they are biased, particularly in small samples. To eliminate bias in small samples, the method of weighted least squares is used to develop the *best linear unbiased estimator* (BLUE) of the mean and variance of a censored normal distribution (for a review, see Cohen, 1991, Chap. 4). The problem has been considered in detail by Gupta (1952) and Sarhan and Greenberg (1962). The estimates are calculated as sums of products of the uncensored observations and the appropriate normal order statistics as

$$\hat{\mu} = \sum_{i=n_0+1}^{n} a_{1i} x_{i|n} \tag{13.8}$$

$$\hat{\sigma} = \sum_{i=n_0+1}^{n} a_{2i} x_{i|n} \tag{13.9}$$

where the x_i, $i = n_0 + 1, \ldots, n$ represent the $k = n - n_0$ uncensored observations. Gupta (1952) has tabulated the coefficients (a_{1i} and a_{2i}) for sample sizes up to $n = 10$ (reproduced here in Tables 13.3 and 13.4), and Sarhan and Greenberg (1962) have tabulated the coefficients for sample sizes up to $n = 20$. For sample sizes greater than 20, maximum likelihood estimators are preferred. The tables are designed principally for right-censored distributions. Gupta (1952) notes that for left-censored distributions, the observations should be ranked from largest to smallest and coefficients for the standard deviates should be of reversed sign.

Illustration Again using the TOC data in Table 13.2 and coefficients in Tables 13.3 and 13.4, we begin by arranging the data from largest to smallest (i.e., $x_{1|n} > x_{2|n} > \cdots > x_{k|n}$, where $k = n - n_0$) as 7, 6, 6, 5, 5, 4, 3. Then

$$\hat{\mu} = 0.02441(7) + 0.06362(6) + 0.08176(6) + 0.09612(5) \\ + 0.10887(5) + 0.12075(4) + 0.50448(3) = 4.07$$

$$\hat{\sigma} = 0.32526(7) + 0.17569(6) + 0.10582(6) + 0.05017(5) \\ + 0.00067(5) - 0.04699(4) - 0.61063(3) = 2.20$$

which are in fact more similar to the simple substitution estimates than the MLE or RMLE estimates, at least for this example. Note that for this estimator, there is no information regarding the value of the censoring point as in the ML or RMLE estimators, only knowledge that $n - k$ observations have been censored. This distribution has sometimes been termed *Type II censoring* to distinguish it from those cases in which there is a known censoring point (i.e., *Type I censoring*).

13.7 ALTERNATIVE LINEAR ESTIMATORS

Gupta (1952) suggested an alternative linear estimator for the mean and standard deviation of the censored normal distribution which is only slightly less efficient than the best linear estimators. The estimators are somewhat easier to compute since they require only the expected values of the order statistics from a standard normal distribution (see Table 13.5). The estimators are

$$\hat{\mu} = \sum_{i=n_0+1}^{n} b_i x_{(i)} \quad (13.10)$$

$$\hat{\sigma} = \sum_{i=n_0+1}^{n} c_i x_{(i)} \quad (13.11)$$

where

$$b_i = \frac{1}{n - n_0} - \frac{\bar{u}_k (u_i - \bar{u}_k)}{\sum_{j=n_0+1}^{n} (u_j - \bar{u}_k)^2} \quad (13.12)$$

Table 13.3 Coefficients for BLUE of the mean

n	n_1	n_r	$x(1)$	$x(2)$	$x(3)$	$x(4)$	$x(5)$	$x(6)$	$x(7)$	$x(8)$	$x(9)$
3	0	1	.0000	1.0000	—	—	—	—	—	—	—
4	0	1	.1161	.2408	.6431	—	—	—	—	—	—
	0	2	−.4056	1.4056	—	—	—	—	—	—	—
	1	1	—	−1.6834	1.6834	—	—	—	—	—	—
5	0	1	.1252	.1830	.2147	.4771	—	—	—	—	—
	0	2	−.0638	.1498	.9139	—	—	—	—	—	—
	0	3	−.7411	1.7411	—	—	—	—	—	—	—
	1	1	—	−1.0101	.0000	1.0101	—	—	—	—	—
	1	2	—	−2.0201	2.0201	—	—	—	—	—	—
6	0	1	.1183	.1510	.1680	.1828	.3799	—	—	—	—
	0	2	.0185	.1226	.1761	.6828	—	—	—	—	—
	0	3	−.2159	.0649	1.1511	—	—	—	—	—	—
	0	4	−1.0261	2.0261	—	—	—	—	—	—	—
	1	1	—	.3198	.1802	.1802	.3198	—	—	—	—
	1	2	—	.1539	.1781	.6680	—	—	—	—	—
	1	3	—	−.4578	1.4578	—	—	—	—	—	—
	2	2	—	—	.5000	.5000	—	—	—	—	—
7	0	1	.1088	.1295	.1400	.1487	.1571	.3159	—	—	—
	0	2	.0465	.1072	.1375	.1626	.5462	—	—	—	—
	0	3	−.0738	.0677	.1375	.8686	—	—	—	—	—
	0	4	−.3474	−.0135	1.3609	—	—	—	—	—	—
	0	5	−1.2733	2.2733	—	—	—	—	—	—	—
	1	1	—	.2718	.1520	.1524	.1520	.2718	—	—	—
	1	2	—	.1748	.1432	.1634	.5186	—	—	—	—
	1	3	—	−.0592	.1270	.9321	—	—	—	—	—
	1	4	—	−.8716	1.8716	—	—	—	—	—	—
	2	2	—	—	.4157	.1686	.4157	—	—	—	—
	2	3	—	—	.0000	1.0000	—	—	—	—	—
8	0	1	.0997	.1139	.1208	.1265	.1318	.1370	.2704	—	—
	0	2	.0569	.0962	.1153	.1309	.1451	.4555	—	—	—
	0	3	−.0167	.0677	.1084	.1413	.6993	—	—	—	—
	0	4	−.1549	.0176	.1001	1.0372	—	—	—	—	—
	0	5	−.4632	−.0855	1.5487	—	—	—	—	—	—
	0	6	−1.4915	2.4915	—	—	—	—	—	—	—
	1	1	—	.2367	.1315	.1319	.1319	.1315	.2367	—	—
	1	2	—	.1716	.1222	.1338	.1442	.4282	—	—	—
	1	3	—	.0431	.1061	.1406	.7102	—	—	—	—
	1	4	—	−.2519	.0741	1.1778	—	—	—	—	—
	1	5	—	−1.2462	2.2462	—	—	—	—	—	—
	2	2	—	—	.3569	.1431	.1431	.3569	—	—	—
	2	3	—	—	.1742	.1429	.6829	—	—	—	—
	2	4	—	—	−.4761	1.4761	—	—	—	—	—
	3	3	—	—	—	.5000	.5000	—	—	—	—
9	0	1	.0915	.1018	.1067	.1106	.1142	.1177	.1212	.2365	—
	0	2	.0602	.0876	.1006	.1110	.1204	.1294	.3909	—	—
	0	3	.0104	.0660	.0923	.1133	.1320	.5860	—	—	—
	0	4	−.0731	.0316	.0809	.1199	.8408	—	—	—	—
	0	5	−.2272	−.0284	.0644	1.1912	—	—	—	—	—
	0	6	−.5664	−.1521	1.7185	—	—	—	—	—	—
	0	7	−1.6868	2.6868	—	—	—	—	—	—	—

13.7 ALTERNATIVE LINEAR ESTIMATORS

Table 13.3 (Continued)

n	n_1	n_r	$x(1)$	$x(2)$	$x(3)$	$x(4)$	$x(5)$	$x(6)$	$x(7)$	$x(8)$	$x(9)$
	1	1	—	.2097	.1159	.1162	.1163	.1162	.1159	.2097	—
	1	2	—	.1626	.1074	.1148	.1214	.1275	.3663	—	—
	1	3	—	.0799	.0936	.1140	.1321	.5804	—	—	—
	1	4	—	−.0768	.0699	.1153	.8916	—	—	—	—
	1	5	—	−.4272	.0218	1.4054	—	—	—	—	—
	1	6	—	−1.5874	2.5874	—	—	—	—	—	—
	2	2	—	—	.3134	.1243	.1246	.1243	.3134	—	—
	2	3	—	—	.2040	.1191	.1330	.5440	—	—	—
	2	4	—	—	−.0527	.1098	.9429	—	—	—	—
	2	5	—	—	−.9229	1.9229	—	—	—	—	—
	3	3	—	—	—	.4315	.1370	.4315	—	—	—
	3	4	—	—	—	.0000	1.0000	—	—	—	—
10	0	1	.0843	.0921	.0957	.0986	.1011	.1036	.1060	.1085	.2101
	0	2	.0605	.0804	.0898	.0972	.1037	.1099	.1161	.3424	—
	0	3	.0244	.0636	.0818	.0962	.1089	.1207	.5045	—	—
	0	4	−.0316	.0383	.0707	.0962	.1185	.7078	—	—	—
	0	5	−.1240	−.0016	.0549	.0990	.9718	—	—	—	—
	0	6	−.2923	−.0709	.0305	1.3327	—	—	—	—	—
	0	7	−.6596	−.2138	1.8734	—	—	—	—	—	—
	0	8	−1.8634	2.8634	—	—	—	—	—	—	—
	1	1	—	.1884	.1036	.1040	.1041	.1041	.1040	.1036	.1884
	1	2	—	.1525	.0961	.1013	.1057	.1098	.1138	.3209	—
	1	3	—	.0942	.0846	.0979	.1095	.1204	.4933	—	—
	1	4	—	−.0043	.0665	.0938	.1179	.7261	—	—	—
	1	5	—	−.1866	.0351	.0892	1.0623	—	—	—	—
	1	6	—	−.5877	−.0289	1.6166	—	—	—	—	—
	1	7	—	−1.9000	2.9000	—	—	—	—	—	—
	2	2	—	—	.2798	.1099	.1103	.1103	.1099	.2798	—
	2	3	—	—	.2050	.1038	.1122	.1198	.4592	—	—
	2	4	—	—	.0606	.0935	.1178	.7281	—	—	—
	2	5	—	—	−.2648	.0735	1.1914	—	—	—	—
	2	6	—	—	−1.3406	2.3406	—	—	—	—	—
	3	3	—	—	—	.3807	.1193	.1193	.3807	—	—
	3	4	—	—	—	.1871	.1198	.6930	—	—	—
	3	5	—	—	—	−.4847	1.4847	—	—	—	—
	4	4	—	—	—	—	.5000	.5000	—	—	—

Source: Reproduced with permission from A.E. Sarhan and B.G. Greenberg, *Ann. Math. Stat.* **27**(1956):427–457.

$$c_i = \frac{u_i - \bar{u}_k}{\sum_{j=n_0+1}^{n} (u_j - \bar{u}_k)^2} \qquad (13.13)$$

$$\bar{u}_k = \frac{1}{n - n_0} \sum_{j=n_0+1}^{n} u_j \qquad (13.14)$$

or the arithmetic mean of the expected values of the uncensored sample elements. The values u_i are the expected values of the order statistics from a standard normal

148 CENSORED DATA

Table 13.4 Coefficients for BLUE of the variance

n	n_1	n_r	x(1)	x(2)	x(3)	x(4)	x(5)	x(6)	x(7)	x(8)	x(9)
3	0	1	−1.1816	1.1816	—	—	—	—	—	—	—
4	0	1	−.6971	−.1268	.8239	—	—	—	—	—	—
	0	2	−1.3654	1.3654	—	—	—	—	—	—	—
	1	1	—	−1.6834	1.6834	—	—	—	—	—	—
5	0	1	−.5117	−.1668	.0274	.6511	—	—	—	—	—
	0	2	−.7696	−.2121	.9817	—	—	—	—	—	—
	0	3	−1.4971	1.4971	—	—	—	—	—	—	—
	1	1	—	−1.0101	.0000	1.0101	—	—	—	—	—
	1	2	—	−2.0201	2.0201	—	—	—	—	—	—
6	0	1	−.4097	−.1685	−.0406	.0740	.5448	—	—	—	—
	0	2	−.5528	−.2091	−.0290	.7909	—	—	—	—	—
	0	3	−.8244	−.2760	1.1004	—	—	—	—	—	—
	0	4	−1.5988	1.5988	—	—	—	—	—	—	—
	1	1	—	−.7531	−.0829	.0829	.7531	—	—	—	—
	1	2	—	−1.1438	−.0878	1.2317	—	—	—	—	—
	1	3	—	−2.2717	2.2717	—	—	—	—	—	—
7	0	1	−.3440	−.1610	−.0681	.0114	.0901	.4716	—	—	—
	0	2	−.4370	−.1943	−.0718	.0321	.6709	—	—	—	—
	0	3	−.5848	−.2428	−.0717	.8994	—	—	—	—	—
	0	4	−.8682	−.3269	1.1951	—	—	—	—	—	—
	0	5	−1.6812	1.6812	—	—	—	—	—	—	—
	1	1	—	−.6108	−.1061	.0000	.1061	.6108	—	—	—
	1	2	—	−.8288	−.1258	.0248	.9298	—	—	—	—
	1	3	—	−1.2483	−.1548	1.4030	—	—	—	—	—
	1	4	—	−2.4712	2.4712	—	—	—	—	—	—
	2	2	—	—	−1.4176	.0000	1.4176	—	—	—	—
	2	3	—	—	−2.8352	2.8352	—	—	—	—	—
8	0	1	−.2978	−.1515	−.0796	−.0200	.0364	.0951	.4175	—	—
	0	2	−.3638	−.1788	−.0881	−.0132	.0570	.5868	—	—	—
	0	3	−.4586	−.2156	−.0970	.0002	.7709	—	—	—	—
	0	4	−.6110	−.2707	−.1061	.9878	—	—	—	—	—
	0	5	−.9045	−.3690	1.2735	—	—	—	—	—	—
	0	6	−1.7502	1.7502	—	—	—	—	—	—	—
	1	1	—	−.5184	−.1115	−.0361	.0361	.1115	.5184	—	—
	1	2	—	−.6608	−.1319	−.0318	.0630	.7615	—	—	—
	1	3	—	−.8894	−.1605	−.0197	1.0696	—	—	—	—
	1	4	—	−1.3337	−.2086	1.5423	—	—	—	—	—
	1	5	—	−2.6357	2.6357	—	—	—	—	—	—
	2	2	—	—	−1.0357	−.0674	.0674	1.0357	—	—	—
	2	3	—	—	−1.5661	−.0678	1.6338	—	—	—	—
	2	4	—	—	−3.1220	3.1220	—	—	—	—	—
	3	3	—	—	—	−3.2784	3.2784	—	—	—	—
9	0	1	−.2633	−.1421	−.0841	−.0370	.0062	.0492	.0954	.3757	—
	0	2	−.3129	−.1647	−.0938	−.0364	.0160	.0678	.5239	—	—
	0	3	−.3797	−.1936	−.1048	−.0333	.0317	.6797	—	—	—
	0	4	−.4766	−.2335	−.1181	−.0256	.8537	—	—	—	—
	0	5	−.6330	−.2944	−.1348	1.0622	—	—	—	—	—
	0	6	−.9355	−.4047	1.3402	—	—	—	—	—	—
	0	7	−1.8092	1.8092	—	—	—	—	—	—	—
	1	1	—	−.4527	−.1107	−.0532	.0000	.0532	.1107	.4527	—

Table 13.4 (Continued)

n	n_1	n_r	x(1)	x(2)	x(3)	x(4)	x(5)	x(6)	x(7)	x(8)	x(9)
	1	2	—	−.5544	−.1291	−.0563	.0109	.0775	.6514	—	—
	1	3	—	−.7015	−.1535	−.0578	.0299	.8828	—	—	—
	1	4	—	−.9399	−.1896	−.0558	1.1852	—	—	—	—
	1	5	—	−1.4057	−.2534	1.6591	—	—	—	—	—
	1	6	—	−2.7753	2.7753	—	—	—	—	—	—
	2	2	—	—	−.8817	−.0885	.0000	.0885	.8317	—	—
	2	3	—	—	−1.1222	−.1023	.0223	1.2022	—	—	—
	2	4	—	—	−1.6894	−.1227	1.8122	—	—	—	—
	2	5	—	—	−3.3620	3.3620	—	—	—	—	—
	3	3	—	—	—	−1.8213	.0000	1.8213	—	—	—
	3	4	—	—	—	−3.6426	3.6426	—	—	—	—
10	0	1	−.2364	−.1334	−.0851	−.0465	−.0119	.0215	.0559	.0937	.3423
	0	2	−.2753	−.1523	−.0947	−.0488	−.0077	.0319	.0722	.4746	—
	0	3	−.3252	−.1758	−.1058	−.0502	−.0006	.0469	.6107	—	—
	0	4	−.3930	−.2063	−.1192	−.0501	.0111	.7576	—	—	—
	0	5	−.4919	−.2491	−.1362	−.0472	.9243	—	—	—	—
	0	6	−.6520	−.3150	−.1593	1.1263	—	—	—	—	—
	0	7	−.9625	−.4357	1.3981	—	—	—	—	—	—
	0	8	−1.8608	1.8608	—	—	—	—	—	—	—
	1	1	—	−.4034	−.1074	−.0616	−.0201	.0201	.0616	.1074	.4034
	1	2	—	−.4803	−.1235	−.0674	−.0166	.0325	.0827	.5726	—
	1	3	—	−.5842	−.1440	−.0734	−.0097	.0514	.7599	—	—
	1	4	—	−.7359	−.1719	−.0797	.0031	.9844	—	—	—
	1	5	—	−.9831	−.2145	−.0859	1.2835	—	—	—	—
	1	6	—	−1.4678	−.2918	1.7595	—	—	—	—	—
	1	7	—	−2.8960	2.8960	—	—	—	—	—	—
	2	2	—	—	−.7021	−.0947	−.0310	.0310	.0947	.7021	—
	2	3	—	—	−.8898	−.1101	−.0262	.0549	.9711	—	—
	2	4	—	—	−1.1952	−.1318	−.0144	1.3415	—	—	—
	2	5	—	—	−1.7947	−.1688	1.9635	—	—	—	—
	2	6	—	—	−3.5677	3.5677	—	—	—	—	—
	3	3	—	—	—	−1.2832	−.0559	.0559	1.2832	—	—
	3	4	—	—	—	−1.9791	−.0553	2.0344	—	—	—
	3	5	—	—	—	−3.9511	3.9511	—	—	—	—
	4	4	—	—	—	—	−4.0761	4.0761	—	—	—

Source: Reproduced with permission from A.E. Sarhan and B.G. Greenberg, *Ann. Math. Stat.* **27**(1956): 427–457.

distribution [i.e., $N(0, 1)$]. In Table 13.3 expected values of normal order statistics (u_j) are tabulated for sample sizes of $n = 2$ to $n = 50$, so that this alternative linear estimator can be computed for most practical problems.

Illustration Returning to the TOC example and ranking from smallest to largest: 3, 4, 5, 5, 6, 6, 7,

Table 13.5 Expected values of normal order statistics

Rank	2	3	4	5	6	7	8
1	0.5642	0.8463	1.0294	1.1630	1.2672	1.3522	1.4236
2	−0.5642	0.0000	0.2970	0.4950	0.6418	0.7574	0.8522
3	—	−0.8463	−0.2970	0.0000	0.2015	0.3527	0.4728
4	—	—	−1.0294	−0.4950	−0.2015	0.0000	0.1525

Rank	9	10	11	12	13	14	15
1	1.4850	1.5388	1.5864	1.6292	1.6680	1.7034	1.7359
2	0.9323	1.0014	1.0619	1.1157	1.1641	1.2079	1.2479
3	0.5720	0.6561	0.7288	0.7928	0.8498	0.9011	0.9477
4	0.2745	0.3758	0.4620	0.5368	0.6028	0.6618	0.7149
5	0.0000	0.1227	0.2249	0.3122	0.3883	0.4556	0.5157
6	−0.2745	−0.1227	0.0000	0.1026	0.1905	0.2673	0.3353
7	−0.5720	−0.3758	−0.2249	−0.1026	0.0000	0.0882	0.1653

Rank	16	17	18	19	20	21	22
1	1.7660	1.7939	1.8200	1.8445	1.8675	1.8892	1.9097
2	1.2847	1.3188	1.3504	1.3799	1.4076	1.4336	1.4582
3	0.9903	1.0295	1.0657	1.0995	1.1309	1.1605	1.1882
4	0.7632	0.8074	0.8481	0.8859	0.9210	0.9538	0.9846
5	0.5700	0.6195	0.6648	0.7066	0.7454	0.7815	0.8153
6	0.3962	0.4513	0.5016	0.5477	0.5903	0.6298	0.6667
7	0.2338	0.2952	0.3508	0.4016	0.4483	0.4915	0.5316
8	0.0773	0.1460	0.2077	0.2637	0.3149	0.3620	0.4056
9	−0.0773	0.0000	0.0688	0.1307	0.1870	0.2384	0.2858
10	−0.2338	−0.1460	−0.0688	0.0000	0.0620	0.1184	0.1700
11	−0.3962	−0.2952	−0.2077	−0.1307	−0.0620	0.0000	0.0564

Rank	23	24	25	26	27	28	29
1	1.9292	1.9477	1.9653	1.9822	1.9983	2.0137	2.0285
2	1.4814	1.5034	1.5243	1.5442	1.5633	1.5815	1.5989
3	1.2144	1.2392	1.2628	1.2851	1.3064	1.3267	1.3462
4	1.0136	1.0409	1.0668	1.0914	1.1147	1.1370	1.1582
5	0.8470	0.8768	0.9050	0.9317	0.9570	0.9812	1.0041
6	0.7012	0.7335	0.7641	0.7929	0.8202	0.8461	0.8708
7	0.5690	0.6040	0.6369	0.6679	0.6973	0.7251	0.7515
8	0.4461	0.4839	0.5193	0.5527	0.5841	0.6138	0.6420
9	0.3297	0.3705	0.4086	0.4444	0.4780	0.5098	0.5398
10	0.2175	0.2616	0.3027	0.3410	0.3771	0.4110	0.4430
11	0.1081	0.1558	0.2001	0.2413	0.2798	0.3160	0.3501
12	0.0000	0.0518	0.0995	0.1439	0.1852	0.2239	0.2602
13	−0.1081	−0.0518	0.0000	0.0478	0.0922	0.1336	0.1724
14	−0.2175	−0.1558	−0.0995	−0.0478	0.0000	0.0444	0.0859

13.7 ALTERNATIVE LINEAR ESTIMATORS

Table 13.5 (Continued)

Rank	30	31	32	33	34	35	36
1	2.0428	2.0565	2.0697	2.0824	2.0947	2.1066	2.1181
2	1.6156	1.6317	1.6471	1.6620	1.6764	1.6902	1.7036
3	1.3648	1.3827	1.3999	1.4164	1.4323	1.4476	1.4624
4	1.1786	1.1980	1.2167	1.2347	1.2520	1.2686	1.2847
5	1.0261	1.0471	1.0672	1.0865	1.1051	1.1229	1.1402
6	0.8944	0.9169	0.9384	0.9590	0.9789	0.9979	1.0162
7	0.7767	0.8007	0.8236	0.8455	0.8666	0.8868	0.9062
8	0.6688	0.6944	0.7187	0.7420	0.7643	0.7857	0.8063
9	0.5683	0.5955	0.6213	0.6460	0.6695	0.6921	0.7138
10	0.4733	0.5021	0.5294	0.5555	0.5804	0.6043	0.6271
11	0.3824	0.4129	0.4418	0.4694	0.4957	0.5208	0.5449
12	0.2945	0.3269	0.3575	0.3867	0.4144	0.4409	0.4662
13	0.2088	0.2432	0.2757	0.3065	0.3358	0.3637	0.3903
14	0.1247	0.1613	0.1957	0.2283	0.2592	0.2886	0.3166
15	0.0415	0.0804	0.1169	0.1515	0.1842	0.2152	0.2446
16	−0.0415	0.0000	0.0389	0.0755	0.1101	0.1428	0.1739
17	−0.1247	−0.0804	−0.0389	0.0000	0.0366	0.0712	0.1040
18	−0.2088	−0.1613	−0.1169	−0.0755	−0.0366	0.0000	0.0346

Rank	37	38	39	40	41	42	43
1	2.1293	2.1401	2.1506	2.1608	2.1707	2.1803	2.1897
2	1.7166	1.7291	1.7413	1.7351	1.7646	1.7757	1.7865
3	1.4768	1.4906	1.5040	1.5170	1.5296	1.5419	1.5538
4	1.3002	1.3151	1.3296	1.3437	1.3573	1.3705	1.3833
5	1.1568	1.1728	1.1883	1.2033	1.2178	1.2319	1.2456
6	1.0339	1.0509	1.0674	1.0833	1.0987	1.1136	1.1281
7	0.9250	0.9430	0.9604	0.9772	0.9935	1.0092	1.0245
8	0.8260	0.8451	0.8634	0.8811	0.8983	0.9148	0.9308
9	0.7346	0.7547	0.7740	0.7926	0.8106	0.8279	0.8447
10	0.6490	0.6701	0.6904	0.7099	0.7287	0.7469	0.7645
11	0.5679	0.5900	0.6113	0.6318	0.6515	0.6705	0.6889
12	0.4904	0.5136	0.5359	0.5574	0.5780	0.5979	0.6171
13	0.4158	0.4401	0.4635	0.4859	0.5075	0.5283	0.5483
14	0.3434	0.3689	0.3934	0.4169	0.4394	0.4611	0.4820
15	0.2727	0.2995	0.3252	0.3498	0.3734	0.3960	0.4178
16	0.2034	0.2316	0.2585	0.2842	0.3089	0.3326	0.3553
17	0.1351	0.1647	0.1929	0.2199	0.2457	0.2704	0.2942
18	0.0674	0.0985	0.1282	0.1564	0.1835	0.2093	0.2341
19	0.0000	0.0328	0.0640	0.0936	0.1219	0.1490	0.1749
20	−0.0674	−0.0328	0.0000	0.0312	0.0608	0.0892	0.1163
21	−0.1351	−0.0985	−0.0640	−0.0312	0.0000	0.0297	0.0580

(Continued)

Table 13.5 (Continued)

Rank	44	45	46	47	48	49	50
1	2.1988	2.2077	2.2164	2.2249	2.2331	2.2412	2.2491
2	1.7971	1.8073	1.8173	1.8271	1.8366	1.8458	1.8549
3	1.5653	1.5766	1.5875	1.5982	1.6086	1.6187	1.6286
4	1.3957	1.4078	1.4196	1.4311	1.4422	1.4531	1.4637
5	1.2588	1.2717	1.2842	1.2964	1.3083	1.3198	1.3311
6	1.1421	1.1558	1.1690	1.1819	1.1944	1.2066	1.2185
7	1.0392	1.0536	1.0675	1.0810	1.0942	1.1070	1.1195
8	0.9463	0.9614	0.9760	0.9902	1.0040	1.0174	1.0304
9	0.8610	0.8767	0.8920	0.9068	0.9212	0.9353	0.9489
10	0.7815	0.7979	0.8139	0.8294	0.8444	0.8590	0.8732
11	0.7067	0.7238	0.7405	0.7566	0.7723	0.7875	0.8022
12	0.6356	0.6535	0.6709	0.6877	0.7040	0.7198	0.7351
13	0.5676	0.5863	0.6044	0.6219	0.6388	0.6552	0.6712
14	0.5022	0.5217	0.5405	0.5586	0.5763	0.5933	0.6099
15	0.4389	0.4591	0.4787	0.4976	0.5159	0.5336	0.5508
16	0.3772	0.3983	0.4187	0.4383	0.4573	0.4757	0.4935
17	0.3170	0.3390	0.3602	0.3806	0.4003	0.4194	0.4379
18	0.2579	0.2808	0.3029	0.3241	0.3446	0.3644	0.3836
19	0.1997	0.2236	0.2465	0.2686	0.2899	0.3105	0.3304
20	0.1422	0.1671	0.1910	0.2140	0.2361	0.2575	0.2781
21	0.0851	0.1111	0.1360	0.1599	0.1830	0.2051	0.2265
22	0.0283	0.0555	0.0814	0.1064	0.1303	0.1534	0.1756
23	−0.0283	0.0000	0.0271	0.0531	0.0781	0.1020	0.1251
24	−0.0851	−0.0555	−0.0271	0.0000	0.0260	0.0509	0.0749
25	−0.1422	−0.1111	−0.0814	−0.0531	−0.0260	0.0000	0.0250

Source: Reproduced with permission from D.B. Owen, *Handbook of Statistical Tables,* Addison-Wesley, Reading, Mass., 1962, pp. 152–154.

we obtain the equation for the mean as

$$\hat{\mu} = 0.2838(3) + 0.2413(4) + 0.2001(5) + 0.1576(5) \\ + 0.1105(6) + 0.0525(6) - 0.0457(7) = 4.26$$

and standard deviation as

$$\hat{\sigma} = -0.3041(3) - 0.2124(4) - 0.1235(5) - 0.0318(5) \\ + 0.0698(6) + 0.1950(6) + 0.4070(7) = 1.90$$

which are similar to the BLU estimators of 4.07 and 2.20, respectively. Again, application of these Type II censoring estimators to a Type I censoring problem yields different results than ML or RML estimators, because the linear estimator does not incorporate knowledge of the censoring point. As the distance between the censoring point and the smallest measured value increases, as in this example, the discrepancy between Type I and Type II estimators will increase. In general, the linear estimators will have higher mean and lower standard deviation than the ML or RML estimators.

13.8 DELTA DISTRIBUTIONS

An alternative approach to the censored data problem is the *delta distribution approach* (Aitchison, 1955), in which the parameters of a continuous probability distribution with some probability mass at zero are estimated. Owen and DeRouen (1980) have shown that the lognormal delta distribution is optimal for measuring exposure to air contaminants. This distribution is well suited to the current problem because it accommodates both the problem of nondetects and the lognormality of the constituent concentrations detected. Usually, the concentration of contaminants in environmental media are lognormally distributed (Ott, 1990). If detection and/or quantitation limits are close to zero, there is typically little loss of information in assuming that the censored portion of the distribution is at zero.

Statistically, the delta distribution is a two-parameter lognormal distribution in which some proportion of the probability mass is located at zero. Computationally, the mean and variance of the delta distribution are obtained as follows.

Denoting the number of observations that are not detected as n_0, the number of detected measurements as n_1, and the total number of measurements as n, the mean and variance of the lognormal delta distribution are given by

$$\hat{\mu} = \frac{n_1}{n} \exp(\bar{y}) \Delta_{n_1}\left(\frac{s_y^2}{2}\right) \tag{13.15}$$

$$\hat{\sigma}^2 = \frac{n_1}{n} \exp(2\bar{y}) \left[\Delta_{n_1}(2s_y^2) - \frac{n_1 - 1}{n - 1} \Delta_{n_1}\left(\frac{n_1 - 2}{n_1 - 1} s_y^2\right) \right] \tag{13.16}$$

where

$$\bar{y} = \sum_{i=1}^{n_1} \frac{\log_e x_i}{n}$$

$$s_y = \sqrt{\sum_{i=1}^{n_1} \frac{(\log_e x_i - \bar{y})^2}{n - 1}}$$

that is, the mean and standard deviation of the natural logarithms of the detected values, and

$$\Delta_{n_1}(z) = 1 + \frac{n_1 - 1}{n_1} z + \frac{(n_1 - 1)^3}{n_1^2 2!} \frac{z^2}{n_1 + 1} + \frac{(n_1 - 1)^5}{n_1^3 3!} \frac{z^3}{(n_1 + 1)(n_1 + 3)} + \cdots \tag{13.17}$$

is a Bessel function with argument z, that can take on the values $z = 2s_y^2$ or $z = s_y^2/2$ or $z = s_y^2[(n_1 - 2)/(n_1 - 1)]$ as shown in equations (13.15) to (13.17). In practice,

154 CENSORED DATA

this function generally converges in fewer than 10 steps, the first three of which are shown above.

Aitchison's (1955) results also apply to a normal distribution with some probability mass at zero. In this case, the adjusted mean concentration is computed as

$$\hat{\mu} = \left(1 - \frac{n_0}{n}\right)\bar{x}' \qquad (13.18)$$

where $\bar{x}'$ is the average of the detected values, n the total number of samples, and n_0 the number of samples in which the compound is not present. The standard deviation is

$$\hat{\sigma} = \sqrt{\left(1 - \frac{n_0}{n}\right)s^{2\prime} + \frac{n_0}{n}\left(1 - \frac{n_0 - 1}{n - 1}\right)\bar{x}^{2\prime}} \qquad (13.19)$$

where s' is the standard deviation of the detected measurements. In the case of a normal distribution, however, these results are identical to substituting zero for the nondetects in the usual calculation for the mean and variance.

Illustration For the TOC example, we have $n = 10$, $n_0 = 3$, $n_1 = 7$, $\bar{y} = 1.60$, and $s_y = 0.29$. After 10 iterations,

$$\Delta_{n_1}\left(\frac{s_y^2}{2}\right) = 1.0365$$

$$\Delta_{n_1}(2s_y^2) = 1.1522$$

$$\Delta_{n_1}\left(\frac{n_1 - 2}{n_1 - 1}s_y^2\right) = 1.0614$$

The mean is therefore

$$\hat{\mu} = \frac{7}{10}(e^{1.6})1.0365 = 3.59$$

and standard deviation

$$\hat{\sigma} = \frac{7}{10}(e^{2(1.6)})\left[1.1522 - \frac{6}{9}(1.0614)\right] = 2.76$$

These estimates are, in fact, quite similar to the ML estimates for the censored normal distribution, despite the fact that they assume an underlying lognormal distribution and place probability mass at zero rather than the censoring point, which in this case is 1.0 mg/L.

Using Aitchison's (1955) estimator for the normal distribution, we obtain

$$\hat{\mu} = \left(1 - \frac{3}{10}\right) 5.14 = 3.60$$

$$\hat{\sigma} = \left[\left(1 - \frac{3}{10}\right) 1.35^2 + \frac{3}{10}\left(1 - \frac{3-1}{10-1}\right) 5.14^2\right]^{1/2} = 2.73$$

which is remarkably similar to the lognormal results. Note that these estimates are identical to what would have been obtained had we simply substituted zero for the censored values. Recall that substitution of $L_D/2 = 0.5$ instead of zero yielded 3.75 and 2.50 for the mean and standard deviation, respectively.

13.9 REGRESSION METHODS

Hashimoto and Trussell (1983) and Gilliom and Helsel (1986) have suggested a method by which values for the censored observations can be imputed based on a linear regression of order statistics on measured concentrations for the uncensored data. Following Gilliom and Helsel (1986), normal scores are computed as

$$z = \Phi^{-1}\left(\frac{r}{n+1}\right) \qquad (13.20)$$

where r is the rank of the measurement ($r = n_0 + 1, \ldots, n$), n the total number of measurements, and Φ^{-1} the inverse normal cumulative distribution function. A least-squares regression of concentration on normal scores for all uncensored data can be used to extrapolate values to the censored observations (ranks $r = 1, \ldots, n_0$). Estimated values below zero are set equal to zero. The estimated values are treated as observed and the mean and variance of all measurements (i.e., estimated and observed concentrations) are used.

The least-squares estimates of the intercept and slope of the regression line for the uncensored values are the usual

$$b_1 = \frac{\sum_{i=n_0+1}^{n}(x_i - \bar{x})(z_i - \bar{z})}{\sum_{i=n_0+1}^{n}(z_i - \bar{z})^2} \qquad (13.21)$$

$$b_0 = \bar{x} - b_1 \bar{z} \qquad (13.22)$$

The prediction equation for a new concentration given the inverse normal probability (z) associated with rank r is therefore

$$\hat{x} = b_0 + b_1 z \qquad (13.23)$$

CENSORED DATA

which is computed for ranks 1 through n_0 (i.e., the n_0 lowest values). It is important to note that nothing prevents the estimated concentration for a censored value (i.e., $\hat{x}$) from being less than zero or greater than the censoring value. Estimated values less than zero should be set to zero and values greater than the censoring point (i.e., L_Q) should be set to that value.

Illustration For the TOC data, the ordered values and inverse normal values associated with their rank are as follows:

Rank	Value	z	Rank	Value	z
1	<1	−1.34	6	5	0.11
2	<1	−0.91	7	5	0.35
3	<1	−0.60	8	6	0.60
4	3	−0.35	9	6	0.91
5	4	−0.11	10	7	1.34

The mean uncensored values are $\bar{x} = 5.14$ and $\bar{z} = .41$. The least-squares estimates are

$$b_1 = \frac{4.59}{2.09} = 2.20$$

$$b_0 = 5.14 - 2.20(0.41) = 4.24$$

The estimated values for the three censored values are:

Rank	Estimated Concentration
1	1.30
2	2.25
3	2.92

which are all greater than the L_Q of 1.0 mg/L. In light of this, we would substitute the L_Q for all three censored values and obtain

$$\hat{\mu} = 3.9$$

$$\hat{\sigma} = 2.28$$

However, had we used the imputed values, the estimates would have been $\hat{\mu} = 4.25$ and $\hat{\sigma} = 1.85$, which are quite similar to the linear estimates. Gilliom and Helsel (1986) have also indicated that best overall success is obtained by first transforming

13.10 SUBSTITUTION OF EXPECTED VALUES OF ORDER STATISTICS

the uncensored measurements to a natural log scale, which in our case yields

$$1.10, 1.39, 1.61, 1.61, 1.79, 1.79, 1.95$$

and a censoring value of $\log_e(1) = 0$. Here $b_0 = 1.42$, $b_1 = 0.45$, and the three uncensored values are:

Rank	$\log_e$ Estimated Censored Value	Estimated Raw Concentration
1	0.82	2.27
2	1.01	2.75
3	1.15	3.16

which again are all above the censoring limit of $\log_e(1) = 0$. If, however, we use these estimated values, we obtain $\hat{\mu}_{\log_e} = 1.42$ and $\hat{\sigma}_{\log_e} = 0.38$. These estimates may then be used to estimate the mean and standard deviation of the lognormal distribution as

$$\hat{\mu} = \exp\left(\hat{\mu}_{\log_e} + \frac{\hat{\sigma}^2_{\log_e}}{z}\right) = 4.45$$

$$\hat{\sigma} = [\hat{\mu}^2 (\exp(\hat{\sigma}^2_{\log_e}) - 1)]^{1/2} = 1.75$$

which are quite similar to the normal regression results.

13.10 SUBSTITUTION OF EXPECTED VALUES OF ORDER STATISTICS

Gleit (1985) suggested an interactive procedure for obtaining improved estimators based on simple substitution methods. The basic idea is to replace the censored observations with the expected values of the order statistics for the n_0 censored observations conditional on a provisional estimate of the mean and standard deviation (e.g., begin by substituting $L_Q/2$ for the censored values). The algorithm is as follows:

1. Compute a provisional estimate of the mean and standard deviation $\hat{\mu}_1$ and $\hat{\sigma}_1$ by substituting $L_Q/2$ for the censored observations.
2. On the basis of $\hat{\mu}_1$ and $\hat{\sigma}_1$, calculate the expected values for the first n_0 order statistics. To do this, select the expected values of the first n_0 order statistics for a sample of size n from a standard normal distribution [i.e., $N(0, 1)$] from Table 13.5 and transform them to the current $N(\hat{\mu}_1, \hat{\sigma}_1)$ scale by

$$u'_i = u_i \hat{\sigma}_1 + \hat{\mu}_1 \tag{13.24}$$

where u_i is the expected value of the order statistic on the $N(0, 1)$ scale and u_i' is the expected value of the normal order statistic on the $N(\hat{\mu}_1, \hat{\sigma}_1)$ scale.

3. Now that all n data points are "known," compute the usual estimate of the mean and standard deviation and call them $\hat{\mu}_2$ and $\hat{\sigma}_2$.

4. Continue steps 2 and 3 until the difference between $\hat{\mu}_t$ and $\hat{\mu}_{t-1}$ and $\hat{\sigma}_t$ and $\hat{\sigma}_{t-1}$ are less than 10^{-6}.

Illustration In the TOC example, the first $n_0 = 3$ samples were censored and the corresponding $N(0, 1)$ order statistics are

$$u_1' = -1.5388$$
$$u_2' = -1.0014$$
$$u_3' = -0.6561$$

Beginning with the simple substitution estimates (i.e., substituting $L_Q/2$), we have

$$\hat{\mu}_1 = 3.75$$
$$\hat{\sigma}_1 = 2.50$$

Therefore,

$$\hat{\mu}_1' = -1.5388(2.50) + 3.75 = -0.0970$$
$$\hat{\mu}_2' = -1.0014(2.50) + 3.75 = 1.2465$$
$$\hat{\mu}_3' = -0.6561(2.50) + 3.55 = 1.9098$$

Repeated iterations yielded

$$\hat{\mu} = 4.37$$
$$\hat{\sigma} = 1.70$$

The final values for the three censored measurements were 1.743, 2.658, and 3.247 mg/L. Note that all imputed values are greater than the detection limit but consistent with the detected concentrations. As such, this approach yields values similar to the linear estimates that ignore information regarding the actual censoring point (i.e., Type II censoring). The imputed values are somewhat larger than those obtained by the normal regression method, probably due to the normality assumption, which is not present in the unweighted least-squares interpolation. However, estimates of the mean and standard deviation of both methods are virtually identical.

13.11 COMPARISON OF ESTIMATORS

There have been several studies comparing the statistical properties of the various estimators, both analytically and via Monte Carlo simulation. In general, the studies have compared the ability of these estimators to recover the true mean and standard deviation of the uncensored parent distribution. In general, the analytical studies have found that the modified maximum likelihood estimators perform nearly as well as the maximum likelihood estimator and that the alternative linear estimators perform nearly as well as the best linear unbiased estimator (Schneider, 1986). As expected, the efficiency of the estimators decreases with the degree of censoring and the effect is more pronounced on the estimate of the variance versus the mean (Sarhan and Greenberg, 1962).

Somewhat more relevant are Monte Carlo simulation studies which compared various estimators in the context of singly left censored samples that are observed in environmental monitoring applications. Again, these studies focus on the ability to recover the mean and variance of the total distribution. The two most relevant studies to this area were conducted by Gilliom and Helsel (1986) and Haas and Scheff (1990). Gilliom and Helsel (1986) considered the following eight estimators:

1. *ZE*: censored observations set to zero
2. *DL*: censored observations set equal to the L_D
3. *UN*: censored observations set equal to $L_D/2$
4. *NR*: censored observations were imputed using the normal regression method
5. *LR*: same as NR but data log transformed
6. *NM*: maximum likelihood estimates
7. *LM*: maximum likelihood estimates based on log transformed data followed by reverse transformation due to Aitchison and Brown (1957)
8. *DT*: the lognormal delta distribution

Evaluation of the reliability of the methods was based on root-mean-square errors (RMSEs) computed from the actual parameters of the underlying distribution used to generate the simulated data: for example,

$$\text{RMSE} = \left[\sum_{i=1}^{N} \left(\frac{\bar{x}_i - \mu}{\mu} \right)^2 / N \right]^{1/2} \tag{13.25}$$

where $\bar{x}_i$ is the estimate of the mean for the ith of N data sets and μ is the true mean value used to generate the data.

Results of their study revealed that overall, the lognormal regression method (LR) of imputing the censored values performed best. The maximum likelihood method computed on log-transformed data performed best for estimating the median and interquartile range.

Haas and Scheff (1990) compared maximum likelihood, bias-corrected maximum likelihood, restricted maximum likelihood, one-half the detection limit, and normal regression methods in terms of recovering the mean of the parent distribution. The parent distributions were standard normal, a normal mixture with common variance, and a normal mixture with different variances. The authors noted that all methods yield biased results as the censoring level approaches 0.5. The normal regression method is positively biased (i.e., overestimates the true mean) over the entire range of censoring values, and for small amounts of censoring was the most strongly biased estimator. Overall, the bias-corrected restricted maximum likelihood estimator performed the best. Both one-half the detection limit and normal regression methods were shown to have "substantial deficiencies" with respect to bias and/or RMSE. The restricted maximum likelihood method also performed best for the normal mixture distributions which are more characteristic of the bimodal, heavy-tailed, and skewed distributions commonly found in environmental data.

The discrepancy between the two studies in terms of the utility of normal regression models is puzzling. Perhaps the emphasis on lognormally distributed generating distributions in the Gilliom and Helsel (1986) study versus the normal or normal mixture generating distributions used in the Haas and Scheff (1990) study may account for some of the difference.

Several other relevant studies have been conducted and have been summarized by Haas and Scheff (1990). Perhaps most interesting is a study by Gleit (1985) in which substituting expected values of the normal order statistics for the censored values greatly outperformed both maximum likelihood and constant fill-in procedures (e.g., $L_D/2$) for samples drawn from normal distributions. Other studies (El-Shaarawi, 1989; Hashimoto and Trussell, 1983; Helsel and Cohn, 1988) generally found similar results for regression- and maximum likelihood–based methods.

McNichols and Davis (1988) performed a limited study on the effect of type of censored data estimator on overall false positive rates for prediction limits of Gibbons (1987) and Davis and McNichols (1987). The prediction limit factors were selected to provide an overall significance level of 5% across eight downgradient monitoring wells. Background sample sizes of 4 and 12 were considered with censoring levels ranging from 20 to 90%. Data were generated from normal and skewed normal distributions. Four methods were compared:

1. Nondetects = 0
2. Nondetects = $L_D/2$
3. Nondetects = L_D
4. Maximum likelihood estimators

Results of the study revealed that for high levels of censoring (i.e., >80%) none of methods achieved their intended nominal false positive rates; however, the Davis and McNichols (1987) prediction limits, which incorporate a verification resample, dramatically decreased the overall false positive rates for all methods, but not to the nominal level. The effect of high degrees of censoring were worse for the maximum likelihood methods than the simple substitution methods; however, this may in part be due to the limited background sample sizes (i.e., 4 and 12) and the way in which they were drawn (i.e., each sample consisted of four aliquots and the maximum likelihood estimator was applied first to the aliquots and then to the set of 4 or 12 samples).

In terms of the TOC example, Table 13.6 displays a summary of the estimates. Inspection of the table reveals the following. First, MLE and RMLE typically have higher standard deviations and lower means, due to their dependence on the

Table 13.6 Summary of estimates for TOC example

Estimator	Mean	Standard Deviation
MLE (adjusted)	3.50	3.22
RMLE (adjusted)	2.91	3.60
DELTA normal	3.60	2.73
Delta lognormal	3.59	2.76
BLUE	4.07	2.20
Alternative linear estimator	4.26	1.90
Regression method	4.25	1.85
Substitution of order statistics	4.37	1.70
Substitution of $L_Q/2$	3.75	2.50

censoring point, which in this example is relatively far from the measured values. The simple substitution method (i.e., $L_Q/2$) and both normal and lognormal delta distributions are quite similar, probably because the censoring point is close to zero. The linear estimators, and the two more sophisticated substitutions methods (i.e., based on linear regression and expected values of order statistics), all yield quite similar results, due to the fact that they all treat the problem as Type II censoring; that is, they are not dependent on the value of the censoring point.

13.12 FURTHER SIMULATION RESULTS

Although far from conclusive, Gibbons (1994) performed a limited Monte Carlo simulation to compare the methods described previously for a typical groundwater monitoring problem. As in the examples, $n = 10$ background measurements were simulated. First, root-mean-square errors (RMSEs) were compared for the mean and standard deviation of the various estimators over 1000 replications. Second, the new monitoring measurements were drawn from the same distribution and compared to a normal prediction limit for one of two samples in bounds at each of 10 monitoring wells. The resulting limit was

$$\hat{\mu} + \hat{\sigma}1.925$$

In the event that the new sample exceeded the limit, a second sample was generated and failure was indicated if both samples exceeded the prediction limit (i.e., false positive result). The generating distribution was normal with $\mu = 5$ and $\sigma = 1$ [i.e., $N(5, 1)$]. With only 10 background measurements, censoring was restricted in probability to 20 and 50%. Background samples with fewer than three detected measurements were discarded.

Results of the simulation are reproduced here in Tables 13.7 and 13.8. Results of this very limited study indicate that the MLE is the best overall estimator even in a sample of only 10 measurements, in terms of minimizing both false positive rates and recovering the true population parameters. Interestingly, the normal and lognormal

Table 13.7 Comparison of estimators for 20% type I censoring

Estimator	False Positive Rate	RMSE $\hat{\mu}$	RMSE $\hat{\sigma}$
MLE	0.074	0.074	0.292
RMLE	0.292	0.065	0.297
Linear	0.385	0.072	0.312
Normal delta	0.046	0.213	1.325
Lognormal delta	0.075	0.213	1.329
Regression	0.343	0.073	0.298
Order statistics	0.377	0.066	0.290
$L_Q/2$	0.132	0.119	0.559

forms of the delta distribution were most effective at minimizing false positive rates, yet they were the worst at recovering the true population mean and standard deviation. The linear estimator and the two more sophisticated substitution methods (i.e., regression and expected value of order statistics) were not effective at controlling the overall false positive rate. The simple substitution method (i.e., $L_Q/2$) actually outperformed the more sophisticated substitution methods in terms of false positive rates (but not recovering the population parameters), but it still had an overall false positive rate almost three times the nominal level. As pointed out by McNichols and Davis (1988), use of verification resampling seems to help ensure that at least some methods can achieve their intended nominal Type I error rates.

13.13 SUMMARY

There are a great many methods for handling nondetectable or nonquantifiable samples in environmental data. Historically, these estimators have been compared on

Table 13.8 Comparison of estimators for 50% type I censoring

Estimator	False Positive Rate	RMSE $\hat{\mu}$	RMSE $\hat{\sigma}$
MLE	0.057	0.074	0.390
RMLE	0.284	0.055	0.385
Linear	0.356	0.085	0.441
Normal delta	0.032	0.438	1.972
Lognormal delta	0.023	0.438	1.971
Regression	0.367	0.088	0.434
Order statistics	0.436	0.079	0.419
$L_Q/2$	0.132	0.188	0.739

the basis of their ability to recover the true population parameters. In many environmental monitoring applications, a more relevant criterion is the overall false positive rate that results from use of a particular method. In many cases, the method that best recovers the parameters of the distribution fails miserably at predicting future individual measurements from that distribution. Overall, the MLE estimator appears to work best for small normally distributed samples, and lognormal versions of the estimator can be obtained simply by taking natural logarithms of the data and censoring point. The delta distributions also performed well in this simple example in terms of minimizing overall false positive rates. None of the other approaches provided adequate protection from false positive results, even with verification resampling. A distinct advantage of the delta distribution over the MLE is that it easily accommodates varying censoring points (i.e., L_Q) which are common in environmental data.

14

TESTING DISTRIBUTIONAL ASSUMPTIONS

14.1 INTRODUCTION

An assumption of many of the methods described in this book is that the measurements are continuous and normally distributed, or can be suitably transformed to approximate a normal distribution [e.g., $\log_e(x) \sim N(\mu_x, \sigma_x^2)$]. There are several approaches to testing this assumption, varying from graphical methods such as normal probability plots to inferential statistical approaches based on normal order statistics (e.g., Shapiro and Wilk, 1965). In this chapter, attention is focused on statistical tests of the hypothesis that the data are normally distributed in the population of constituent measurements. In the context of environmental monitoring applications there are two special problems. First, measurements are nested within locations (e.g., monitoring wells). Due to spatial variability, the within-location measurements may all be normally distributed; however, each sampling location may have a different mean, offsetting the measurements from one location to another. A test of normality for the composite will generally yield a rejection of the null hypothesis that the data are normally distributed when no such rejection is warranted.

Second, the presence of censored measurements (i.e., measurements not quantifiable) will generally produce rejection of the normality hypothesis regardless of whether or not the quantifiable measurements are normally distributed. One solution is simply to ignore the nonquantifiable measurements and test the assumption of normality in the samples measured. When detection frequency is high (i.e., 90% or more), this may produce reasonable results. Alternatively, modifications of some normality tests, which incorporate the censored observations, have also been proposed and are preferable to simply ignoring the censored observations.

In the following sections, several commonly used tests of normality are described and generalizations to joint assessment of normality in several locations and extensions to censored normal distributions are discussed.

14.2 SIMPLE GRAPHICAL APPROACH

In Chapter 13, the expected values of normal order statistics were used to impute quantitative values for the censored measurements. These same expected values can be used to produce a simple graphical test of normality. The measurements observed are first ordered from lowest to highest value and then plotted against the expected values of the normal order statistics (i.e., x-axis). If the data are normally distributed, the points should lie close to a straight line, except for chance sampling fluctuations. The disadvantage of this type of graphical examination is that it provides no means of judging the significance of departure from linearity. It is often good to plot the ordered measurements both in their original metric and transformed [e.g., $x_{(i)}$ and $\log_{10}(x_{(i)})$].

Illustration Consider the ordered chloride measurements in Table 14.1, obtained from a single upgradient well over 15 quarterly monitoring events. The normal probability plots are displayed graphically in Figures 14.1 and 14.2. Inspection of

Table 14.1 Ordered chloride measurements (mg/L) with and without transformation

Ordered Measurement, i	Original Value $x_{(i)}$	Transformed Value, $\log_{10}(10x_{(i)})$	Expected Value of Order Statistic, $E(i\,|\,15)$
1	0.200	0.301	−1.736
2	0.330	0.519	−1.248
3	0.450	0.653	−0.948
4	0.490	0.690	−0.715
5	0.780	0.892	−0.516
6	0.920	0.964	−0.335
7	0.950	0.978	−0.165
8	0.970	0.987	0.000
9	1.040	1.017	0.165
10	1.710	1.233	0.335
11	2.220	1.346	0.516
12	2.275	1.357	0.715
13	3.650	1.562	0.948
14	7.000	1.845	1.248
15	8.800	1.944	1.736

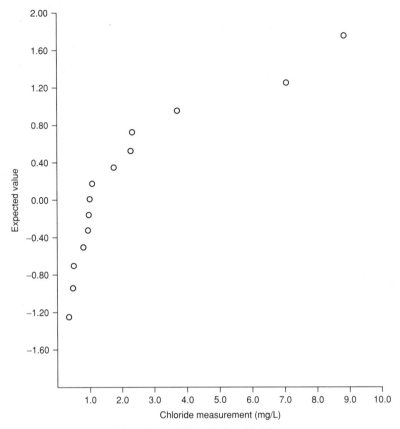

Figure 14.1 Normal probability plot for chloride measurements.

the figures reveals that the log-transformed values are more nearly normally distributed than the original values. The statistical significance of the departure from linearity observed in Figure 14.2 is, however, unknown using the graphical approach.

14.3 SHAPIRO–WILK TEST

In an attempt to summarize formally information contained in normal probability plots, Shapiro and Wilk (1965) proposed a test of normality based on normal order statistics. Their statistic W is proportional to the ratio of the slope of the normal probability plot to the usual mean-square estimate:

$$W = \frac{\left(\sum_{i=1}^{n} a_{i,n} x_{(i)}\right)^2}{\sum_{i=1}^{n} (x_i - \bar{x})^2} \quad (14.1)$$

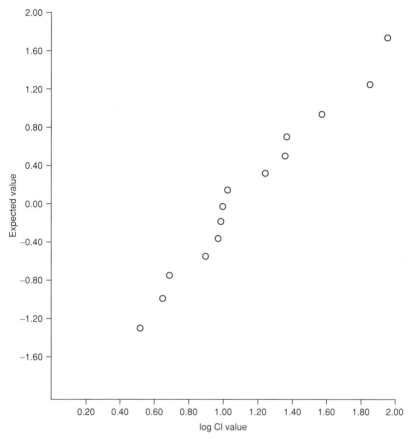

Figure 14.2 Normal probability plot for log transformed chloride values: $\log_{10}(10x)$.

The coefficients $a_{1,n}$ are given in Table 14.2 for $n = 2$ to 50. Note that since the distribution is symmetric,

$$a_{n-1+1,n} = -a_{i,n} \tag{14.2}$$

The numerator in equation (14.1) can be written as

$$A^2 = \left[\sum_{i=1}^{h} a_{i,n} \left(x_{(n-i+1)} - x_i \right) \right]^2 \tag{14.3}$$

where $h = n/2$ if n is even or $(n-1)/2$ if n is odd. For ties (i.e., $x_{(i)} = x_{(i+1)}$) the authors suggest multiplying both by

$$\tfrac{1}{2}(a_{i,n} + a_{i+1,n})$$

which can be extended to cover ties of more than two. Critical values of the W statistic are given in Table 14.3. The larger the value of W (i.e., closer to 1), the greater

Table 14.2 Coefficients a_i for the Shapiro–Wilk W test of normality

	2	3	4	5	6	7	8	9	10
1	.7071	.7071	.6872	.6646	.6431	.6233	.6052	.5888	.5739
2	—	.0000	.1677	.2413	.2806	.3031	.3164	.3244	.3291
3	—	—	—	.0000	.0875	.1401	.1743	.1976	.2141
4	—	—	—	—	—	.0000	.0561	.0947	.1224
5	—	—	—	—	—	—	—	.0000	.0399

	11	12	13	14	15	16	17	18	19	20
1	.5601	.5475	.5359	.5251	.5150	.5056	.4968	.4886	.4808	.4734
2	.3315	.3325	.3325	.3318	.3306	.3290	.3273	.3253	.3232	.3211
3	.2260	.2347	.2412	.2460	.2495	.2521	.2540	.2553	.2561	.2565
4	.1429	.1586	.1707	.1802	.1878	.1939	.1988	.2027	.2059	.2085
5	.0695	.0922	.1099	.1240	.1353	.1447	.1524	.1587	.1641	.1686
6	.0000	.0303	.0539	.0727	.0880	.1005	.1109	.1197	.1271	.1334
7	—	—	.0000	.0240	.0433	.0593	.0725	.0837	.0932	.1013
8	—	—	—	—	.0000	.0196	.0359	.0496	.0612	.0711
9	—	—	—	—	—	—	.0000	.0163	.0303	.0422
10	—	—	—	—	—	—	—	—	.0000	.0140

	21	22	23	24	25	26	27	28	29	30
1	.4643	.4590	.4542	.4493	.4450	.4407	.4366	.4328	.4291	.4254
2	.3185	.3156	.3126	.3098	.3069	.3043	.3018	.2992	.2968	.2944
3	.2578	.2571	.2563	.2554	.2543	.2533	.2522	.2510	.2499	.2487
4	.2119	.2131	.2139	.2145	.2148	.2151	.2152	.2151	.2150	.2148
5	.1736	.1764	.1787	.1807	.1822	.1836	.1848	.1857	.1864	.1870
6	.1399	.1443	.1480	.1512	.1539	.1563	.1584	.1601	.1616	.1630
7	.1092	.1150	.1201	.1245	.1283	.1316	.1346	.1372	.1395	.1415
8	.0804	.0878	.0941	.0997	.1046	.1089	.1128	.1162	.1192	.1219
9	.0530	.0618	.0696	.0764	.0823	.0876	.0923	.0965	.1002	.1036
10	.0263	.0368	.0459	.0539	.0610	.0672	.0728	.0778	.0822	.0862
11	.0000	.0122	.0228	.0321	.0403	.0476	.0540	.0598	.0650	.0697
12	—	—	.0000	.1007	.0200	.0284	.0358	.0424	.0483	.0537
13	—	—	—	—	.0000	.0094	.0178	.0253	.0320	.0381
14	—	—	—	—	—	—	.0000	.0084	.0159	.0227
15	—	—	—	—	—	—	—	—	.0000	.0076

	31	32	33	34	35	36	37	38	39	40
1	.4220	.4188	.4156	.4127	.4096	.4068	.4040	.4015	.3989	.3964
2	.2921	.2898	.2876	.2854	.2834	.2813	.2794	.2774	.2755	.2737
3	.2475	.2463	.2451	.2439	.2427	.2415	.2403	.2391	.2380	.2368
4	.2145	.2141	.2137	.2132	.2127	.2121	.2116	.2110	.2104	.2098
5	.1874	.1878	.1880	.1882	.1883	.1883	.1883	.1881	.1880	.1878
6	.1641	.1651	.1660	.1667	.1673	.1678	.1683	.1686	.1689	.1691
7	.1433	.1449	.1463	.1475	.1487	.1496	.1505	.1513	.1520	.1526
8	.1243	.1265	.1284	.1301	.1317	.1331	.1344	.1356	.1366	.1376

Table 14.2 (Continued)

	31	32	33	34	35	36	37	38	39	40
9	.1066	.1093	.1118	.1140	.1160	.1179	.1196	.1211	.1225	.1237
10	.0899	.0931	.0961	.0988	.1013	.1036	.1056	.1075	.1092	.1108
11	.0739	.0777	.0812	.0844	.0873	.0900	.0924	.0947	.0967	.0986
12	.0585	.0629	.0669	.0706	.0739	.0770	.0798	.0824	.0848	.0870
13	.0435	.0485	.0530	.0572	.0610	.0645	.0677	.0706	.0733	.0759
14	.0289	.0344	.0395	.0441	.0484	.0523	.0559	.0592	.0622	.0651
15	.0144	.0206	.0262	.0314	.0361	.0404	.0444	.0481	.0515	.0546
16	.0000	.0068	.0131	.0187	.0239	.0287	.0331	.0372	.0409	.0444
17	—	—	.0000	.0062	.0119	.0172	.0220	.0264	.0305	.0343
18	—	—	—	—	.0000	.0057	.0110	.0158	.0203	.0244
19	—	—	—	—	—	—	.0000	.0053	.0101	.0146
20	—	—	—	—	—	—	—	—	.0000	.0049

	41	42	43	44	45	46	47	48	49	50
1	.3940	.3917	.3894	.3872	.3850	.3830	.3808	.3789	.3770	.3751
2	.2719	.2701	.2684	.2667	.2651	.2635	.2620	.2604	.2589	.2574
3	.2357	.2345	.2334	.2323	.2313	.2302	.2291	.2281	.2271	.2260
4	.2091	.2085	.2078	.2072	.2065	.2058	.2052	.2045	.2038	.2032
5	.1876	.1874	.1871	.1868	.1865	.1862	.1859	.1855	.1851	.1847
6	.1693	.1694	.1695	.1695	.1695	.1695	.1695	.1693	.1692	.1691
7	.1531	.1535	.1539	.1542	.1545	.1548	.1550	.1551	.1553	.1554
8	.1384	.1392	.1398	.1405	.1410	.1415	.1420	.1423	.1427	.1430
9	.1249	.1259	.1269	.1278	.1286	.1293	.1300	.1306	.1312	.1317
10	.1123	.1136	.1149	.1160	.1170	.1180	.1189	.1197	.1205	.1212
11	.1004	.1020	.1035	.1049	.1062	.1073	.1085	.1095	.1105	.1113
12	.0891	.0909	.0927	.0943	.0959	.0972	.0986	.0998	.1010	.1020
13	.0782	.0804	.0824	.0842	.0860	.0876	.0892	.0906	.0919	.0932
14	.0677	.0701	.0724	.0745	.0765	.0783	.0801	.0817	.0832	.0846
15	.0575	.0602	.0628	.0651	.0673	.0694	.0713	.0731	.0748	.0764
16	.0476	.0506	.0534	.0560	.0584	.0607	.0628	.0648	.0667	.0685
17	.0379	.0411	.0442	.0471	.0497	.0522	.0546	.0568	.0588	.0608
18	.0283	.0318	.0352	.0383	.0412	.0439	.0465	.0489	.0511	.0532
19	.0188	.0227	.0263	.0296	.0328	.0357	.0385	.0411	.0436	.0459
20	.0094	.0136	.0175	.0211	.0245	.0277	.0307	.0335	.0361	.0386
21	.0000	.0045	.0087	.0126	.0163	.0197	.0229	.0259	.0288	.0314
22	—	—	.0000	.0042	.0081	.0118	.0153	.0185	.0215	.0244
23	—	—	—	—	.0000	.0039	.0076	.0111	.0143	.0174
24	—	—	—	—	—	—	.0000	.0037	.0071	.0104
25	—	—	—	—	—	—	—	—	.0000	.0035

Source: Reproduced with permission from S.S. Shapiro and M.B. Wilk, *Biometrika* **52**(1965):591–611.

Table 14.3 Lower 1% and 5% critical values for the Shapiro–Wilk test statistic W used in testing normality

Sample Size	$W_{0.01}$	$W_{0.05}$	Sample Size	$W_{0.01}$	$W_{0.05}$
3	0.753	0.767	27	0.894	0.923
4	0.687	0.748	28	0.896	0.924
5	0.686	0.762	29	0.898	0.926
6	0.713	0.788	30	0.900	0.927
7	0.730	0.803	31	0.902	0.929
8	0.749	0.818	32	0.904	0.930
9	0.764	0.829	33	0.906	0.931
10	0.781	0.842	34	0.908	0.933
11	0.792	0.850	35	0.910	0.934
12	0.805	0.859	36	0.912	0.935
13	0.814	0.866	37	0.914	0.936
14	0.825	0.874	38	0.916	0.938
15	0.835	0.881	39	0.917	0.939
16	0.844	0.887	40	0.919	0.940
17	0.851	0.892	41	0.920	0.941
18	0.858	0.897	42	0.922	0.942
19	0.863	0.901	43	0.923	0.943
20	0.868	0.905	44	0.924	0.944
21	0.873	0.908	45	0.926	0.945
22	0.878	0.911	46	0.927	0.945
23	0.881	0.914	47	0.928	0.946
24	0.884	0.916	48	0.929	0.947
25	0.886	0.918	49	0.929	0.947
26	0.891	0.920	50	0.930	0.947

Source: Reproduced with permission from S.S. Shapiro and M.B. Wilk, *Biometrika*, **52**(1965):591–611.

the support for the normality assumption. The assumption of normality is rejected if the computed value of W is less than the critical value.

Illustration   Using the chloride data in Table 14.1, we obtain

$$\sum (x_i - \bar{x})^2 = 90.41 \qquad A = 8.036 \qquad W(x) = 0.714$$

for the original data, and letting $y = \log_{10}(10x)$,

$$\sum (y_i - \bar{y})^2 = 3.062 \qquad A = 1.723 \qquad W(y) = 0.970$$

Table 14.3 reveals that for $n = 15$, the lower one percentage point is 0.835; therefore, we reject normality for the original data but not for the log-transformed data.

14.4 SHAPIRO–FRANCIA TEST

When $n > 50$, the Shapiro–Francia test can be used in place of the Shapiro–Wilks test (Shapiro and Francia, 1972). The modified W statistic is computed as

$$W' = \frac{[\sum_{i=1}^{n} b_{i,n} x_{(i)}]^2}{\sum_{i=1}^{n} (x_{(i)} - \bar{x})^2 \sum_{i=1}^{n} b_{i,n}^2} \qquad (14.4)$$

where the $x_{(i)}$ represent the ordered observations, the b_i represent the expected values of the normal order statistics $b_i = E(i \mid n)$. For large n (i.e., $n > 50$), the expected value of the normal order statistic can be approximated as

$$b_i = \Phi^{-1}\left(\frac{i}{n+1}\right) \qquad (14.5)$$

where Φ^{-1} is the inverse standard normal distribution function described in Chapter 13. Critical values of W' are provided in Table 14.4.

14.5 D'AGOSTINO'S TEST

An alternative and even computationally easier test to compute when $n > 50$ was developed by D'Agostino (1971). To test if the underlying distribution is normal,

Table 14.4 Lower 1% and 5% critical values of the Shapiro–Francia test of normality for $n > 50$

Sample Size	Lower 1%	Lower 5%	Sample Size	Lower 1%	Lower 5%
35	.919	.943	75	.956	.969
50	.935	.953	77	.957	.969
51	.935	.954	79	.957	.970
53	.938	.957	81	.958	.970
55	.940	.958	83	.960	.971
57	.944	.961	85	.961	.972
59	.945	.962	87	.961	.972
61	.947	.963	89	.961	.972
63	.947	.964	91	.962	.973
65	.948	.965	93	.963	.973
67	.950	.966	95	.965	.974
69	.951	.966	97	.965	.975
71	.953	.967	99	.967	.976
73	.956	.968			

Source: Reproduced with permission from S.S. Shapiro and R.S. Francia, *J. Am. Stat. Assoc.* **67**(1972): 215–216. Copyright 1972 by the American Statistical Association. All rights reserved.

compute the statistic

$$D = \frac{\sum_{i=1}^{n}[i - \frac{1}{2}(n+1)]x_{(i)}}{n^2 s} \quad (14.6)$$

where s is the sample standard deviation and the $x_{(i)}$ are the sample order statistics (i.e., the observed measurements ranked from lowest to highest). To test the significance of the statistic, compute

$$Y = \frac{D - 0.28209479}{0.02998598/\sqrt{n}} \quad (14.7)$$

The values of Y can be compared to critical values computed by D'Agostino (1971) and provided here in Table 14.5, for selected values of $n = 50$ to 1000 and percentage points of 0.5 to 99.5%. The null hypothesis of normality is rejected if Y is less than $Y_{\alpha/2}$ or greater than $Y_{1-\alpha/2}$.

Table 14.5 Critical values of the D'Agostino test of normality for $n > 50$

Sample Size	$Y_{0.01}$	$Y_{0.05}$	$Y_{0.95}$	$Y_{0.95}$
50	−3.442	−2.220	0.923	1.140
60	−3.360	−2.179	0.986	1.236
70	−3.293	−2.146	1.036	1.312
80	−3.237	−2.118	1.076	1.374
90	−3.100	−2.095	1.109	1.426
100	−3.150	−2.075	1.137	1.470
150	−3.009	−2.004	1.233	1.623
200	−2.922	−1.960	1.290	1.715
250	−2.861	−1.926	1.328	1.779
300	−2.816	−1.906	1.357	1.826
350	−2.781	−1.888	1.379	1.863
400	−2.753	−1.873	1.396	1.893
450	−2.729	−1.861	1.411	1.918
500	−2.709	−1.850	1.423	1.938
550	−2.691	−1.841	1.434	1.957
600	−2.676	−1.833	1.443	1.972
650	−2.663	−1.826	1.451	1.986
700	−2.651	−1.820	1.458	1.999
750	−2.640	−1.814	1.465	2.010
800	−2.630	−1.809	1.471	2.020
850	−2.621	−1.804	1.476	2.029
900	−2.613	−1.800	1.481	2.037
950	−2.605	−1.796	1.485	2.045
1000	−2.599	−1.792	1.489	2.052

Source: Reproduced with permission from R.B. D'Agostino, *Biometrika* **58**(1971):341–348.

14.6 METHODS BASED ON MOMENTS OF A NORMAL DISTRIBUTION

There are three general methods for testing departure from normality based on sample moments. The sample moments are defined as

$$\bar{x} = \sum_{i=1}^{n} \frac{x_i}{n} \tag{14.8}$$

$$m_r = \sum_{i=1}^{n} \frac{(x_i - \bar{x})^r}{n} \quad \text{for} \quad r \geq 2 \tag{14.9}$$

The relevant moment ratios are

$$\sqrt{b_1} = \frac{m_3}{m_2^{3/2}} \tag{14.10}$$

$$b_2 = \frac{m_4}{m_2^2} \tag{14.11}$$

Departure of $\sqrt{b_1}$ from the normal value of zero is an indication of skewness in the frequency distribution, whereas departure of b_2 from the normal value of 3 is an indication of kurtosis. Pearson and Hartley (1976) suggest that for large samples (e.g., $n > 50$), a "rough" test of normality can be obtained by comparing $\sqrt{b_1}$ and $b_2 - 3$ with the approximate values of their standard errors, which are $\sqrt{6/n}$ and $\sqrt{24/n}$, respectively. The ratio of the estimate to its standard error is approximately normal, therefore, rejection is indicated by comparing the ratio to the corresponding normal tail probability. Fisher (1929, 1930) first developed methods of calculating higher sample moments of $\sqrt{b_1}$ and b_2, which led to the critical values in Tables 14.6 and 14.7. Shapiro

Table 14.6 Critical values for the distribution of $\sqrt{b_1} = m_3/m_2^{3/2}$ in samples from a normal distribution

Sample Size	Upper (or Lower) 5%	Upper (or Lower) 1%	Sample Size	Upper (or Lower) 5%	Upper (or Lower) 1%
20	.772	1.150	100	.390	.567
25	.711	1.059	125	.351	.508
30	.662	.986	150	.322	.465
35	.621	.923	175	.299	.430
40	.588	.871	200	.280	.403
45	.559	.826	250	.251	.361
50	.534	.788	300	.230	.329
60	.492	.724	350	.213	.305
70	.459	.673	400	.200	.285
80	.432	.632	450	.188	.269
90	.409	.597	500	.179	.255

Source: Reproduced with permission from E.S. Pearson and H.O. Hartley, *Biometrika Tables for Statisticians*, Vol. 1, Oxford University Press, Oxford, 1976, pp. 207–208.

Table 14.7 Critical values for the distribution of $b_2 = m_4/m_2^2$ in samples from a normal population

Sample Size	Upper 1.0%	Upper 5.0%	Lower 5.0%	Lower 1.0%
20	5.38	4.18	1.83	1.64
30	5.20	4.12	1.98	1.79
40	5.04	4.06	2.07	1.89
50	4.88	4.00	2.15	1.95
75	4.59	3.87	2.27	2.08
100	4.39	3.77	2.35	2.18
125	4.24	3.70	2.40	2.24
150	4.13	3.65	2.45	2.29
175	4.04	3.61	2.48	2.34
200	3.98	3.57	2.51	2.37
250	3.87	3.52	2.55	2.42
300	3.79	3.47	2.59	2.46
400	3.67	3.41	2.64	2.52
500	3.60	3.37	2.67	2.57
600	3.54	3.34	2.70	2.60
700	3.50	3.31	2.72	2.62
800	3.46	3.29	2.74	2.65
900	3.43	3.28	2.75	2.66
1000	3.41	3.26	2.76	2.68
2000	3.28	3.18	2.83	2.77

Source: **Reproduced** with permission from E.S. Pearson and H.O. Hartley, *Biometrika Tables for Statisticians*, Vol. 1, Oxford University Press, Oxford, 1976, pp. 207–208.

et al. (1968) have shown that combined use of $\sqrt{b_1}$ and b_2 is slightly less powerful than the W statistic.

An alternative statistic better suited to small samples was proposed by Geary (1935, 1936). The statistic is the ratio of mean deviation to standard deviation,

$$a = \frac{\sum_{i=1}^{n} |x_i - \bar{x}|}{[n \sum_{i=1}^{n} (x_i - \bar{x})^2]^{1/2}} \tag{14.12}$$

For a normally distributed random variable, the ratio has value $\sqrt{2/\pi} = 0.7979$. For platykurtic distributions the ratio will be higher, and for leptokurtic distributions the ratio will be lower. Table 14.8 gives upper and lower 5% and 1% points for a.

Illustration Consider the following 20 chloride measurements from four monitoring locations listed in Table 14.9. Assuming that samples from all four upgradient wells were drawn from the same population, we obtain the following three statistics:

$$\sqrt{b_1} = \frac{m_3}{m_2^{3/2}} = \frac{0.0040}{1.3475^{3/2}} = 0.0025$$

Table 14.8 Critical values for the distribution of a = (mean deviation)/(standard deviation)

Sample Size	Upper 1%	Upper 5%	Lower 5%	Lower 1%
11	.9359	.9073	.7153	.6675
16	.9137	.8884	.7236	.6829
21	.9001	.8768	.7304	.6950
26	.8901	.8686	.7360	.7040
31	.8827	.8625	.7404	.7110
36	.8769	.8578	.7440	.7167
41	.8722	.8540	.7470	.7216
46	.8682	.8508	.7496	.7256
51	.8648	.8481	.7518	.7291
61	.8592	.8434	.7554	.7347
71	.8549	.8403	.7583	.7393
81	.8515	.8376	.7607	.7430
91	.8484	.8353	.7626	.7460
101	.8460	.8344	.7644	.7487
201	.8322	.8229	.7738	.7629
301	.8260	.8183	.7781	.7693
401	.8223	.8155	.7807	.7731
501	.8198	.8136	.7825	.7757
601	.8179	.8123	.7838	.7776
701	.8164	.8112	.7848	.7791
801	.8152	.8103	.7857	.7803
901	.8142	.8096	.7864	.7814
1001	.8134	.8090	.7869	.7822

Source: Reproduced with permission from E.S. Pearson and H.O. Hartley, *Biometrika Tables for Statisticians*, 1(1976): 207–208.

$$b_2 = \frac{m_4}{m_2^2} = \frac{5.200}{1.3475^2} = 2.8640$$

$$a = \frac{\sum_{i=1}^{n} |x_i - \bar{x}|}{[n \sum_{i=1}^{n} (x_i - \bar{x})^2]^{1/2}} = \frac{19.0000}{[20(26.9500)]^{1/2}} = 0.8184$$

Comparison of these estimates to the critical values in Tables 14.6 to 14.8 reveal that in all three cases, normality is not rejected.

14.7 MULTIPLE INDEPENDENT SAMPLES

In the context of environmental monitoring, it is hard to imagine a case in which normality would be assessed in a single sample. More typical is the case described in Section 14.6, in which the collection of sampling locations is used to establish

Table 14.9 Chloride measurements from four upgradient wells

	Chloride Measurement (mg/L)			
Sample	Well 1	Well 2	Well 3	Well 4
1	6	4	5	5
2	5	6	6	8
3	6	5	4	5
4	7	6	7	5
5	4	5	3	6

that background and multiple measurements are available at each location. Due to spatial variability, the monitoring locations may have different means and variances, so that the simple pooling of measurements as in Section 14.6 is not justifiable. Wilk and Shapiro (1968) have suggested a generalization of their original test that is suitable for the joint assessment of normality in T independent samples. The idea is to compute individual values of W for each location, denoted W_t, obtain a normal equivalent G_t and obtain an overall test by referring the normalized mean

$$G = \sum_{i=1}^{T} \frac{G_t}{\sqrt{T}} \qquad (14.13)$$

to a standard table of the normal integral. Values of G_t are given by

$$G_t = \gamma(n) + \delta(n) \log \frac{W_t - \varepsilon(n)}{1 - W_t} \qquad (14.14)$$

The coefficients γ, δ, and ε as functions of n are given in Table 14.10 for $n = 7$ to 50. For values of $n = 3$ to 6, values of G_t can be obtained in terms of the transformed function

$$\gamma_t = \log_e \frac{W_t - \varepsilon(n)}{1 - W_t} \qquad (14.15)$$

where G_t is given as a function of γ_t in Table 14.11 [values of $\varepsilon(n)$ are given in the column headers in Table 14.11].

Illustration Returning to the data in Table 14.9, in which there were five samples in each of four upgradient wells, we have $T = 4$, $n_t = n = 5$. Using the coefficients in Table 14.2, we obtain the following values of W_t, γ_t, and G_t:

Table 14.10 Coeffecients γ, δ, and ϵ as Functions of $n = 7(1)50$ for assessing normality jointly in several independent samples

Sample Size	$\gamma_{(n)}$	$\delta_{(n)}$	$\epsilon_{(n)}$	Sample Size	$\gamma_{(n)}$	$\delta_{(n)}$	$\epsilon_{(n)}$
7	−2.356	1.245	.4533	29	−6.074	1.934	.1907
8	−2.696	1.333	.4186	30	−6.150	1.949	.1872
9	−2.968	1.400	.3900	31	−6.248	1.965	.1840
10	−3.262	1.471	.3660	32	−6.324	1.976	.1811
11	−3.485	1.515	.3451	33	−6.402	1.988	.1781
12	−3.731	1.571	.3270	34	−6.480	2.000	.1755
13	−3.936	1.613	.3111	35	−6.559	2.012	.1727
14	−4.155	1.655	.2969	36	−6.640	2.024	.1702
15	−4.373	1.695	.2842	37	−6.721	2.037	.1677
16	−4.567	1.724	.2727	38	−6.803	2.049	.1656
17	−4.713	1.739	.2622	39	−6.887	2.062	.1633
18	−4.885	1.770	.2528	40	−6.961	2.075	.1612
19	−5.018	1.786	.2440	41	−7.035	2.088	.1591
20	−5.153	1.802	.2359	42	−7.111	2.101	.1572
21	−5.291	1.818	.2264	43	−7.188	2.114	.1552
22	−5.413	1.835	.2207	44	−7.266	2.128	.1534
23	−5.508	1.848	.2157	45	−7.345	2.141	.1516
24	−5.605	1.862	.2106	46	−7.414	2.155	.1499
25	−5.704	1.876	.2063	47	−7.484	2.169	.1482
26	−5.803	1.890	.2020	48	−7.555	2.183	.1466
27	−5.905	1.905	.1980	49	−7.615	2.198	.1451
28	−5.988	1.919	.1943	50	−7.677	2.212	.1436

Source: Reproduced with permission from E.S. Pearson and H.O. Hartley, *Biometrika Tables for Statisticians*, Vol. 2, Oxford University Press, Oxford, 1976, p. 221.

	Well			
	1	2	3	4
w_t	0.961	0.881	0.987	0.833
γ_t	2.342	1.015	3.478	0.516
G_t	0.869	−0.493	1.717	−1.058

From the last row of figures we find that $G = \Sigma G_t / \sqrt{4} = 0.518$, which has an associated probability of 0.302, indicating that we cannot reject the assumption of normality.

14.8 TESTING NORMALITY IN CENSORED SAMPLES

When data are censored (e.g., reported as less than the detection limit), the distributional tests described previously do not apply. Ignoring the nondetects can be misleading because it eliminates the lower tail of the distribution and can therefore

Table 14.11 Values of G_t for argument γ_t for normal conversion of W [$n = 3(1)6$]

γ_t	$n = 3$ (0.7500)	$n = 4$ (0.6297)	$n = 5$ (0.5521)	$n = 6$ (0.4963)	γ_t	$n = 3$ (0.7500)	$n = 4$ (0.6297)	$n = 5$ (0.5521)	$n = 6$ (0.4963)
−7.0	−3.29	—	—	—	2.2	0.52	0.74	0.75	0.64
−5.4	−2.81	—	—	—	2.6	0.67	1.00	1.09	1.06
−5.0	−2.68	—	—	—	3.0	0.81	1.23	1.40	1.45
−4.6	−2.54	—	—	—	3.4	0.95	1.44	1.67	1.83
−4.2	−2.40	—	—	—	3.8	1.07	1.65	1.91	2.17
−3.8	−2.25	−3.50	—	—	4.2	1.19	1.85	2.15	2.50
−3.4	−2.10	−3.27	—	—	4.6	1.31	2.03	2.47	2.77
−3.0	−1.94	−3.05	−4.01	—	5.0	1.42	2.19	2.85	3.09
−2.6	−1.77	−2.84	−3.70	—	5.4	1.52	2.34	3.24	3.54
−2.2	−1.59	−2.64	−3.38	—	5.8	1.62	2.48	3.64	—
−1.8	−1.40	−2.44	−3.11	—	6.2	1.72	2.62	—	—
−1.4	−1.21	−2.22	−2.87	—	6.6	1.81	2.75	—	—
−1.0	−1.01	−1.96	−2.56	−3.72	7.0	1.90	2.87	—	—
−0.6	−0.80	−1.66	−2.20	−2.88	7.4	1.98	2.97	—	—
−0.2	−0.60	−1.31	−1.81	−2.27	7.8	2.07	3.08	—	—
0.2	−0.39	−0.94	−1.41	−1.85	8.2	2.15	3.22	—	—
0.6	−0.19	−0.57	−0.97	−1.38	8.6	2.23	3.36	—	—
1.0	−0.00	−0.19	−0.51	−0.84	9.0	2.31	—	—	—
1.4	0.18	0.15	−0.06	−0.33	9.4	2.38	—	—	—
1.8	0.35	0.45	0.37	0.18	9.8	2.45	—	—	—

Source: Reproduced with permission from E.S. Pearson and H.O. Hartley, *Biometrika Tables for Statisticians*, Vol. 2, Oxford University Press, Oxford, 1976, pp. 207–208.

falsely reject the null hypothesis of normality. On the other hand, including the non-quantifiable values at L_Q introduces a spike in the distribution that can also, incorrectly, cause the rejection of normality.

To develop a test of normality, we begin by noting that an alternative view of the Shapiro–Wilk test is as a correlation coefficient. Ryan and Joiner (1973) noted that the Shapiro–Francia test could be written as the square of the correlation coefficient Z:

$$Z^2 = \frac{\sum_{i=1}^{n}[(x_{(i)} - \bar{x})(z_{i,n} - \bar{z})]}{[\sum_{i=1}^{n}(x_i - \bar{x})^2 \sum_{i=1}^{n}(z_{i,n} - \bar{z})^2]^{1/2}}$$

where the $z_{i,n}$ are approximately $\Phi^{-1}[i/(n + 1)]$. Filliben (1975) suggested substituting the inverse normal transform of the median of the ith order statistic (m_i) from a sample of n standard normal random variables. The median order statistics (m_i) are given by

$$m_i = \begin{cases} 1 - m_n & i = 1 \\ \dfrac{i - 0.3175}{n + 0.365} & i = 2, 3, \ldots, n-1 \\ 0.5^{1/n} & i = n \end{cases}$$

Therefore, $z_i = \Phi^{-1}(m_i)$. Filliben (1975) developed percentage points of the normal probability plot correlation coefficient z for $n = 3$ to 100 which are reproduced in Table 14.12 for the 1% and 5% points. A major advantage of this interpretation is that it is more general in terms of sample size than any of the other tests, easier to compute, and essentially indistinguishable in terms of power.

Smith and Bain (1976) obtained percentage points of the statistic $1 - z^2$ for complete and censored samples for $n = 8$ to 80 and censoring of 0%, 25%, and 50%. The critical 1% and 5% values of $1 - z^2$ are displayed in Table 14.13.

Smith and Bain (1976) note that the choice of z_i is not critical and suggest the approximation

$$z_i = t - \frac{2.30753 + 0.27061t}{1.0 + 0.99229t + 0.04481t^2}$$

where

$$t = \left\{\log_e\left[\frac{(n+1)^2}{i^2}\right]\right\}^{1/2}$$

Note that this is an approximation for $\Phi^{-1}[i/(n + 1)]$; therefore, there are slight differences between Tables 14.12 and 14.13 for the complete data case. To apply their test, the correlation of the uncensored data with the corresponding largest order statistics is computed.

Table 14.12 1% and 5% critical values for the normal probability plot correlation coefficient Z

Sample Size	Lower 1%	Lower 5%	Sample Size	Lower 1%	Lower 5%	Sample Size	Lower 1%	Lower 5%
3	.869	.879	23	.933	.955	43	.959	.973
4	.822	.868	24	.936	.957	44	.960	.973
5	.822	.879	25	.937	.958	45	.961	.974
6	.835	.890	26	.939	.959	46	.962	.974
7	.847	.899	27	.941	.960	47	.963	.974
8	.859	.905	28	.943	.962	48	.963	.975
9	.868	.912	29	.945	.962	49	.964	.975
10	.876	.917	30	.947	.964	50	.965	.977
11	.883	.922	31	.948	.965	55	.967	.978
12	.889	.926	32	.949	.966	60	.970	.980
13	.895	.931	33	.950	.967	65	.972	.981
14	.901	.934	34	.951	.967	70	.974	.982
15	.907	.937	35	.952	.968	75	.975	.983
16	.912	.940	36	.953	.968	80	.976	.984
17	.916	.942	37	.955	.969	85	.977	.985
18	.919	.945	38	.956	.970	90	.978	.985
19	.923	.947	39	.957	.971	95	.979	.986
20	.925	.950	40	.958	.972	100	.981	.987
21	.928	.952	41	.958	.972			
22	.930	.954	42	.959	.973			

Source: Reproduced with permission from J.J. Filliben, *Technometrics* **17**, 1(1975):113. Copyright (1975) by the American Statistical Asscociation. All rights reserved.

Table 14.13 1% and 5% critical values for the censored normal probability plot correlation coefficient Z Tabled values are $T = 1 - Z^2$

r/n	Sample Size	Upper 95%	Upper 99%
.5	8	.268	.369
	20	.170	.268
	40	.115	.175
	60	.085	.128
	80	.066	.101
.75	8	.227	.315
	20	.128	.197
	40	.078	.125
	60	.052	.078
	80	.043	.062
1.0	8	.187	.286
	20	.100	.146
	40	.059	.086
	60	.040	.060
	80	.033	.045

Source: Reproduced with permission from R.M. Smith and L.J. Bain, *Commun. Stat. Theory Methods* A5, 2(1976):119–132 by courtesy of Marcel Dekker, Inc.

Illustration Using the chloride data from Table 14.1, we obtain the various estimates of z_i in Table 14.14. The three estimates of z_i lead to values of $Z = 0.917$, 0.915, and 0.917, respectively, indicating further that the correlation estimate is robust to the choice of z_i. Comparison of these estimates to the critical values in Table 14.12 reveals a 1% critical value of 0.907 and a 5% critical value of 0.937, indicating that the data fit a normal distribution marginally at best. Recall that for the Shapiro–Wilk test, normality was rejected at both the 5% and 1% levels.

To illustrate application to censored data, assume that the seven lowest values were not detected. In this case, summation was from 8 to 15 and the resulting values of $Z = 0.766$, 0.757, and 0.765 for expected values of normal order statistics, $\Phi^{-1}[i/(n+1)]$, and $\Phi^{-1}(m_i)$, respectively. For 50% censoring and $n = 15$, the 1% critical value found by interpolation in Table 14.13 is 0.310. Given correlation of approximately 0.76, $T = 1 - 0.76^2 = 0.422$ and normality is rejected at the 1% level.

14.9 KOLMOGOROV–SMIRNOV TEST

The Kolmogorov–Smirnov (KS) test is a nonparametric test that can be used to evaluate the fit of any hypothesized distribution. These tests are described in detail by Conover (1980) and discussed by Gilbert (1987) for environmental applications. In general, the KS test is considered more powerful than alternative goodness-of-fit chi-square tests. The three general limitations are: (1) the method

182 TESTING DISTRIBUTIONAL ASSUMPTIONS

Table 14.14 Three methods for computing z_i using chloride data from Table 14.1

Ordered Measurement	Value	Expected Value of Order Statistics	$\Phi^{-1}[i/(n+1)]$	$\Phi^{-1}(m_i)$
1	0.200	−1.736	−1.538	−1.698
2	0.330	−1.248	−1.152	−1.232
3	0.450	−0.948	−0.887	−0.937
4	0.490	−0.715	−0.673	−0.706
5	0.780	−0.516	−0.487	−0.509
6	0.920	−0.335	−0.317	−0.330
7	0.950	−0.165	−0.156	−0.162
8	0.970	0.000	0.000	0.000
9	1.040	0.165	0.156	0.162
10	1.710	0.335	0.317	0.330
11	2.220	0.516	0.487	0.509
12	2.275	0.715	0.673	0.706
13	3.650	0.948	0.887	0.937
14	7.000	1.248	1.152	1.232
15	8.800	1.736	1.538	1.698

is computationally complex, (2) it requires large sample sizes (i.e., 50 or more), and (3) the parameters of the hypothesized distribution (e.g., mean and variance of a normal distribution) are assumed to be known. Lilliefors (1967, 1969) generalized the test to the case of a normal or lognormal distribution with unknown mean and variance, although the method is still computationally complex and requires large samples.

14.10 SUMMARY

Although several methods for testing normality have been described in this chapter, no single method is suitable for generic use because none of the methods incorporate both multiple groups and censoring. From the perspective of testing normality in multiple independent samples, the test described by Wilk and Shapiro (1968) appears to be most rigorous. From the perspectives of computational simplicity, generality with respect to sample size and censoring, tests based on normal probability plot correlation coefficients appear to be most useful. Gibbons (1994) suggested that one possibility is to compute censored versions of Z separately in each location and compute the average value of Z weighted by the number of measurements available in each location (in the event that the number of location-specific measurements are unequal). This average value of Z could then be compared to the corresponding critical value corresponding to the total number of measurements. Although not exact, this approach, which combines both multiple locations and censoring, should be sufficiently accurate for most practical purposes.

15

TESTING FOR OUTLIERS

15.1 INTRODUCTION

It is common to find outliers in environmental data. Outliers, those values that do not conform to the pattern established by other observations (see Gilbert, 1987; Hunt et al., 1981), can arise from errors in transcription, data coding, analytical instrument failure, and calibration errors, or underestimation of inherent spatial or temporal variability in constituent concentrations. Despite the cause, including outliers to establish statistical limit estimates will often lead to biased and misleading results.

As an example, consider the TOC data in Table 15.1 from a single monitoring location. The mean concentration is 4.13 mg/L and the standard deviation is 2.03 mg/L. A 99% normal upper prediction limit for the next single measurement from that population is

$$\bar{x} + st_{7,0.01}\left(1 + \frac{1}{n}\right)^{1/2} = 10.59 \text{ mg/L}$$

Now assume that the 1992 fourth quarter observation was recorded incorrectly as 50 mg/L. In this case the mean is 9.75 mg/L, the standard deviation is 16.39 mg/L, and the 99% normal upper prediction limit is 61.87 mg/L. Inclusion of the outlier will have a profound effect on the false negative rate of the statistical test. Including the outlier will cause us to accept potentially contaminated measurements as consistent with background levels. This is also true for low outliers since the increase in s increases the prediction limit $t\sqrt{1 + 1/n}$ times more than the decrease in $\bar{x}$.

In practice, however, we rarely know the source of an outlying observation. Is the TOC value of 50 mg/L really due to error, or is it simply natural variability that one might observe in nature? Deleting the observation from the background

184   TESTING FOR OUTLIERS

Table 15.1 Hypothetical TOC data

Year	Quarter	TOC (mg/L)
1991	1	1
	2	7
	3	3
	4	6
1992	1	5
	2	4
	3	2
	4	5

database may cause us incorrectly to reject a new monitoring measurement at that level (i.e., false positive). Should we remove the outlier, or not? In the following sections several methods for detecting outliers in background data are described. There are numerous approaches to this problem and the interested reader is referred to Barnett and Lewis (1984) and Beckman and Cook (1983) for an excellent reviews. Since there is no reason to assume a specific number of outliers (e.g., one or two) or direction (i.e., outliers can be either too small or too large, although censoring may prevent identification of outliers on the low side), attention is focused on two-tailed outlier detection procedures for up to m outliers.

15.2 ROSNER'S TEST

Gilbert (1987) suggests use of Rosner's test (Rosner, 1983) for use with environmental data. Rosner's test is a generalization of the extreme Studentized deviate (ESD) originally suggested by Grubbs (1969) and tabulated by Grubbs and Beck (1972) for testing a single outlier. Rosner's generalization is for multiple outliers (i.e., $m = 2$ to 10). To use Rosner's test, an upper limit m must be specified on the number of potential outliers present. The test is valid for samples of size 25 or more.

Let $\bar{x}^{(i)}$ and $s^{(i)}$ be the mean and standard deviation of the $m - i$ measurements that remain after the i most extreme observations have been deleted:

$$\bar{x}^{(i)} = \frac{1}{n-i} \sum_{j=1}^{n-i} x_j \qquad (15.1)$$

$$s^{(i)} = \left[\frac{1}{n-i} \sum_{j=1}^{n-i} (x_j - \bar{x}^{(i)})^2 \right]^{1/2} \qquad (15.2)$$

Let $x^{(i)}$ denote the farthest remaining observation from the mean $\bar{x}^{(i)}$ after i more extreme values (large or small) have been detected. Then

$$R_{i+1} = \frac{|x^{(i)} - \bar{x}^{(i)}|}{s^{(i)}} \qquad (15.3)$$

is a test statistic for deciding whether the $i + 1$ most extreme values in the complete dataset are outliers from a normal distribution. Critical values for this statistic are provided in Table 15.2.

Table 15.2 Critical values for Rosner's test statistic

n	i + 1	0.05	0.01	n	i + 1	0.05	0.01	n	i + 1	0.05	0.01
25	1	2.82	3.14	32	1	2.94	3.27	39	1	3.03	3.37
	2	2.80	3.11		2	2.92	3.25		2	3.01	3.36
	3	2.78	3.09		3	2.91	3.24		3	3.00	3.34
	4	2.76	3.06		4	2.89	3.22		4	2.99	3.33
	5	2.73	3.03		5	2.88	3.20		5	2.98	3.32
	10	2.59	2.85		10	2.78	3.09		10	2.91	3.24
26	1	2.84	3.16	33	1	2.95	3.29	40	1	3.04	3.38
	2	2.82	3.14		2	2.94	3.27		2	3.03	3.37
	3	2.80	3.11		3	2.92	3.25		3	3.01	3.36
	4	2.78	3.09		4	2.91	3.24		4	3.00	3.34
	5	2.76	3.06		5	2.89	3.22		5	2.99	3.33
	10	2.62	2.89		10	2.80	3.11		10	2.92	3.25
27	1	2.86	3.18	34	1	2.97	3.30	41	1	3.05	3.39
	2	2.84	3.16		2	2.95	3.29		2	3.04	3.38
	3	2.82	3.14		3	2.94	3.27		3	3.03	3.37
	4	2.80	3.11		4	2.92	3.25		4	3.01	3.36
	5	2.78	3.09		5	2.91	3.24		5	3.00	3.34
	10	2.65	2.93		10	2.82	3.14		10	2.94	3.27
28	1	2.88	3.20	35	1	2.98	3.32	42	1	3.06	3.40
	2	2.86	3.18		2	2.97	3.30		2	3.05	3.39
	3	2.84	3.16		3	2.95	3.29		3	3.04	3.38
	4	2.82	3.14		4	2.94	3.27		4	3.03	3.37
	5	2.80	3.11		5	2.92	3.25		5	3.01	3.36
	10	2.68	2.97		10	2.84	3.16		10	2.95	3.29
29	1	2.89	3.22	36	1	2.99	3.33	43	1	3.07	3.41
	2	2.88	3.20		2	2.98	3.32		2	3.06	3.40
	3	2.86	3.18		3	2.97	3.30		3	3.05	3.39
	4	2.84	3.16		4	2.95	3.29		4	3.04	3.38
	5	2.82	3.14		5	2.94	3.27		5	3.03	3.37
	10	2.71	3.00		10	2.86	3.18		10	2.97	3.30
30	1	2.91	3.24	37	1	3.00	3.34	44	1	3.08	3.43
	2	2.89	3.22		2	2.99	3.33		2	3.07	3.41
	3	2.88	3.20		3	2.98	3.32		3	3.06	3.40
	4	2.86	3.18		4	2.97	3.30		4	3.05	3.39
	5	2.84	3.16		5	2.95	3.29		5	3.04	3.38
	10	2.73	3.03		10	2.88	3.20		10	2.98	3.32
31	1	2.92	3.25	38	1	3.01	3.36	45	1	3.09	3.44
	2	2.91	3.24		2	3.00	3.34		2	3.08	3.43
	3	2.89	3.22		3	2.99	3.33		3	3.07	3.41
	4	2.88	3.20		4	2.98	3.32		4	3.06	3.40
	5	2.86	3.18		5	2.97	3.30		5	3.05	3.39
	10	2.76	3.06		10	2.89	3.22		10	2.99	3.33

(*Continued*)

Table 15.2 (Continued)

n	i + 1	0.05	0.01	n	i + 1	0.05	0.01	n	i + 1	0.05	0.01
46	1	3.09	3.45	70	1	3.26	3.62	250	1	3.67	4.04
	2	3.09	3.44		2	3.25	3.62		5	3.67	4.04
	3	3.08	3.43		3	3.25	3.61		10	3.66	4.03
	4	3.07	3.41		4	3.24	3.60	300	1	3.72	4.09
	5	3.06	3.40		5	3.24	3.60		5	3.72	4.09
	10	3.00	3.34		10	3.21	3.57		10	3.71	4.09
47	1	3.10	3.46	80	1	3.31	3.67	350	1	3.77	4.14
	2	3.09	3.45		2	3.30	3.67		5	3.76	4.13
	3	3.09	3.44		3	3.30	3.66		10	3.76	4.13
	4	3.08	3.43		4	3.29	3.66	400	1	3.80	4.17
	5	3.07	3.41		5	3.29	3.65		5	3.80	4.17
	10	3.01	3.36		10	3.26	3.63		10	3.80	4.16
48	1	3.11	3.46	90	1	3.35	3.72	450	1	3.84	4.20
	2	3.10	3.46		2	3.34	3.71		5	3.83	4.20
	3	3.09	3.45		3	3.34	3.71		10	3.83	4.20
	4	3.09	3.44		4	3.34	3.70	500	1	3.86	4.23
	5	3.08	3.43		5	3.33	3.70		5	3.86	4.23
	10	3.03	3.37		10	3.31	3.68		10	3.86	4.22
49	1	3.12	3.47	100	1	3.38	3.75	750	1–10	3.95	4.30
	2	3.11	3.46		2	3.38	3.75	1000	1–10	4.02	4.37
	3	3.10	3.46		3	3.38	3.75	2000	1–10	4.20	4.54
	4	3.09	3.45		4	3.37	3.74	3000	1–10	4.29	4.63
	5	3.09	3.44		5	3.37	3.74	4000	1–10	4.36	4.70
	10	3.04	3.38		10	3.35	3.72	5000	1–10	4.41	4.75
50	1	3.13	3.48	150	1	3.52	3.89				
	2	3.12	3.47		2	3.51	3.89				
	3	3.11	3.46		3	3.51	3.89				
	4	3.10	3.46		4	3.51	3.88				
	5	3.09	3.45		5	3.51	3.88				
	10	3.05	3.39		10	3.50	3.87				
60	1	3.20	3.56	200	1	3.61	3.98				
	2	3.19	3.55		2	3.60	3.98				
	3	3.19	3.55		3	3.60	3.97				
	4	3.18	3.54		4	3.60	3.97				
	5	3.17	3.53		5	3.60	3.97				
	10	3.14	3.49		10	3.59	3.96				

Source: Reproduced with permission from R. O. Gilbert, *Statistical Methods for Environmental Pollution Monitoring*, Van Nostrand Reinhold, New York, 1987, pp. 268–270.

Gilbert (1989) suggests that test for outliers from a lognormal distribution can be constructed simply by log transforming the original data (e.g., $y_i = \log_e[x_i]$). Fortran IV source code that iteratively rejects up to m outliers is given by Rosner (1983) and reproduced by Gilbert (1987). Illustrative examples are given by Gilbert (1987).

Table 15.3 Critical values for extreme studentized deviate

Sample Size	Upper 5%	Upper 1%
5	.858	.882
6	.844	.882
7	.825	.873
8	.804	.860
9	.783	.844
10	.763	.827
11	.745	.811
12	.727	.795
13	.711	.779
14	.695	.764
15	.681	.750
16	.668	.737
17	.655	.724
18	.643	.711
19	.632	.700
20	.621	.688
22	.602	.668
24	.584	.649
26	.568	.632
28	.554	.616
30	.540	.601

Source: Reproduced with permission from W. von Türk (formerly Stefansky), *Technometrics* **14**(1972):475–476. Copyright 1972 by the American Statistical Association. All rights reserved.

For samples of less than 25, critical values for the ESD test for a single outlier can be used (see Table 15.3). When testing outliers, a conservative approximation can be obtained by assuming independence and setting $\alpha^* = \alpha/m$ and interpolating to find α^* in Table 15.3.

15.3 SKEWNESS TEST

The skewness test can be used to test discordance for one or more upper or lower outliers in a normal sample with μ and σ^2 unknown. The sample skewness statistic is

$$\left[\frac{\sum_{i=1}^{n}(x_i - \bar{x})^3}{ns^3}\right]^{1/2} \tag{15.4}$$

To use the test consecutively, observations are ordered from lowest to highest and the value of the statistic is compared to the critical value in Table 14.6. If the sample skewness statistic exceeds the critical value, the lowest or highest value is rejected, depending on the sign of the statistic (note that the square root must be taken on the absolute value of the statistic). The procedure is repeated until the sample skewness statistic does not exceed the critical value. Note that this test is essentially a data screening procedure that ends when the data are consistent with a normal distribution. If the data are not normally distributed, it is possible to reject values falsely. Comparisons of results for raw and log transformed values are strongly recommended.

15.4 KURTOSIS TEST

An analogous test to the skewness test is based on the sample kurtosis statistic

$$\frac{\sum_{i=1}^{n}(x_i - \bar{x})^4}{ns^4} \tag{15.5}$$

Discrepancy is indicated by high values of the statistic. The smallest or largest value is rejected based on whichever is farther from $\bar{x}$. Critical values for this statistic are provided in Table 14.7. As in the skewness test, the kurtosis test can be applied repeatedly until the sample value of the statistic is less than the critical value.

15.5 SHAPIRO–WILK TEST

As in the two prior outlier tests, the Shapiro–Wilk test of normality (W-statistic)

$$W = \frac{\{\sum_{i=1}^{n/2} a_{(n,n-i+1)}[x_{(n-i+1)} - x_{(i)}]\}^2}{s^2} \tag{15.6}$$

can be used to reject outliers consecutively until the fit of a normal distribution can no longer be rejected. Critical values of W are given in Table 14.2, and the coefficients (a) are given in Table 14.3. Note that for the skewness test, kurtosis test, and Shapiro–Wilk test, critical values underestimate their intended nominal levels as number of repeated evaluations increase. For example, with five outlier tests, the 1% critical value of these tests approximates the 5% critical value. It is only approximate, because the repeated tests are not independent.

15.6 E_m-STATISTIC

Tietjen and Moore (1972) developed a two-sided test for m outliers in a normal sample with μ and σ^2 unknown. Their statistic is

$$E_m = \frac{\sum_{i=1}^{n-m}(r_{(i)} - \bar{r}_{n-m})^2}{\sum_{i=1}^{n}(r_{(i)} - \bar{r})^2} \tag{15.7}$$

Table 15.4 Critical values for E_m test statistic $\alpha = .05$

n	1	2	3	4	5	6	7	8	9	10
3	.001	—	—	—	—	—	—	—	—	—
4	.025	.001	—	—	—	—	—	—	—	—
5	.081	.010	—	—	—	—	—	—	—	—
6	.146	.034	.004	—	—	—	—	—	—	—
7	.208	.065	.016	—	—	—	—	—	—	—
8	.265	.099	.034	.010	—	—	—	—	—	—
9	.314	.137	.057	.021	—	—	—	—	—	—
10	.356	.172	.083	.037	.014	—	—	—	—	—
11	.386	.204	.107	.055	.026	—	—	—	—	—
12	.424	.234	.133	.073	.039	.018	—	—	—	—
13	.455	.262	.156	.092	.053	.028	—	—	—	—
14	.484	.293	.179	.112	.068	.039	.021	—	—	—
15	.509	.317	.206	.134	.084	.052	.030	—	—	—
16	.526	.340	.227	.153	.102	.067	.041	.024	—	—
17	.544	.362	.248	.170	.116	.078	.050	.032	—	—
18	.562	.382	.267	.187	.132	.091	.062	.041	.026	—
19	.581	.398	.287	.203	.146	.105	.074	.050	.033	—
20	.597	.416	.302	.221	.163	.119	.085	.059	.041	.028
25	.652	.493	.381	.298	.236	.186	.146	.114	.089	.068
30	.698	.549	.443	.364	.298	.246	.203	.166	.137	.112
35	.732	.596	.495	.417	.351	.298	.254	.214	.181	.154
40	.758	.629	.534	.458	.395	.343	.297	.259	.223	.195
45	.778	.658	.567	.492	.433	.381	.337	.299	.263	.233
50	.797	.684	.599	.529	.468	.417	.373	.334	.299	.268

$\alpha = .01$

n	1	2	3	4	5	6	7	8	9	10
3	.000	—	—	—	—	—	—	—	—	—
4	.004	.000	—	—	—	—	—	—	—	—
5	.029	.002	—	—	—	—	—	—	—	—
6	.068	.012	.001	—	—	—	—	—	—	—
7	.110	.028	.006	—	—	—	—	—	—	—
8	.156	.050	.014	.004	—	—	—	—	—	—
9	.197	.078	.026	.009	—	—	—	—	—	—
10	.235	.101	.018	.006	—	—	—	—	—	—
11	.274	.134	.064	.030	.012	—	—	—	—	—
12	.311	.159	.083	.042	.020	.008	—	—	—	—
13	.337	.181	.103	.056	.031	.014	—	—	—	—
14	.374	.207	.123	.072	.042	.022	.012	—	—	—
15	.404	.238	.146	.090	.054	.032	.018	—	—	—
16	.422	.263	.166	.107	.068	.040	.024	.014	—	—
17	.440	.290	.188	.122	.079	.052	.032	.018	—	—
18	.459	.306	.206	.141	.094	.062	.041	.026	.014	—
19	.484	.323	.219	.156	.108	.074	.050	.032	.020	—
20	.499	.339	.236	.170	.121	.086	.058	.040	.026	.017
25	.571	.438	.320	.245	.188	.146	.110	.087	.066	.050
30	.624	.482	.386	.308	.250	.204	.166	.132	.108	.087
35	.669	.533	.435	.364	.299	.252	.211	.177	.149	.124
40	.704	.574	.480	.408	.347	.298	.258	.220	.190	.164
45	.728	.607	.518	.446	.386	.336	.294	.258	.228	.200
50	.748	.636	.550	.482	.424	.376	.334	.297	.264	.235

Source: Reproduced with permission from G. L. Tietjen and R. H. Moore, *Technometrics* **14**, 3(1972): 583–597. Copyright 1972 by the American Statistical Association. All rights reserved.

where $r_i = |x_i - \bar{x}|$ is the absolute deviation of x_i from the sample mean, $\{r_{(i)}\}$ are the values of r_i in ascending order (i.e., $r_{(1)} < r_{(2)} \cdots < r_{(n)}$), $\bar{r}$ is the mean of all of the r_i, and $\bar{r}_{n-m}$ is the mean of the $(n - m)$ lowest r_i. Critical values for the statistic are provided in Table 15.4. Since E_m is a simultaneous test for m outliers, it is applied only once. Note that if the choice of m is too large (e.g., $m = 3$, when only $m = 2$ true outliers are present), the third observation will often be rejected as well, or even worse, the third may mask the first two outliers.

15.7 DIXON'S TEST

Although not specifically designed for a general number of possible outliers, Dixon's test can be used for applications in which a small number of outliers are suspected. Arranging the sample in ascending order, the best test criterion for varying sample sizes is provided by Dixon (1953) as follows:

n	Highest Value	Lowest Value	n	Highest Value	Lowest Value
3–7	$\dfrac{x_n - x_{n-1}}{x_n - x_1}$	$\dfrac{x_2 - x_1}{x_n - x_1}$	11–13	$\dfrac{x_n - x_{n-2}}{x_n - x_2}$	$\dfrac{x_3 - x_1}{x_{n-1} - x_1}$
8–10	$\dfrac{x_n - x_{n-1}}{x_n - x_2}$	$\dfrac{x_2 - x_1}{x_{n-1} - x_1}$	14–25	$\dfrac{x_n - x_{n-2}}{x_n - x_3}$	$\dfrac{x_3 - x_1}{x_{n-2} - x_1}$

Critical values for Dixon's statistic are provided in Table 15.5. The test may be repeated consecutively by first testing the most extreme value and then testing the next most extreme value in a sample of size $(n - 1)$, found by omitting the most extreme value. If m outliers are suspected, all m tests must be performed no matter what the verdict of the first $m - 1$ test, since the former may mask the latter, particularly if they are on the same side. If the mth test exceeds the critical value,

Table 15.5 Critical values for Dixon's statistic

n	5% Level	1% Level	n	5% Level	1% Level
3	.941	.988	14	.546	.641
4	.765	.889	15	.525	.616
5	.642	.780	16	.507	.595
6	.560	.698	17	.490	.577
7	.507	.637	18	.475	.561
8	.554	.683	19	.462	.547
9	.512	.635	20	.450	.535
10	.477	.597	21	.440	.524
11	.576	.679	23	.421	.505
12	.546	.642	24	.413	.497
13	.521	.615	25	.406	.489

Source: Reproduced with permission from W. J. Dixon, *Biometrics* **9**(1953):89.

15.8 ILLUSTRATION

Gibbons (1994) considered the following 12 TOC measurements obtained from a single background monitoring location over a period of three years (see Table 15.6). The values on the first quarter of 1990 and second quarter of 1991 appear to be possible outliers. Table 15.7 lists data sorted by maximum deviation from the mean and test statistics for tests of $m = 2$ outliers. Table 15.8 lists the same results for $\log_e(x)$.

In general, each test rejected the second largest deviation for both raw and log transformed data. Each method would have resulted in rejection of the two suspected outliers. Some variation in result was found for the most extreme deviation, particularly for raw data. For the skewness test and E_m, the most extreme value was masked by the second most extreme value. For the kurtosis test and Shapiro–Wilk test, the most extreme value was significant at the 5% level, but the second most extreme value was significant at the 1% level, suggesting a small masking effect. The most consistent results in the example appeared to have been for Rosner's extension of the ESD test and for Dixon's test. Also note that the ESD test and E_m test are designed for simultaneous tests of all m outliers and should not be interpreted as sequential tests. Nonetheless, the fact that the largest outlier was masked by the second largest for the E_m test is of little consequence since with $m = 2$ we would have computed the second test only. However, if we had, instead, failed to recognize the second outlier and set $m = 1$, we would have obtained a misleading result.

Table 15.6 Hypothetical TOC data

Year	Quarter	TOC (mg/L)	$\log_e(x_i) + 1$
1990	1	<1	0.693
	2	21	3.091
	3	22	3.135
	4	18	2.944
1991	1	19	2.996
	2	40	3.714
	3	21	3.091
	4	25	3.258
1992	1	17	2.890
	2	18	2.944
	3	19	2.996
	4	22	3.135

Table 15.7 Raw TOC data sorted by $|x - \bar{x}|$ and associated test statistics for $m = 2$ outliers

Raw Data	Sorted	ESD	Skew	Kurt	SWT	E_m	Dixon
<1	40	3.456**	0.287	4.413*	0.819*	0.542	0.783**
21	<1	7.868**	−1.358**	5.719**	0.706**	0.026**	0.708**
22	25						
18	17						
19	18						
40	18						
21	22						
25	22						
17	19						
18	19						
19	21						
22	21						

*$p < 0.05$.
**$p < 0.01$.

Finally, it is of some interest to investigate what would have happened if we had set $m = 3$. The third most deviant value is 25 mg/L, which is not particularly deviant. Nevertheless, this value would have been rejected by Rosner's test for $m = 3$ and E_m for $m = 3$. These results indicate that the two simultaneous tests are influenced by having a subset of the m outliers as true outliers. Setting m too large will result in false rejection of valid measurements for the simultaneous tests.

Table 15.8 $\log_e(x) + 1$ transformed data sorted by $|x - \bar{x}|$ and associated test statistics for $m = 2$ outliers

$\log_e(x) = y$	Sorted	ESD	Skew	Kurt	SWT	E_m	Dixon
0.693	0.693	10.619**	−1.481**	7.175**	0.592**	0.120**	0.878**
3.091	3.714	5.888**	−1.251**	4.701**	0.779**	0.025**	0.702**
3.135	3.258						
2.944	3.135						
2.996	3.135						
3.714	3.091						
3.091	3.091						
3.258	2.996						
2.890	2.996						
2.944	2.944						
2.996	2.944						
3.135	2.890						

*$p < 0.05$.
**$p < 0.01$.

15.9 SUMMARY

Statistical approaches to outlier detection broadly classify into simultaneous and sequential methods. Simultaneous methods are attractive because they control the false positive rate for tests of all m suspected outliers; however, presence of some outliers can lead to rejection of all m candidate outliers. In contrast, sequential methods appear to work reasonably well despite lack of explicitly controlled multiple comparisons through adjustment of critical values. Skewness and kurtosis tests should probably be used jointly to eliminate potential masking produced by the presence of high and low outliers. The Shapiro–Wilk test is computationally complex in that separate sets of coefficients must be used at each step (i.e., for each suspected outlier) and the test is also sensitive to nonnormality (i.e., the test may confuse nonnormality with the presence of outliers from a normal distribution). We are then left with the one test really not designed for testing multiple outliers, namely Dixon's test. As described here, the test can be applied easily to the two highest and two lowest values in a sample, which may be sufficient for most practical applications where historical measurements from each location are examined separately.

16

DETECTING TREND

16.1 INTRODUCTION

Detecting trend in environmental data is a broad area that could easily encompass a separate volume. Issues of seasonality, autocorrelation, and corrections for flow substantially complicate trend detection application. Good sources of information include Box and Jenkins (1976), Chatfield (1984), Gilbert (1987), and McCleary and Hay (1980). Environmental applications can be found Carlson et al. (1970), Fuller and Tsokos (1971), Hsu and Hunter (1976), McCallister and Wilson (1975) and McMichael and Hunter (1972). These papers generally focus on groundwater streams, rivers and air monitoring applications. Typically, they require more historical measurements than are available in many environmental monitoring applications. For example, time-series methods such as Box–Jenkins (Montgomery and Johnson, 1976) are often used, which generally require at least 50 to 100 measurements at equally spaced intervals to evaluate the autocorrelation function properly. Additionally, trace-level or nondetected measurements must be treated as missing data, complicating application of traditional estimation methods, and often invalidating parametric methods. For example, if a series of measurements is reported at the quantification limit, deviations from the trend line will not be normally distributed and the standard error of the usual least-squares trend estimator will no longer apply, thus invalidating the test of the null hypothesis that the trend is zero. In many cases, outliers in the data will produce biased estimates of the least-squares estimated slope itself. As such, nonparametric trend estimators are generally most useful for environmental monitoring problems. Fortunately, several estimators are available, which are described in the following.

16.2 SEN'S TEST

Sen (1968) developed a simple nonparametric estimator of trend which is particularly useful for environmental monitoring applications. Gilbert (1987) points out that the method is an extension of an earlier work by Theil (1950). The method is robust to outliers, missing data, and nondetects.

To compute Sen's trend estimator, begin by obtaining the N' slope estimates, Q for each location as

$$Q = \frac{x'_i - x_i}{i' - i} \tag{16.1}$$

where x'_i and x_i are the measured concentrations on monitoring events i' and i, where $i' > i$ and N' is the number of data pairs for which $i' > i$. The median value of the N' values of Q is Sen's estimator of trend (i.e., slope of the time by concentration regression line). With a single measurement per monitoring event (which is recommended),

$$N' = \frac{n(n-1)}{2} \tag{16.2}$$

where n is the number of monitoring events. For nondetects, the quantification limit may be used for x_i since it represents the lowest quantifiable concentration. To obtain the median value of Q, denoted as S, the N' values of Q are ranked from smallest to largest (i.e., $Q_1 \leq Q_2 \cdots \leq Q_{n-1} \leq Q_n$) and the median slope is computed as

$$S = Q_{[(N'+1)/2]} \quad \text{if } N' \text{ is odd} \tag{16.3}$$

or

$$S = \frac{Q_{[N'/2]} + Q_{[(N'+2)/2]}}{2} \quad \text{if } N' \text{ is even} \tag{16.4}$$

To test the null hypothesis of zero slope (i.e., no trend), the lower $100(1 - \alpha)\%$ confidence limit for true slope is computed. To compute the confidence limit, an estimate of the variance of S is required. Gilbert (1987) suggests employing the estimator due to Kendall (1975), which can be used with small samples (i.e., $n > 8$) and in the presence of ties (i.e., measurements with the same value or nondetects). The variance estimate is given by

$$\text{Var}(S) = \tfrac{1}{18}\left[n(n-1)(2n+5) - \sum_{p=1}^{q} t_p(t_p - 1)(2t_p + 5)\right] \tag{16.5}$$

where q is the number of values for which there are ties and t_p is the number of tied measurements for a particular value. The approximate normal theoretical lower confidence limit is the M_1th largest value of Q, where

196 DETECTING TREND

Table 16.1 Monitoring data for TOC

Year	Quarter	TOC (mg/L)
1991	1	1
	2	5
	3	4
	4	8
1992	1	8
	2	12
	3	16
	4	30

$$M_1 = \frac{N' - Z_{1-\alpha}[\text{Var}(S)]^{1/2}}{2} \qquad (16.6)$$

and $Z_{1-\alpha}$ is the $(1 - \alpha)100\%$ point of the normal distribution (e.g., $Z_{0.95} = 1.645$). If $Q_{M1} > 0$, the null hypothesis that the trend is zero can be rejected.

The upper confidence limit for S is given by the $(M_2 + 1)$th largest value of Q, where

$$M_2 = \frac{N' + Z_{1-\alpha}[\text{Var}(S)]^{1/2}}{2} \qquad (16.7)$$

Note that to compute a 95% two-sided confidence interval for S, $Z_{1-\alpha/2} = 1.965$ is required.

Illustration Consider the data in Table 16.1 for TOC sampled quarterly for the past two years. There are $N' = 28$ pairs for which $i' > i$. Individual slope estimates for these pairs are obtained by dividing the differences by $i' - i$, as in Table 16.2. Ranking these values for smallest to largest gives

$-1, 0, 1, 1.5, 1.5, 1.75, 1.75, 2.00, 2.00, 2.20, 2.20, 2.33, 2.50, 2.67, 2.67, 3,$
$4, 4, 4, 4, 4.14, 4.17, 5.20, 5.50, 7.33, 9, 14$

Table 16.2 Individual slope estimates for illustration

Time Period:	1	2	3	4	5	6	7	8
Concentration:	1	5	4	8	8	12	16	30
		4.00	1.50	2.33	1.75	2.20	2.50	4.14
			−1.00	1.50	1.00	1.75	2.20	4.17
				4.00	2.00	2.67	3.00	5.20
					0.00	2.00	2.67	5.50
						4.00	4.00	7.33
							4.00	9.00
								14.00

Since $N' = 28$ is even, the median is the average of the 14th and 15th largest values (2.67), which is the Sen slope estimate. Since there is one pair of tied values, the variance estimate is

$$\text{Var}(S) = \tfrac{1}{18}[8(7)(21) - 2(1)(9)] = 64.33$$

The 95% confidence interval is therefore given by the values corresponding to the order statistics:

$$M_1 = \frac{28 - 1.96\sqrt{64.33}}{2} = 6.14$$

$$M_2 = \frac{28 + 1.96\sqrt{64.33}}{2} = 21.86$$

which correspond to values of 1.75 and 4.12, obtained by interpolating between the sixth and seventh and 21st and 22nd ordered values. This interval does not contain the value zero; therefore, the null hypothesis of no trend (i.e., slope equals zero) can be rejected. Note that if the concern had been with increasing trends, a one-sided interval or limit should have been used. In this case, the one-sided lower confidence limit is computed as

$$M_1 = \frac{28 - 1.65\sqrt{64.33}}{2} = 7.38$$

which has the value 1.85 mg/L, which is also greater than zero.

16.3 MANN–KENDALL TEST

Another trend estimator well suited to environmental data is the Mann–Kendall test (or Kendall's test) for trend (Kendall, 1975; Mann, 1945; Theil, 1950). As in Sen's test, there are no distributional assumptions and missing data (i.e., nondetects) or irregularly spaced measurement periods are permitted. Nondetects are assigned a value smaller than the smallest measured value, typically L_Q, since no quantitative value can or should be reported below the L_Q.

The version of the Mann–Kendall test presented here is recommended for 40 or fewer measurements (Gilbert, 1987). First, order the data by sampling date $x_1, x_2, \ldots, x_n$ where x_i is the measured value on occasion i. Second, record the signs of each of the N' possible differences $x_{i'} - x_i$, where $i' > i$; for example, let

$$\text{sgn}(x_{i'} - x_i) = \begin{cases} 1 & \text{if } x_{i'} - x_i > 0 \\ 0 & \text{if } x_{i'} - x_i = 0 \\ -1 & \text{if } x_{i'} - x_i < 0 \end{cases} \qquad (16.8)$$

Table 16.3 Mann–Kendall test

\multicolumn{5}{c	}{Measurements Ordered by Time}	Number of + signs	Number of − signs
x_1	x_2	x_3	... x_{n-1}	x_n		
	$x_2 - x_1$	$x_3 - x_1$	$x_{n-1} - x_1$	$x_n - x_1$		
		$x_3 - x_2$	$x_{n-1} - x_2$	$x_n - x_2$		
			$x_{n-1} - x_3$	$x_n - x_3$		
			.	.		
			.	.		
			.	.		
			$x_{n-1} - x_{n-2}$	$x_n - x_{n-2}$		
				$x_n - x_{n-1}$		
					Sum of + signs	Sum of − signs

The Mann–Kendall statistic is then computed as

$$S = \sum_{i=1}^{n-1} \sum_{i'=k+1}^{n} \operatorname{sgn}(x_{i'} - x_i) \qquad (16.9)$$

which is the number of positive differences minus the number of negative differences, easily obtained from the last two columns of Table 16.3. Values of S, n, and associated probability for the test of $S = 0$ (i.e., no increasing trend) are given in Table 16.4 (see Kendall, 1975). Negative trends can also be tested (e.g., to determine if effects of remediation are significant) by reversing the sign of S (i.e., for decreasing trends $S < 0$). The two-sided test (i.e., either increasing or decreasing trend) can also be obtained by doubling the probability values listed in Table 16.4. For $n > 10$, Gilbert (1987) suggests that the normal approximation

$$Z = \begin{cases} \dfrac{S - 1}{[\operatorname{Var}(S)]^{1/2}} & \text{if } S > 0 \\ 0 & \text{if } S > 0 \\ \dfrac{S + 1}{[\operatorname{Var}(S)]^{1/2}} & \text{if } S < 0 \end{cases} \qquad (16.10)$$

works adequately for most purposes.

Illustration Using the example data in Table 16.1, the Mann–Kendall test statistic is computed as shown in Table 16.5. The value of S is therefore $26 - 1 = 25$. From Table 16.4, entering at $n = 8$ and $S = 25$ yields a probability of 0.00053, indicating a significant positive trend. Using the normal approximation,

$$Z = \frac{25 - 1}{\sqrt{64.33}} = 2.99$$

16.3 MANN–KENDALL TEST

Table 16.4 Values of S, n, and associated probability for Mann–Kendall test[a]

		n					n	
S	4	5	8	9	S	6	7	10
0	0.625	0.592	0.548	0.540	1	0.500	0.500	0.500
2	0.375	0.408	0.452	0.460	3	0.360	0.386	0.431
4	0.167	0.242	0.360	0.381	5	0.235	0.281	0.364
6	0.042	0.117	0.274	0.306	7	0.136	0.191	0.300
8	—	0.042	0.199	0.238	9	0.068	0.119	0.242
10	—	0.0^283	0.138	0.179	11	0.028	0.068	0.190
12	—	—	0.089	0.130	13	0.0^283	0.035	0.146
14	—	—	0.054	0.090	15	0.0^214	0.015	0.108
16	—	—	0.031	0.060	17	—	0.0^254	0.078
18	—	—	0.016	0.038	19	—	0.0^214	0.054
20	—	—	0.0^271	0.022	21	—	0.0^320	0.036
22	—	—	0.0^228	0.012	23	—	—	0.023
24	—	—	0.0^387	0.0^263	25	—	—	0.014
26	—	—	0.0^319	0.0^229	27	—	—	0.0^283
28	—	—	0.0^425	0.0^212	29	—	—	0.0^246
30	—	—	—	0.0^343	31	—	—	0.0^223
32	—	—	—	0.0^312	33	—	—	0.0^211
34	—	—	—	0.0^425	35	—	—	0.0^347
36	—	—	—	0.0^528	37	—	—	0.0^318
					39	—	—	0.0^458
					41	—	—	0.0^415
					43	—	—	0.0^528
					45	—	—	0.0^628

[a]Repeated zeros are indicated by powers; for example, 0.0^347 stands for 0.00047.

Source: Reproduced with permission from M.G. Kendall, *Rank Correlation Methods*, Charles Griffin, London, 1975.

Table 16.5 Computation of the Mann–Kendall test

Time Period:	1	2	3	4	5	6	7	8	Number of + signs	Number of − signs
Concentration:	1	5	4	8	8	12	16	30		
			4	3	7	7	11	15	29	7	0
				−1	3	3	7	11	25	5	1
					4	4	8	12	26	5	0
						0	4	8	22	3	0
							4	8	22	3	0
								4	18	2	0
									14	1	0
										26	1

16.4 SEASONAL KENDALL TEST

When data are influenced by seasonal effects, the estimators described previously will yield biased results, requiring a trend estimator adjustable for seasonal variation. The seasonal Kendall test was developed by Smith et al. (1982) and is discussed further by Gilbert (1987). As in the tests of sections 16.2 and 16.3, the seasonal Kendall test is free of distributional assumptions and will work with a mixture of quantifiable and nonquantifiable measurements.

Let x_{il} be the measurement for the ith season in the lth year, K the number of seasons, and L the number of years as in Table 16.6. In this illustration, $K = 4$ seasons; however, there is no reason that the test could not consider monthly effects (i.e., $K = 12$) as long as independent measurements are possible.

For each season, data are accumulated over the L years to compute the Mann–Kendall test statistic S. For season i, compute

$$S_i = \sum_{k=1}^{n_i - 1} \sum_{l=k+1}^{n_i} \text{sgn}(x_{il} - x_{ik}) \qquad (16.11)$$

where $l > k$, n_i is the number of measurements available for season i (i.e., over the L years), and sgn $(x_{il} - x_{ik})$ is defined as in equation (16.8).

The variance of S_i is computed as

$$\begin{aligned}\text{Var}(S_i) = &\tfrac{1}{18}\left[n_i(n_i - 1)(2n_i + 5) - \sum_{p=1}^{g_i} t_{ip}(t_{ip} - 1)(2t_{ip} + 5) \right.\\ &\left. - \sum_{q=1}^{h_i} \mu_{iq}(\mu_{iq} - 1)(2\mu_{iq} + 5)\right]\\ &+ \frac{\sum_{p=1}^{g_i} t_{ip}(t_{ip} - 1)(t_{ip} - 2) \sum_{q=1}^{h_i} \mu_{iq}(\mu_{iq} - 1)(\mu_{iq} - 2)}{9 n_i(n_i - 1)(n_i - 2)}\\ &+ \frac{\sum_{p=1}^{g_i} t_{ip}(t_{ip} - 1) \sum_{q=1}^{h_i} \mu_{iq}(\mu_{iq} - 1)}{2 n_i(n_i - 1)}\end{aligned} \qquad (16.12)$$

Table 16.6 Data Set for computing seasonal Kendall test

	Season			
Year	Spring	Summer	Fall	Winter
1	x_{11}	x_{21}	x_{31}	x_{41}
2	x_{12}	x_{22}	x_{32}	x_{42}
⋮				
L	x_{1L}	x_{2L}	x_{3L}	x_{4L}

where g_i is the number of groups of values tied in season i, t_{ip} the number of ties in the pth group for season i, h_i the number of sampling times in season i that contains more than one measurement, and μ_{iq} the number of measurements in the qth time period in season i. Note that in many cases there is a single measurement per sampling event (e.g., quarterly measurements); therefore, $h_i = 0$ for $i = 1, \ldots, 4$ and $\mu_{iq} = 0$ for all i and q. In this case the variance estimator is simply

$$\text{Var}(S_i) = \frac{1}{18} \left[n_i(n_{i-1})(2n_i + 5) - \sum_{p=1}^{g_i} t_{ip}(t_{ip} - 1)(2t_{ip} + 5) \right] \quad (16.13)$$

which is the usual estimator of variance of the Mann–Kendall test computed separately for the data in season i. Pooling across the K seasons,

$$S' = \sum_{i=1}^{K} S_i \quad (16.14)$$

$$\text{Var}(S') = \sum_{i=1}^{K} \text{Var}(S_i) \quad (16.15)$$

The quantity

$$Z = \begin{cases} \dfrac{S' - 1}{[\text{Var}(S')]^{1/2}} & \text{if } S' > 0 \\ 0 & \text{if } S' = 0 \\ \dfrac{S' + 1}{[\text{Var}(S')]^{1/2}} & \text{if } S' < 0 \end{cases} \quad (16.16)$$

can be compared to standard normal cumulative distribution probabilities to test the null hypothesis of no trend. For 95% confidence, the critical one-tailed value is 1.65 (i.e., test of an upward trend) or for a two-tailed test, the value is 1.96 (i.e., test of either an upward or downward trend). The corresponding 99% confidence values are 2.33 and 2.57, respectively.

Although the previous method provides a test of the null hypothesis of no trend, it does not provide an estimate of the slope of the trend line or corresponding interval estimates. The seasonal Kendall slope estimator is obtained by computing the individual N_i' slope estimates for season i as

$$Q_i = \frac{x_{il} - x_{ik}}{l - k} \quad (16.17)$$

where x_{il} is the concentration for the ith season of the lth year and x_{ik} is the concentration for the ith season of the kth year, where $l > k$. This computation is repeated for each season. The median of the $N' = N_1' + N_2' + \cdots + N_K'$ individual slope estimates is then the Kendall slope estimator. Confidence limits are obtained in the same way as for Sen's estimator, substituting S' for S.

Illustration To illustrate the seasonal Kendall test, consider the data in Table 16.7, covering three years and four seasons. In this example,

$$n_1 = n_2 = n_3 = n_4$$
$$g_1 = g_2 = g_3 = g_4$$
$$t_{ip} = 0 \quad \text{for all } i \text{ and } p$$
$$N'_1 = N'_2 = N'_3 = N'_4 = 3$$
$$N' = 12$$

Individual slope estimates are computed as follows:

	Spring			Summer			Fall			Winter		
Year	1	2	3	1	2	3	1	2	3	1	2	3
TOC	3	4	5	4	5	6	5	6	7	6	7	8
		+1	+1		+1	+1		+1	+1		+1	+1
			+1			+1			+1			+1
S_i:		3			3			3			3	

The seasonal Kendall test statistic is therefore

$$S' = S_1 + S_2 + S_3 + S_4 = 3 + 3 + 3 + 3 = 12$$

with seasonal variances

$$\text{Var}(S_1) = \text{Var}(S_2) = \text{Var}(S_3) = \text{Var}(S_4) = \tfrac{1}{18}[3(2)(11) - 0] = 3.67$$

and overall variance

$$\text{Var}(S') = 3.67 + 3.67 + 3.67 + 3.67 = 14.68$$

The test statistic is therefore

$$Z = \frac{12 - 1}{\sqrt{14.68}} = 2.87$$

which has probability 0.002, indicating rejection of the null hypothesis of no trend.

Table 16.7 Seasonal intrawell monitoring data for TOC

Year	Season	TOC (mg/L)
1990	Spring	3
	Summer	4
	Fall	5
	Winter	6
1991	Spring	4
	Summer	5
	Fall	6
	Winter	7
1992	Spring	5
	Summer	6
	Fall	7
	Winter	8

The seasonal Kendall slope estimator is the median of the 12 individual slope estimates, which is 1.0 mg/L per quarter, with a 95% lower confidence limit,

$$M_1 = \frac{12 - 1.65\sqrt{14.68}}{2} = 2.84$$

which also corresponds to a concentration of 1.0 mg/L.

16.5 STATISTICAL PROPERTIES

El-Shaarawi and Niculescu (1992) have considered the statistical properties of the Mann–Kendall test of trend in environmental time-series data. They showed that (1) the statistic is asymptotically normal, thus justifying use of the approximate normal theory confidence limits derived in this chapter, and (2) the statistic is strongly affected by statistical dependence among the reported measurements. They derive the appropriate variance estimators for moving average correlated time-series data for both seasonal and nonseasonal models. The interested reader is referred to their paper, particularly equations (15) and (40). In their illustration, using monthly chloride data in the South Saskatchewan River, they show that the amount of dependence is sufficient to generate incorrect conclusions using the original test (which assumes independence).

16.6 SUMMARY

The detection of trend in environmental data is a field unto itself, and in this chapter we presented a small selection of possible approaches that are potentially useful in this area. In general, trend analysis is useful as a global screening tool. The presence of a significant trend is an indication of a problem that should be investigated fully prior to establishing a routine environmental monitoring program.

17

DETECTION MONITORING

17.1 INTRODUCTION

Protection of our nation's natural resources is a major theme in public policy and social thought. New regulations rely on statistical methods to detect the earliest release of pollutants into the air and to reduce and control the impact of industrial discharges on air, surface-water, and groundwater resources. Statistical decision rules often are the first consideration in examining the repercussions of environmental pollutants. However, these decision rules have not always been accompanied by statistically rigorous methodology.

In this chapter we focus on statistical determination of environmental impact of hazardous and municipal solid waste disposal, specifically the consequences to groundwater beneath landfills. The EPA has promulgated regulations for disposal of hazardous waste (Subtitle C Regulation, U.S. EPA, 1988) and municipal solid waste (Subtitle D Regulation, U.S. EPA, 1991) and associated guidance (U.S. EPA, 1989, 1992), which often use statistical decision rules unsuited to the problems encountered in environmental monitoring. In the following sections, general conceptual and statistical features of decision rules are described and statistical approaches are compared and contrasted. Although the focus of this chapter is primarily on groundwater monitoring problems, the paradigm of detection monitoring has relevance to environmental monitoring in general.

17.2 GROUNDWATER DETECTION MONITORING

Groundwater detection monitoring is used to determine the earliest possible release of pollutants from a waste disposal facility into groundwater underneath the facility.

New waste disposal facilities are required to limit leakage by using clay and synthetic liners. However, older facilities were not required to use liners, and the likelihood that pollutants would be released into groundwater was high. As landfill waste biodegrades, liquid (termed *leachate*), which may contain hazardous constituents in varying concentrations, forms at the bottom of the facility. If a liner is absent or leaky, the leachate can escape, contaminate groundwater beneath the facility, migrate off-site, and affect drinking water supplies negatively, since some leachate constituents are carcinogenic initiators or promoters (Gibbons 1994; Gibbons et al., 1992).

Groundwater detection monitoring typically involves a series of monitoring wells hydraulically upgradient and downgradient of the facility to compare concentrations of chemical constituents between the upgradient and downgradient locations (see Figure 17.1), assuming that any difference in groundwater quality is caused by leachate released from the facility.

However, this assumption is often false because widespread spatial variability exists in groundwater chemistry. In the worst case (often the most typical circumstance), regulations (U.S. EPA, 1988, 1991) require only one upgradient well and a minimum of three monitoring wells located downgradient from the facility. When a single upgradient well is used to characterize natural variability in background, spatial variability and contamination are confounded (i.e., differences between the upgradient and downgradient wells could be due to natural differences between any two locations, regardless of their relation to the waste disposal facility). Even with two upgradient wells, characterization of natural background variability may not be

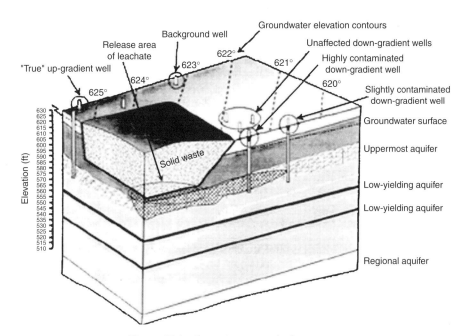

Figure 17.1 Groundwater monitoring system.

possible; that is, two upgradient wells may not display the same amount of variability observed in downgradient wells, which often number between 10 and 100.

Additionally, regulations require each downgradient monitoring well and constituent to be separately tested because releases from a waste disposal facility into groundwater are "plume" shaped (see Figure 17.1), which may influence only a single downgradient well. Pooling data over downgradient wells might mask a release that affected only a single well. In addition, chemical constituents travel at different rates in groundwater; the leading edge of the plume may contain only a small number of highly mobile chemical constituents. In many parts of the country, groundwater flows quite slowly, in some cases only a foot or two per year. Hydrogeologically independent observations from a given monitoring well may be available only quarterly, semiannually, or annually. Pooling data may be impractical since it may result in mixing contaminated and uncontaminated measurements, masking early stage release. Therefore, each new datum must be evaluated individually. The two most critical problems are that (1) numerous statistical evaluations must be performed on each monitoring event (typically, 100 to 1000), and (2) environmental data are often censored (i.e., the analyte may or may not be detected when it is present at a level below the capability of the analytical instrument). These two problems complicate analysis of groundwater monitoring data as detailed in the following sections.

As shown in the following sections, solution of these problems leads to the construction of prediction limits adapted to the case of simultaneous statistical inference and sequential testing. Simultaneous statistical inference refers to the construction of limits or bounds that apply simultaneously to all comparisons made on a given monitoring event. In this context the number of comparisons, which is denoted as k, is the set of all downgradient monitoring wells for all constituents for which statistical evaluation must be performed. Using these methods, we can therefore control the overall site-wide false positive rate (i.e., concluding that there has been an impact when there has not been) at a nominal level (e.g., 5%). As the number of comparisons on any given monitoring event becomes large, however, the associated false negative rate (i.e., the failure to detect contamination when it is present) also becomes large. To minimize the false negative rate, we use a sequential testing strategy in which an initial exceedance is then verified by one or more independent verification resamples. In this way, a smaller prediction limit can be used repeatedly, achieving the same site-wide false positive rate but greatly minimizing the false negative rate. These ideas are now more fully developed in the following sections. The reader is also referred to ASTM Standard D6312 and Gibbons (1994).

17.3 STATISTICAL PREDICTION INTERVALS

17.3.1 Single Location and Constituent

If the problem were to set a $(1 - \alpha)100\%$ confidence limit on the next single measurement for one location and one normally distributed constituent, a β-expectation tolerance limit (i.e., a prediction limit; Guttman, 1970; Hahn, 1970) could be computed from n independent background measurements as

$$\bar{x} + t_{[n-1, 1-\alpha]} s \sqrt{1 + \frac{1}{n}} \tag{17.1}$$

where concern is that the concentration is elevated above background, $\bar{x}$ and s are the background sample mean and standard deviation, and t is the $100(1 - \alpha)$ percentile of Student's t-distribution on $n - 1$ degrees of freedom. If upgradient versus downgradient comparisons are to be performed, a minimum of two background locations (e.g., wells) should be sampled repeatedly at a time interval sufficient to ensure independence (e.g., quarterly or semiannually). The background time period must include at least one year to ensure that the same seasonal variation present in downgradient locations is reflected in the upgradient background. The reader should note that with multiple upgradient locations, s^2, the traditional estimator of σ^2 is biased (i.e., it is too small) because measurements are nested within upgradient monitoring locations. Alternative estimators for σ^2 based on variance components models have been proposed and should be used where appropriate (Gibbons, 1987, 1994).

17.3.2 Multiple Locations

In practice, multiple comparisons are performed, one for each downgradient monitoring location and constituent. Using the Bonferroni inequality (Miller, 1966), a conservative prediction bound (i.e., the probability of at least one false rejection is at most α) for all kq comparisons (i.e., k locations each tested for q constituents) is

$$\bar{x} + t_{[n-1, 1-\alpha/(kq)]} s \sqrt{1 + \frac{1}{n}} \tag{17.2}$$

In the present context, the comparisons are dependent because (1) constituents may be correlated, and (2) all downgradient locations are compared to a common background. In this case, the Bonferroni adjustment may be unnecessarily conservative. Some improvement may be gained by adapting the approach of Dunnett and Sobel (1955) originally developed to compare multiple treatment groups to a common control group. The resulting correlation between the comparison of locations i and j to a common background is

$$\rho_{ij} = 1 \bigg/ \sqrt{\left(\frac{n_0}{n_i} + 1\right)\left(\frac{n_0}{n_j} + 1\right)} \tag{17.3}$$

where n_0 is the number of background measurements, n_j the number of measurements in monitoring location j, and n_i the number of measurements in monitoring location i. In the measurement of groundwater the correlation is constant with value $\rho = 1/(n + 1)$, since the number of background measurements $n_0 = n$ and

$n_i = n_j = 1$ for all i and j (i.e., we are comparing a single new value in each monitoring location to n background measurements). Dunnett (1955) has shown how required values from the multivariate t-distribution can be reduced to evaluation of the equa-correlated multivariate normal distribution for which the required probabilities are easily obtained. These critical points have been tabulated by a number of authors (Gibbons, 1994; Gupta and Panchpakesan, 1979). As we show in the following section, increasing statistical power can be achieved by generalizing the single-stage Dunnett procedure described here to the case of multistage sampling using verification resampling. Alternative stagewise comparison procedures have also been considered (Hochberg and Tamhane, 1987) in the context of multiple comparisons to a common control.

17.3.3 Verification Resampling

As the number of future comparisons increases, the prediction limit increases and false negative rates can become unacceptably large. Gibbons (1987) and Davis and McNichols (1987) noted this problem and suggested sequential testing of new groundwater monitoring measurements such that the presence of an initial exceedance in a downgradient location requires obtaining one or more independent resamples for that constituent. Failure is indicated only if both initial sample and verification resample(s) exceed the prediction limit. In this way, fewer samples are required and both false positive and false negative rates are controlled at minimum levels. Davis and McNichols (1987) derived simultaneous normal prediction limits for the next r of m measurements at each of k monitoring locations, where in the preceding example, $r = 1$ and $m = 2$. Their result is a further generalization to Dunnett's test. The derivation is complicated, but a few key features are described. Again assume that the background observations and new monitoring measurements are drawn from the same normal distribution $N(\mu, \sigma^2)$. Expressing $y_{ij} = x_{ij} - \bar{x}$ (i.e., a mean deviation) for $i = 1, \ldots, k$ locations and $j = 1, \ldots, m$ samples and letting $y_{i(r)}$ denote the rth smallest of the y_{ij} for location i, and $y^* = \max_i (y_{i(r)})$, then having at least r of m future observations below $\bar{x} + Ks$ is equivalent to $y^* < Ks$, where K is the multiplier sought.

Davis and McNichols (1987) have shown that

$$\Pr(y^* < Ks) = \int_{-\infty}^{\infty} T_{n-1,\sqrt{nz^*}}(\sqrt{n}K)$$

$$\times k \left[\int_{-\infty}^{z^*} m \binom{m-1}{r-1} \Phi^{r-1}(t)\phi(t)[1 - \Phi(t)^{m-r}] \, dt \right]^{k-1}$$

$$\times m \binom{m-1}{r-1} \Phi^{r-1}(z^*)\phi(z^*)[1 - \Phi(z^*)]^{m-r} dz^* \qquad (17.4)$$

where $T_{\nu,\delta}(\cdot)$ is the cumulative density function of the noncentral t distribution, z^* the maximum rth order statistic across all k locations, and ϕ and Φ the standard normal probability density and cumulative distribution functions. The equation is then solved for K such that the right-hand side is equal to $1 - \alpha$. Extensive tables of K for varying levels of α, n, k, r, and m are available (Gibbons, 1994). For example, with $n = 8$ background samples $k = 10$ monitoring locations and one verification resample (i.e., $r = 1$, $m = 2$), the prediction limit

$$\bar{x} + 2.03s$$

will include at least one of the next two measurements in each of 10 downgradient locations with 95% confidence.

17.3.4 Multiple Constituents

Little is known about the correlation between monitoring constituents, except that the interrelationship is highly variable and that there are too few background measurements to characterize the correlation matrix precisely or use the matrix to construct accurate multivariate prediction limits. For this reason, the Bonferroni inequality has been used to derive conservative prediction bounds. This practice will produce prediction limits larger than required when positive association is present (which in our experience, appears to be common). For example, in the previous illustration, if we were to monitor 10 constituents, $\alpha = 0.05/10 = 0.005$ and the limit

$$\bar{x} + 3.36s$$

would be applied to each location and constituent with an overall site-wide confidence level of 95%.

Alternatively, there has been some work on multivariate prediction bounds (Bock, 1975; Guttman, 1970) which might apply to those cases where background sample sizes were sufficiently large to obtain a reasonable estimate of the interconstituent covariance matrix. Unfortunately, the presence of nondetects (i.e., left censored distributions) violates the joint normality assumption of the multivariate procedure.

Finally, it should be noted that in all cases, the smallest number of constituents that are indicative of a potential release from the facility should be used. Using fewer constituents will decrease the total number of comparisons and provide more conservative (i.e., smaller) prediction bounds. Also note that some constituents exhibit greater spatial variability than others (e.g., geochemical parameters such as chloride versus metals such as barium) and may be less useful at some facilities with heterogeneous geologic formations.

17.3.5 Problem of Nondetects

In practice, environmental measurements consist of a mixture of detected and nondetected constituents ranging in detection frequency from 0 to 100%. When the detection frequency is high (e.g., >85%), several studies (Gibbons, 1994; Gilliom and Helsel, 1986; Hass and Scheff, 1990) have shown that most estimates of mean and variance of a left censored normal or lognormal distribution yield reasonable results. This is not true when detection frequencies are between 50 and 85% (Gibbons, 1994). In this case, available methods include maximum likelihood estimators (MLE; Cohen, 1959, 1961), restricted maximum likelihood estimators (Persson and Rootzen, 1977), an estimator based on the delta distribution which is a lognormal distribution with probability mass at zero (Aitchison, 1955), best linear unbiased estimators (Gupta, 1952; Sarhan and Greenberg, 1962), alternative linear estimators (Gupta, 1952), regression estimators (Gilliom and Helsel, 1986; Hashimoto and Trussell, 1983), and substitution of expected values of normal order statistics (Gleit, 1985). These methods are described in some detail in Chapter 13. In addition, the EPA has often advocated simple substitution of one-half the method detection limit. Methods that adequately recover the mean and variance of the underlying distribution from the censored data often inadequately recover the tail probabilities used in computing prediction limits (Gibbons, 1994). In a simulation study (Gibbons, 1994) the MLE was the best overall estimator, but the estimator based on the delta distribution was best at preserving confidence levels for prediction limits in the presence of censoring (see Chapter 13).

17.3.6 Nonparametric Prediction Limits

When detection frequency is less than 50%, none of the methods discussed in Section 17.3.5 work well and an alternative strategy must be employed. In practice, an excellent alternative is to compute a nonparametric prediction limit, which is the maximum of n background measurements. The nonparametric limit is attractive because it makes no distributional assumptions and is defined even if only one of the n background measurements is quantifiable. In some cases, however, the number of background measurements is insufficient to provide a reasonable overall confidence level; therefore, the nonparametric prediction limit may not always be an available alternative. Confidence levels for the nonparametric limits are a function of n, kq, and the number of verification resamples similar to the parametric case. For example, let $X_{(\max,n)}$ represent the maximum value obtained out of a sample of size n and $Y_{(\min,m)}$ represent the minimum value out of a sample of size m. In the present context, $X_{(\max,n)}$ is the maximum background concentration and $Y_{(\min,m)}$ is the minimum of the initial sample and verification resample(s) for a constituent in a downgradient monitoring location. The objective is to compare $Y_{(\min,m)}$ to $X_{(\max,n)}$. The confidence level for the simultaneous upper prediction limit defined as $X_{(\max,n)}$ is

$$\Pr(Y_{1(\min,m)} \leq X_{(\max,n)}, Y_{2(\min,m)} \leq X_{(\max,n)},\ldots,Y_{k(\min,m)} \leq X_{(\max,n)}) = 1 - \alpha \quad (17.5)$$

To achieve a desired confidence level (say, $1 - \alpha = 0.95$ for a fixed number of background measurements), m must be adjusted; the more resamples, the greater the confidence. This probability can be evaluated using a variant of the multivariate hypergeometric distribution (Chou and Owen, 1986; Hall et al., 1975) function as

$$1 - \alpha = \frac{n}{km + n} \sum_{j_1=1}^{m} \sum_{j_2=1}^{m} \cdots \sum_{j_k=1}^{m} \frac{\binom{m}{j_1}\binom{m}{j_2}\cdots\binom{m}{j_k}}{\binom{km + n - 1}{\sum_{i=1}^{k} j_i + n - 1}} \qquad (17.6)$$

Based on this result, approximate confidence levels have been derived (Gibbons, 1990) for nonparametric prediction limits defined as the maximum of n background samples in which it is required to pass 1 of m samples (i.e., the initial sample or at least one verification resample) at each of k monitoring locations. To incorporate multiple constituents, the confidence level is adjusted to $1 - \alpha/q$. Exact confidence levels for the previous case and approximate confidence levels for the case in which it is required to pass the first or all of m resamples are now also available (Gibbons, 1991a). Exact confidence levels for the latter case have been derived (Davis and McNichols, 1994, 1999; Willits, 1993) and extensive tables have been prepared (Gibbons, 1994). The case in which the prediction limit is the second largest measurement has also been considered (Davis and McNichols, 1994, 1999; Gibbons, 1994).

17.4 INTRAWELL COMPARISONS

Upgradient versus downgradient comparisons are often inappropriate (e.g., spatial variability may be present) and some form of intrawell comparisons (i.e., each location compared to its own history) must be performed. Note that intrawell comparisons are only appropriate when (1) predisposal data are available or (2) it can be demonstrated that the facility has not affected that well in the past. In this case, there are two good statistical methods available; combined Shewhart–CUSUM (cumulative sum) control charts (Lucas, 1982; Gibbons 1999a) and intrawell prediction limits (Davis, 1994; Gibbons, 1994). The advantage of the combined Shewhart–CUSUM control chart is that the method is sensitive to both immediate and gradual releases, whereas prediction limits are only sensitive to absolute increases over background. For combined Shewhart–CUSUM control charts, Gibbons (1999a) has examined extensively the effects of verification resampling, multiple comparisons, and alternative multipliers on facility-wide false positive and false negative rates.

In the intrawell setting, comparisons are independent since each well is compared to its own history. Gibbons (1994, Table 8.3) provides appropriate factors for computing intrawell prediction limits for up to $kq = 500$ future comparisons under a variety of resampling strategies. These factors apply to normally distributed constituents or constituents that can be suitably transformed to approximate normality. In the nonparametric case, selecting a single future sample and setting the confidence level to $(1 - \alpha)/kq$ is also possible; however, overall confidence levels may be poor due to small numbers of background measurements typically available in individual monitoring wells (i.e., generally eight or fewer). If seasonality is present, adjustments may

17.5 ILLUSTRATION

Consider the data in Table 17.1 for total organic carbon (TOC) measurements from a single well over two years of quarterly monitoring. Inspection of the data reveals no obvious trends, and these data have mean $\bar{x} = 11.0$ and standard deviation $s = 0.61$. The upper 95% point of Student's t-distribution on seven degrees of freedom is $t_{[7,1-0.05]} = 1.895$; therefore, the upper 95% confidence normal prediction limit in equation (17.1) is given by

$$11.0 + 1.895(0.61)\sqrt{1 + \tfrac{1}{8}} = 12.22 \text{ mg/L}$$

which is larger than any of the values observed. This limit provides 95% confidence of including the next single observation from a normal distribution for which eight previous measurements have been obtained with observed mean 11.0 mg/L and standard deviation 0.61 mg/L.

Assuming that spatial variability does not exist (in many cases a demonstrably false assumption) and that values from this single well are representative of values from each of 10 downgradient wells in the absence of contamination, the corresponding Bonferroni adjusted 95% confidence normal prediction limit in equation (17.2) for the next 10 new downgradient measurements is

$$11.0 + 3.50(0.61)\sqrt{1 + \tfrac{1}{8}} = 13.26 \text{ mg/L}$$

In contrast, if the dependence introduced by comparing all 10 downgradient wells to the same background were incorporated as in equation (17.3), the result of

$$11.0 + 3.31(0.61)\sqrt{1 + \tfrac{1}{8}} = 13.14 \text{ mg/L}$$

Table 17.1 Eight quarterly TOC measurements

Year	Quarter	TOC (mg/L)
1992	1	10.0
	2	11.5
	3	11.0
	4	10.6
1993	1	10.9
	2	12.0
	3	11.3
	4	10.7

is obtained (see Gibbons, 1994, Table 1.4). Note that the limit is slightly lower because the multiplier incorporates the dependence introduced by repeated comparison to a common background (i.e., the number of independent comparisons is less than 10 given that they are correlated). Although the Bonferroni-based limit is too conservative, there is little difference in the limits.

Extending this result to include the effects of a verification resample as in equation (17.4) further decreases the limit to

$$11.0 + 2.03(0.61)\sqrt{1 + \tfrac{1}{8}} = 12.31 \text{ mg/L}$$

If each of 10 constituents in each of the 10 downgradient wells had been monitored, $\alpha = 0.05/10 = 0.005$ and the limit would become

$$11.0 + 3.36(0.61)\sqrt{1 + \tfrac{1}{8}} = 13.17 \text{ mg/L}$$

(see Gibbons, 1994, Table 1.5). Note that the verification resample allows application of essentially the same limit derived for 10 wells and one constituent (13.14 mg/L) to a problem of 10 wells and 10 constituents (13.17 mg/L).

Now, consider the nonparametric alternative of taking the maximum of the initial eight background measurements and applying it to the next future monitoring measurement(s). In this example, the nonparametric prediction limit is 12.00 mg/L. For a single future measurement, confidence is 88% without a resample and 98% with a resample (see Gibbons, 1994, Tables 2.5 and 2.6). For a single measurement in each of 10 monitoring wells, confidence is 44% without a resample and 84% with a resample (see Gibbons, 1994, Tables 2.5 and 2.6). With 10 constituents and 10 monitoring wells, an overall 95% confidence level would be obtained with $n = 60$ background samples for one verification resample (see Gibbons, 1994, Table 2.6) or $n = 20$ samples for passage of one of two verification resamples (see Gibbons, 1994, Table 2.7). Note that if either the initial sample or both of two resamples must be passed, $n = 90$ background measurements must be obtained (see Gibbons, 1994, Table 2.13). Other illustrations and additional statistical details are available (Davis, 1994; Davis and McNichols, 1994; Gibbons, 1994).

17.6 METHODS TO BE AVOIDED

17.6.1 Analysis of Variance

In both EPA Subtitle C and D regulations and associated guidance (U.S. EPA, 1989, 1991, 1992, 1988), ANOVA is suggested as the statistical method of choice. Their specific recommendation is a one-way fixed-effects model where the upgradient wells are pooled as one level, and each downgradient well represents an additional level in the design. A minimum of four samples is obtained from each well within a semiannual period. In the presence of a significant F-statistic, posthoc comparisons (i.e., Fisher's LSD method) between each downgradient well and the pooled upgradient background are performed. Either parametric or nonparametric ANOVA models (i.e., Kruskal–Wallis test) is acceptable. Unfortunately, application of either parametric or nonparametric ANOVA procedures to detection monitoring is inadvisable for the following reasons.

1. Univariate ANOVA procedures do not adjust for multiple comparisons due to multiple constituents. This can be devastating to the site-wide false positive rate. As such, a site with 10 indicator constituents will have as much as a 40% probability of failing for at least one constituent on every monitoring event by chance alone.
2. ANOVA is more sensitive to spatial variability than to contamination. Spatial variability produces systematic differences between locations that are large relative to within-location variation (i.e., small consistent differences due to spatial variation achieve statistical significance). In contrast, contamination increases variability within the affected location(s); therefore, a much larger between-location difference is required to achieve statistical significance. In fact, application of ANOVA methods to predisposal groundwater monitoring data often results in statistically significant differences between upgradient and downgradient wells, even when no waste is present (Gibbons, 1994), as illustrated in the following example.
3. Nonparametric ANOVA is often presented as if it would protect the user from all of the weakness of its parametric counterpart; however, the only assumption relaxed is that of normality. The nonparametric ANOVA still assumes independence, homogeneity of variance, and that each measurement is identically distributed. Violation of any of these assumptions can corrupt the power of detection, or false positive rate.
4. ANOVA requires pooling of downgradient data. Specifically, the EPA suggests that four samples per semiannual monitoring event be collected (i.e., eight samples per year). However, ANOVA cannot rapidly detect a release since only a subset of the required four semiannual samples will initially be affected by a site impact. This heterogeneity will decrease the mean concentration and increase the variance for the affected location, limiting the ability of the statistical test to detect actual contamination.

Illustration Consider the data in Table 17.2 obtained from a facility in which disposal of waste has not yet taken place (Gibbons, 1994). Applying both parametric and nonparametric ANOVA to these predisposal data yielded an effect that approached significance for chemical oxygen demand (COD) ($p < 0.072$ parametric and $p < 0.066$ nonparametric) and a significant difference for alkalinity (ALK) ($p < 0.002$ parametric and $p < 0.009$ nonparametric). Individually compared (using Fisher's LSD), significantly increased COD levels were found for well MW05 ($p < 0.026$), and significantly increased ALK levels were found for wells MW06 ($p < 0.026$) and P14 ($p < 0.003$) relative to upgradient wells. These results represent false positives due to spatial variability since no waste has been deposited at

Table 17.2 Raw data for all detection monitoring wells and constituents (mg/L)[a]

Well	Event	TOC	TKN	COD	ALK
MW01	1	5.2000	0.8000	44.0000	58.0000
	2	6.8500	0.9000	13.0000	49.0000
	3	4.1500	0.5000	13.0000	40.0000
	4	15.1500	0.5000	40.0000	42.0000
MW02	1	1.6000	1.6000	11.0000	59.0000
	2	6.2500	0.3000	10.0000	82.0000
	3	1.4500	0.7000	10.0000	54.0000
	4	1.0000	0.2000	13.0000	51.0000
MW03	1	1.0000	1.8000	28.0000	39.0000
	2	1.9500	0.4000	10.0000	70.0000
	3	1.5000	0.3000	11.0000	42.0000
	4	4.8000	0.5000	26.0000	42.0000
MW04	1	4.1500	1.5000	41.0000	54.0000
	2	1.0000	0.3000	10.0000	40.0000
	3	1.9500	0.3000	24.0000	32.0000
	4	1.2500	0.4000	45.0000	28.0000
MW05	1	2.1500	0.6000	39.0000	51.0000
	2	1.0000	0.4000	26.0000	55.0000
	3	19.6000	0.3000	31.0000	60.0000
	4	1.0000	0.2000	48.0000	52.0000
MW06	1	1.4000	0.8000	22.0000	118.0000
	2	1.0000	0.2000	23.0000	66.0000
	3	1.5000	0.5000	25.0000	59.0000
	4	20.5500	0.4000	28.0000	63.0000
P14	1	2.0500	0.2000	10.0000	79.0000
	2	1.0500	0.3000	10.0000	96.0000
	3	5.1000	0.5000	10.0000	89.0000

[a] This facility has no garbage in it.

this site (i.e., a "greenfield" site). Most remarkable is the absence of significant results for TOC, notwithstanding that some values are as much as 20 times higher than others. These extreme values increase the within-well variance estimate, rendering the ANOVA powerless to detect differences regardless of magnitude. Elevated TOC data are inconsistent with chance expectations (based on analysis using prediction limits) and should be investigated. In this case, elevated TOC data are probably caused by contamination from insects getting into the wells since this greenfield facility is located in the middle of the Mohave desert.

17.6.2 Cochran's Approximation to the Behrens–Fisher *t*-test

For years the EPA Resource Conservation and Recovery Act (RCRA) regulation (U.S. EPA, 1982) was based on application of the Cochran's approximation to the Behrens–Fisher (CABF) *t*-test. The test was incorrectly implemented by requiring that four quarterly upgradient samples from a single well and single samples from a minimum of three downgradient wells each be divided into four aliquots and treated as if there were $4n$ independent measurements. The result was that most hazardous waste disposal facilities regulated under RCRA were declared "leaking." As an illustration, consider the data in Table 17.3.

Note that the aliquots are almost perfectly correlated and add virtually no independent information, yet they are assumed by the statistic to be completely independent. The CABF t-test is computed as

$$t = \frac{\bar{X}_B - \bar{X}_M}{\sqrt{S_B^2/N_B + S_M^2/N_M}} = \frac{7.62 - 7.40}{\sqrt{0.032/16 + 0.0004/4}} = \frac{0.22}{0.05} = 4.82$$

The associated probability of this test statistic is 1 in 10,000, indicating that the chance that the new monitoring measurement came from the same population as the background measurements is remote. Note that, in fact, the mean concentration of the four aliquots for the new monitoring measurement is identical to one of the four mean values for background, suggesting intuitively that probability is closer to 1 in 4 rather than 1 in 10,000. Averaging the aliquots yields the statistic

$$t = \frac{\bar{X}_B - \bar{X}_M}{S_B\sqrt{1/N_B + 1}} = \frac{7.62 - 7.40}{0.20\sqrt{\frac{1}{4} + 1}} = \frac{0.22}{0.22} = 1.0$$

Table 17.3 pH data used in computing the CABF t-test

Date	Replicate				Average
	1	2	3	4	
Background					
Nov. 1981	7.77	7.76	7.78	7.78	7.77
Feb. 1982	7.74	7.80	7.82	7.85	7.80
May 1982	7.40	7.40	7.40	7.40	7.40
Aug. 1982	7.50	7.50	7.50	7.50	7.50
$\bar{X}_B$			7.62		7.62
SD_B			0.18		0.20
N_B			16		4
Monitoring					
Sept. 1983	7.39	7.40	7.38	7.42	7.40
$\bar{X}_B$			7.40		7.40
SD_B			0.02		
N_B			4		1

which has an associated probability of 1 in 2. Had the sample size been increased to $N_B = 20$, the probability would have decreased to 1 in 3. The EPA eliminated this method from the regulation (U.S. EPA, 1988).

17.7 SUMMARY

Protection of our natural resources is critical; however, statistical tools used to make environmental impact decisions are limited and often confusing. The problem is not only interesting regarding development of public policy, but it also contains features of statistical interest such as multiple comparisons, sequential testing, and censored distributions. Highlighting the weaknesses of currently mandated regulations may lead to further critical examination of public policy in the field of groundwater monitoring as well as heightened interest in environmental statistics.

18

ASSESSMENT AND CORRECTIVE ACTION MONITORING: OVERVIEW

18.1 INTRODUCTION

As discussed in Chapter 17, statistical methods for detection monitoring have been well studied in recent years [see Gibbons, 1994, 1996; 1998a; U.S. EPA, 1992; and ASTM Standard D6312 (formerly PS 64-96)]. Although equally important, statistical methods for assessment sampling, ongoing monitoring, and corrective action sampling and monitoring have received less attention. In this chapter we present an overview of the general principles and strategies used in assessment sampling and monitoring and corrective action programs for environmental sampling and monitoring (soil, groundwater, air, surface water, and waste streams).

One may ask why statistical analysis is necessary in assessment and corrective action. Why not simply compare each measurement to the corresponding criterion? There are several reasons why statistical methods are essential in assessment and corrective action sampling programs. First, a single measurement indicates very little about the true concentration in the sampling location of interest, and with only one sample there is no way of knowing if the measured concentration is a typical or an extreme value. The objective is to compare the true concentration (or some interval that contains it) to the relevant criterion or standard. Second, in many cases the constituents of interest are naturally occurring (e.g., metals), and the naturally existing concentrations may exceed the relevant criteria. In this case, the relevant comparison is to background (e.g., off-site soil or upgradient groundwater), not to a fixed criterion. As such, background data must be characterized statistically to obtain a statistical estimate of an upper bound for the naturally occurring concentrations so that it can confidently be determined if on-site concentrations are above background levels. Third, there is often a need to compare numerous potential constituents of concern to criteria or background, at numerous sampling locations. By chance alone there will be exceedances as the number of comparisons becomes large. The statistical approach to this problem can ensure that false positive results are minimized.

Of course, there is considerable EPA support for statistical methods applied to detection, assessment and corrective action sampling, and monitoring programs. For example, the 90% upper confidence limit (UCL) is used in SW846 (Chapter 9) for determining if a waste is hazardous. If the UCL is less than the criterion for a particular hazardous waste code, the waste stream is considered to be nonhazardous even if certain individual measurements exceed the criterion. Similarly, in the EPA "Statistical Analysis of Groundwater Monitoring Data at RCRA Facilities, Addendum to the Interim Final Guidance" (1992), confidence intervals for the mean and various upper percentiles of the distribution are advocated for assessment and corrective action sampling. Both the 1989 and 1992 EPA guidance documents suggests use of the lower 95% confidence limit (LCL) as a tool for determining whether a criterion has been exceeded in assessment sampling. The latest EPA guidance in this area (i.e., the draft EPA Unified Statistical Guidance) calls for use of the LCL in assessment monitoring and the UCL in corrective action. In this way, corrective action is triggered only if there is a high degree of confidence that the true concentration has exceeded the criterion or standard, whereas corrective action continues until there is a high degree of confidence that the true concentration is below the criterion or standard. This is the general approach adopted in this chapter as well.

In terms of definitions, *assessment monitoring* refers to investigative monitoring that is initiated after the presence of a contaminant has been detected in a particular media (e.g., groundwater) above a relevant criterion at one or more locations. The objective of the program is to determine if there is a statistically significant exceedance of a standard or criteria at a potential area of concern (PAOC) or at an interface of groundwater venting to surface water and/or to quantify the rate and extent of migration of constituents detected in groundwater above residential criteria.

Corrective action monitoring is instituted when hazardous constituents from a regulated unit have been detected at statistically significant concentrations between the compliance point and the downgradient facility property boundary. Corrective action monitoring is conducted throughout the corrective action program that is implemented to address groundwater contamination. In some cases, corrective action monitoring is conducted throughout the period of corrective action to determine the progress of remediation and to identify statistically significant trends in groundwater contaminant concentrations.

18.2 STRATEGY

In the following, the logic and a corresponding general methodology for statistical comparisons in assessment and corrective action monitoring programs are described.

Specific recommendations are provided that will work well in most cases but may not be optimal in all possible situations.

1. To begin, we must identify relevant constituents for the specific type of facility, media (e.g., soil, groundwater, etc.), and area of interest. A facility is generally comprised of a series of subunits or *source areas* that may have a distinct set of sampling locations and relevant constituents of concern (referred to as a PAOC). The subunit may consist of a single sampling point or collection of sampling points. In some cases, the entire site may comprise the area of interest, and all sampling locations are considered jointly. In all cases, the owner/operator should select the smallest possible list of constituents that adequately characterize the source area in terms of historical use.

2. For each constituent, obtain the appropriate regulatory criterion or standard [e.g., maximum contaminant level (MCL)] if one is available. The appropriate criterion or standard should be selected based on relevant pathways (e.g., direct contact, ingestion, inhalation) and appropriate land use criteria (e.g., commercial, industrial, residential).

3. For each constituent that may have a background concentration higher than the relevant health-based criterion, set "background" to the upper 95% confidence prediction limit (UPL). The prediction limits are computed from all available data collected from background, or outside source areas that are unlikely to be contaminated, upstream, upwind, or upgradient locations only. Henceforth, background refers to any of these types of off-site sources. The background data are first screened for outliers and then tested for normality and lognormality (see Chapters 14 and 15).

    (a) If the test of normality cannot be rejected (e.g., at the 95% confidence level), background is equal to the 95% confidence normal prediction limit.

    (b) If the test of normality is rejected but the test of lognormality cannot be rejected, background is equal to the 95% confidence lognormal prediction limit.

    (c) If the data are neither normally or lognormally distributed, or the detection frequency is less than 50%, background is the nonparametric prediction limit which is computed as a particular order statistic (i.e., ranked observation) of the background measurements (e.g., the maximum). Note that if the detection frequency is zero, background is set equal to the appropriate quantification limit (QL) for that constituent which is the lowest concentration that can be determined reliably within specified limits of precision and accuracy under routine laboratory operating conditions.

4. If the background is greater than the relevant criterion or standard or if there is no criterion or standard, comparisons are made to the background prediction limit. If the criterion is greater than background, compare the

appropriate confidence limit to the criterion. Note that if nothing is detected in background, background is the QL. If the criterion is lower than the QL, the criterion is set to the QL.

5. The number of samples taken depends on whether comparison is to background or a criterion and whether comparisons are made at individual locations or by pooling samples within a source area. If comparison is to background, collect a minimum of one sample from each source area or sampling location. If comparison is to a criterion (i.e., the criterion is greater than background) and interest is in a single location, a minimum of four independent samples from each sampling location should be taken. If the comparison is to a criterion for an entire source area, a minimum of one sample from each of four sampling locations within the source area is required. If there are fewer than four sampling locations within a given source area, the total number of measurements from the source area must be four or more (e.g., two sampling locations each with two independent samples). Note that a minimum of four samples is for statistical purposes only, and may be grossly inadequate to characterize potential contamination at a particular site.

6. If comparison is to a criterion or standard, there are two general approaches. In assessment sampling where interest is in determining if a criterion has been exceeded, compare the 95% lower confidence limit (LCL) for the mean of at least four samples from a single location, source area, or the entire site to the relevant criterion. In corrective action sampling and monitoring, where interest is in demonstrating that the on-site concentration is lower than the criterion, compare the 95% upper confidence limit (UCL) for the mean of at least four samples from a single location, source area, or the entire site to the relevant criterion.

7. If the background prediction limit is larger than the relevant criterion, do one of the following:

    (a) For a single measurement obtained from an individual location, compare this individual measurement to the background prediction limit for the next single measurement from each of k locations.

    (b) For multiple measurements obtained from a given source area or the entire site, compare the mean of the measurements to the background prediction limit for the mean of m measurements based on the best-fitting statistical distribution or nonparametric alternative.

8. Note that if the background UPL and the regulatory criterion are quite similar, it may be possible for the downgradient mean to exceed the background UPL, but the LCL for the downgradient mean may still be less than the regulatory criterion (LCL < Reg < UPL < mean). In this case, an exceedance is generally not determined.

9. Figure 18.1 presents a decision tree that can be used to step through the statistical approach to this problem.

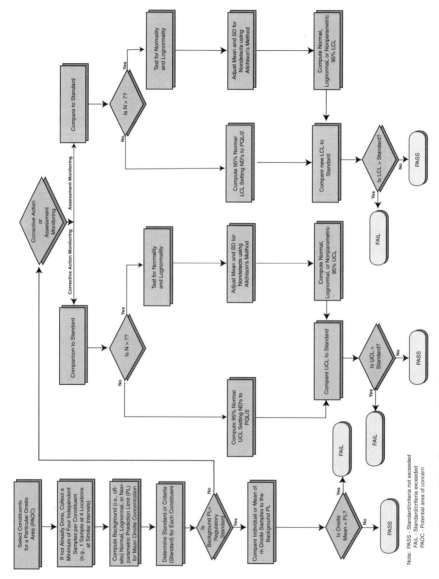

Figure 18.1 Flowchart of statistical decision tree for assessment and corrective action monitoring.

18.3 APPLICATION TO SPECIFIC MEDIA

In the following sections, application to specific media and types of sampling and monitoring programs is described. The areas covered include soil, groundwater, and waste stream sampling; however, similar approaches can be taken for air and surface water monitoring.

18.3.1 Soils: Evaluation of Individual Source Areas (PAOCs)

1. Collect soil samples from the surface to the groundwater table at appropriate intervals (e.g., 2-foot intervals) in the most likely contaminated location in the source area and screen soils to determine the interval with highest concentration(s).
2. At a minimum of three other nearby borings located in the same source area, collect one sample in the same vertical interval (geologic profile) as the previously identified highest concentration interval (i.e., the first boring in the interval of highest screening concentration).
3. Compute the 95% LCL (assessment) or UCL (corrective action) for the mean for the four results to determine if the particular PAOC exceeds the regulatory criterion.
4. If an exceedance is found, assess whether it is naturally occurring (e.g., metals) by obtaining a minimum of eight independent background samples (i.e., off-site soil samples from the same interval) and compute the 95% confidence upper prediction limit (UPL) for the mean of the four (or more) samples, and compare the UPL to the observed mean at each PAOC.

Figures 18.2 and 18.3 illustrate the sampling location approaches for this scenario.

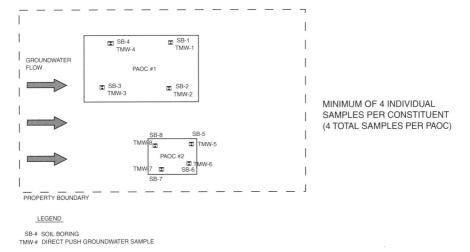

Figure 18.2 Single PAOC example: comparison to a standard/criterion.

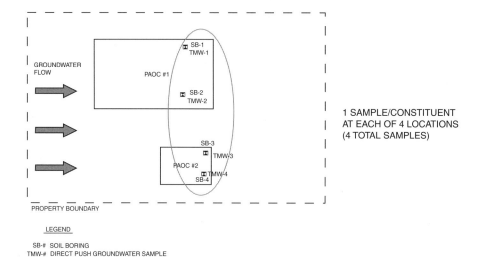

Figure 18.3 Multiple PAOC example: comparison to a standard/criteria.

18.3.2 Soils: Area- or Site-wide Evaluations

1. Collect soil samples to be representative of the entire spatial distribution of constituents of concern (a minimum of four samples).
2. Compute the 95% LCL (assessment) or UCL (corrective action) for the mean of all on-site samples and determine if the area or site as a whole exceeds the regulatory criterion.
3. If an exceedance is found, ensure that it is not naturally occurring by obtaining a minimum of eight independent background samples (i.e., off-site soil samples from the same stratigraphic unit) and compute the 95% confidence UPL.

   (a) If the level of hazardous substance concentrations at the site is relatively homogeneous, compute the UPL for the mean of the m on-site measurements and compare the observed mean to the UPL.

   (b) If the level of hazardous substance concentrations at the site is heterogeneous, compute the UPL for the simultaneous coverage of each of the m individual on-site measurements and compare each measurement to the UPL.

Figure 18.4 illustrates the sampling location approach for this scenario.

18.3.3 Groundwater: Aquifer

As in the soil sampling above, if soil sampling and screening indicate that groundwater may be impacted, one groundwater sample should be obtained in each of the four borings and results should be evaluated statistically to determine if the entire PAOC requires additional assessment. The general methodology previously

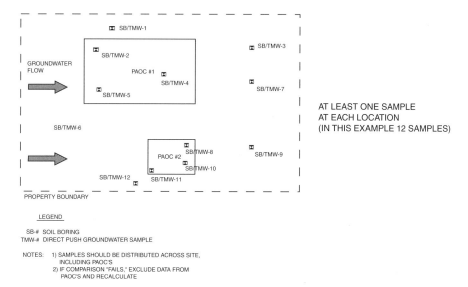

Figure 18.4 Comparison of mean concentrations of an entire site to a standard/criterion.

described for soil PAOCs can be used here as well, as illustrated in Figures 18.2 and 18.3. To characterize background, a minimum of eight independent samples must be collected. This can be four samples from each of two locations, two samples from four locations, or one sample from each of eight locations. A minimum of two locations is required. Statistical independence implies that the same groundwater is not sampled repeatedly and that the background data are representative of the same temporal variation as are the on-site data. This precludes, for example, establishing background in the winter and comparing on-site measurements obtained in the summer.

Figure 18.5 illustrates another approach for evaluating groundwater at a site. Sampling locations are set up as shown and four independent samples are collected from background locations GW-7 and GW-8 (i.e., eight total background samples). If the background UPL exceeds the appropriate regulatory criterion, the mean from downgradient samples GW-1 through GW-6 is compared to the background UPL to determine if an exceedance exists. If the background UPL is less than the appropriate regulatory criterion, the downgradient LCL should be compared to the criterion. Another modification to this approach is if an exceedance exists, data from the PAOCs can be excluded from the downgradient mean and LCL computation to determine conditions for site groundwater, excluding hot spots. As discussed previously, if the background UPL and the regulatory criterion are quite similar, it may be possible for the downgradient mean to exceed the background UPL, but the LCL for the downgradient mean may still be less than the regulatory criterion. In this case, an exceedance is not determined. This applies equally to all media.

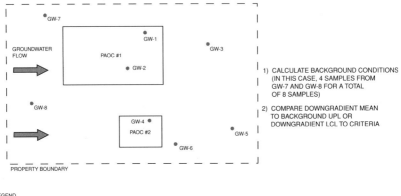

Figure 18.5 Evaluation of groundwater concentrations for an entire site.

18.3.4 Groundwater: Groundwater–Surface Water Interface

1. Characterize background as described previously. As indicated earlier, background is established by obtaining at least eight independent samples from a minimum of two locations (i.e., to incorporate spatial variability). The background limit is established by computing the UPL from these data (i.e., a minimum of eight background samples).
2. If the only comparison is to background, obtain a single sample from each GSI sampling location (i.e., one sample from each compliance point) and compare to the appropriate upgradient UPL.
3. If comparison is to regulatory criteria, obtain a minimum of four independent samples from each GSI sampling location (i.e., four samples from each compliance point) and compare the LCL (assessment) or the UCL (corrective action) to the regulatory criterion. If the upgradient UPL is greater than the regulatory criterion for a particular constituent, compare each GSI sampling location to background.
4. Depending on the application, each GSI sampling location can be compared to background or the appropriate regulatory criterion individually or as a group.

Figure 18.6 illustrates the sampling strategy for this scenario.

18.3.5 Groundwater: Long-Term Monitoring

1. When sampling for long-term monitoring of a plume, compute the 95% confidence normal UCL for the most recent four measurements in each sampling location and compare to the relevant regulatory criteria.

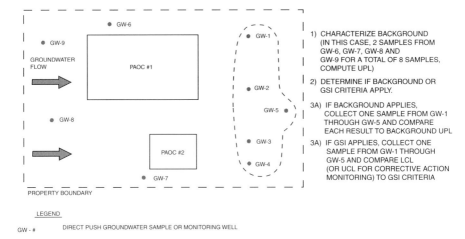

Figure 18.6 Evaluation of groundwater data to determine compliance with GSI criteria.

2. Compute Sen's test to determine if there are increasing or decreasing trends (at a 95% confidence level) at each sampling location (need a minimum of eight measurements per well).

18.3.6 Groundwater: Natural Attenuation Evaluation

1. Here temporal changes are considered in the mean of all wells within a plume or wells in the relatively higher concentration area of a plume.
2. Obtain a minimum of eight independent samples (e.g., one from each of eight monitoring wells or two from each of four monitoring wells). This should be done either for all wells within the plume or the relatively higher concentration area of the plume. Note that if there is seasonal variability in analyte concentrations, four quarterly samples within a period of no less than one year should be obtained from each sampling location.
3. Compute the 95% confidence lower prediction limit (LPL) and the UPL for the mean of all wells in the plume or all wells within the relatively higher concentration area. For example, if there are eight wells, compute the LPL and UPL for the mean of the next eight samples.
4. If the actual mean exceeds the UPL, there is evidence that the plume is getting significantly worse.
5. If the actual mean is less than the LPL, there is evidence that the plume is getting significantly better (i.e., natural attenuation is occurring).
6. Compute Sen's test to determine if there are increasing or decreasing trends (at a 95% confidence level) at each sampling location (need a minimum of eight measurements per well).

18.3.7 Waste Stream Sampling

1. To determine if a particular waste stream is hazardous, obtain a series of $n \geq 4$ representative samples from the waste stream for all relevant characteristically hazardous criteria.
2. Compute the appropriate 90% UCL for the mean concentration.
3. Note that the 90% confidence level is used based on guidance provided in SW846, Chapter 9.
4. If the 90% UCL is less than the regulatory criterion or standard, the waste stream is not hazardous.

18.4 SUMMARY

In this chapter, a general strategy for assessment, compliance, and corrective action monitoring has been outlined for several media and potential applications. The strategy is not all encompassing; however, it provides a general framework for this type of monitoring problem and can easily be adapted to other monitoring problems as well. Of course, no details have been provided regarding what an LCL or UCL or LPL or UPL is, how one should be computed, and what the alternative forms are, based on distributional form and amount of censoring in the data. In Chapter 19, statistical methods for comparison to fixed standards are presented. In Chapter 20, statistical methods for comparison to background are presented. Finally, in Chapter 21, the methods are illustrated through application to a series of three case studies.

19

ASSESSMENT AND CORRECTIVE ACTION MONITORING: COMPARISON TO A STANDARD

19.1 INTRODUCTION

In this chapter we describe general statistical methods for comparisons to fixed regulatory standards. Surprisingly, such methods have received relatively sparse attention in both statistical and environmental literatures.

When comparison is in reference to a fixed regulatory standard, quite different statistical strategies are required than when comparing to a background distribution with unknown parameters. Here we are interested in comparison of the true on-site concentration to the regulatory standard. Of, course, given a finite set of m on-site/downgradient samples, we can never know the true concentration with certainty. We can, however, determine an interval that will include the true mean concentration with a given level of confidence. In assessment monitoring we can then use this interval to determine if the true mean concentration has exceeded the regulatory standard with a reasonable level of confidence. This determination is made if the entire interval exceeds the regulatory standard. More conservatively, we can compute a one-sided lower bound on the true mean concentration as a $100(1 - \alpha)\%$ lower confidence limit (LCL). By contrast, in corrective action monitoring, we want to ensure that the true mean concentration is below the regulatory standard. As such, the entire confidence interval must be below the regulatory standard, or more conservatively, the $100(1 - \alpha)\%$ one-sided upper confidence limit (UCL) for the true mean concentration must be below the standard.

19.2 LCL OR UCL?

One may question the logic involved in using the LCL for assessment monitoring and switching to the UCL for corrective action. In assessment monitoring we are typically interested in determining whether the true concentration has exceeded the standard. In this case the null hypothesis is that the true concentration is less than or equal to the regulatory standard. If we reject the null hypothesis the site is placed under increased regulatory scrutiny, often leading to corrective action. Since the standard is fixed (i.e., we assume that it is known with certainty), and if we assume normality of the measured concentrations, we are led immediately to use of a one-sample t-statistic, which provides a test of the null hypothesis of no difference between the true mean concentration and the regulatory standard at some specified level of confidence, usually 95%. In terms of t, we therefore have

$$t = \frac{\bar{x} - \text{STD}}{s/\sqrt{m}} \tag{19.1}$$

where $\bar{x}$ is the sample mean of the m measurements and s is the observed sample standard deviation. The quantity $s/\sqrt{m}$ is the standard error of the mean, which describes our uncertainty in this sample-based statistic. Rewriting the equation in terms of the standard (STD'), we obtain

$$\text{STD}' = \bar{x} - \frac{ts}{\sqrt{m}} \tag{19.2}$$

where t is the one-tailed $100(1 - \alpha)$ percentage point of Student's t-distribution on $m - 1$ degrees of freedom. It is clear from equation (19.2) that STD' is the lower confidence limit for the mean.

By contrast, in corrective action monitoring, we already know that there is contamination, and now the burden of proof is to demonstrate with reasonable confidence that the contamination is no longer present. As such, we use the UCL so that we may have $100(1 - \alpha)\%$ confidence that the true mean concentration is below the regulatory standard. Whereas in assessment, we were interested in determining if the mean exceeds the standard, in corrective action we are interested in determining if the standard exceeds the mean. In terms of t, we therefore have

$$t = \frac{\text{STD} - \bar{x}}{s/\sqrt{m}} \tag{19.3}$$

Rewriting the equation for the lowest possible standard that would be differentiable from the mean (STD′), we obtain

$$\text{STD}' = \bar{x} + \frac{ts}{\sqrt{m}} \quad (19.4)$$

which is a one-sided $100(1 - \alpha)\%$ UCL.

Another way of expressing the same idea is in terms of specifying null and alternative hypotheses for the two monitoring programs. In assessment monitoring, interest is in testing the null hypothesis that the true mean is less than or equal to the standard (i.e., H_0: $\mu \leq$ STD), with the alternative hypothesis being that the true mean is greater than the standard (i.e., H_1: $\mu >$ STD). In corrective action monitoring interest is in testing the null hypothesis that the true mean is greater than or equal to the standard (i.e., H_0: $\mu \geq$ STD), with the alternative hypothesis being that the true mean is less than the standard (i.e., H_1: $\mu <$ STD). In assessment monitoring we assume that the prior state is uncontaminated and reject the null hypothesis if we have sufficient confidence (e.g., 95%) that the standard has been exceeded (i.e., LCL), whereas in corrective action monitoring we assume that the prior state is contaminated and reject the null hypothesis if we have sufficient confidence (e.g., 95%) that the true mean is less than the standard (i.e., UCL).

19.3 NORMAL CONFIDENCE LIMITS FOR THE MEAN

For a normally distributed constituent that is detected in all cases, the $100(1 - \alpha)\%$ normal LCL (assessment sampling and monitoring) for the mean of m measurements is computed as

$$\bar{x} - t_{[m-1,\alpha]} \frac{s}{\sqrt{m}} \quad (19.5)$$

The $100(1 - \alpha)\%$ normal UCL (corrective action) for the mean of m measurements is computed as

$$\bar{x} + t_{[m-1,\alpha]} \frac{s}{\sqrt{m}}$$

When nondetects are present, several reasonable options are possible. If $m < 8$, nondetects are replaced by one-half of the QL, since with fewer than eight measurements, more sophisticated statistical adjustments are typically not appropriate. Similarly, a normal UCL is typically used because seven or fewer samples are insufficient to determine confidently the distributional form of the data. Use of a lognormal limit with small samples can result in extreme limit estimates; therefore, it is reasonable and conservative to default to normality for cases in which $m < 8$.

If $m \geq 8$, a good choice is to use Aitchison's (1955) method to adjust for nondetects (see Chapter 13) and test for normality and lognormality of the data using the single- or multiple-group version of the Shapiro–Wilk test (see Chapter 14). However, the ability of the Shapiro–Wilk test (and other distributional tests) to detect nonnormality is highly dependent on sample size. The multiple group version of the Shapiro–Wilk test is used when there are multiple measurements from multiple on-site locations. For most applications, 95% confidence is a reasonable choice. Note that alternatives such as Cohen's (1961) method can be used; however, the reporting limit must be constant for each constituent, which is rarely the case.

19.4 LOGNORMAL CONFIDENCE LIMITS FOR THE MEDIAN

For a lognormally distributed constituent [i.e., $y = \log_e(x)$ is distributed $N(\mu_y, \sigma_y^2)$] the $100(1 - \alpha)\%$ LCL for the median or 50th percentile of the distribution is given by

$$\exp\left[\bar{y} - t_{[m-1,\alpha]} \frac{s_y}{\sqrt{m}}\right] \quad (19.6)$$

where $\bar{y}$ and s_y are the mean and standard deviation of the natural log–transformed concentrations. Note that the exponentiated limit is, in fact, an LCL for the median and not the mean concentration. In general, the median and corresponding LCL will be lower than the mean and its corresponding LCL. The $100(1 - \alpha)\%$ UCL for the median or 50th percentile of the distribution is given by

$$\exp\left[\bar{y} + t_{[m-1,\alpha]} \frac{s_y}{\sqrt{m}}\right] \quad (19.7)$$

19.5 LOGNORMAL CONFIDENCE LIMITS FOR THE MEAN

19.5.1 Exact Method

Land (1971) developed an exact method for computing confidence limits for linear functions of the normal mean and variance. The classic example is the normalization of a lognormally distributed random variable x through the transformation $y = \log_e(x)$, where, as noted previously, $y \sim N(\mu_y, \sigma_y^2)$. Using Land's (1975) tabled coefficients H_α, the one-sided $(1 - \alpha)100\%$ lognormal LCL for the mean is

$$\exp\left(\bar{y} + 0.5s_y + \frac{H_\alpha s_y}{\sqrt{m-1}}\right) \quad (19.8)$$

19.5 LOGNORMAL CONFIDENCE LIMITS FOR THE MEAN

Alternatively, using $H_{1-\alpha}$, the one-sided $(1 - \alpha)100\%$ lognormal UCL for the mean is

$$\exp\left(\bar{y} + 0.5 s_y + \frac{H_{1-\alpha} s_y}{\sqrt{m-1}}\right) \qquad (19.9)$$

The factors H are given by Land (1975) and $\bar{y}$ and s_y and the mean and standard deviation of the natural log–transformed data [i.e., $y = \log_e(x)$]. Gilbert (1987) has a small subset of these extensive tables for $m = 3$ through 101, $s_y = 0.1$ through 10.0, and $\alpha = 0.05$ and 0.10 (i.e., upper and lower 90% and 95% confidence limit factors). Unfortunately, Land's 1975 tables are difficult to find; therefore, they are reproduced here in Appendix A. Land (1975) suggests that cubic interpolation (i.e., four-point Lagrangian interpolation; Abramowitz and Stegun, 1964, p. 879) be used when working with these tables. A much easier and quite reasonable alternative is to use logarithmic interpolation.

19.5.2 Approximating Land's Coefficients

In the following, a series of approximations for Land's tables that are amenable to machine computation are provided. The approximations are based on the original work of Hewett and Ganser (1997) and their further unpublished work on this topic (Hewett and Ganser, personal communication, 1999).

90% Confidence A good approximation to the tabulated values in Appendix A for 90% confidence LCLs and UCLs is given by Hewett and Ganser (personal communication, 1999). To compute the 90% LCL factor $(H_{0.10})$, let

$$a = 0.71602171 \qquad f = 0.95597342$$
$$b = 0.33948241 \qquad g = -0.57180834$$
$$c = -0.87496047 \qquad h = 1.581226$$
$$d = 0.80696638 \qquad i = -0.59781156$$
$$e = 2.2317352$$

Next, compute the intermediate equations:

$$F1 = \frac{a}{(m-2)^b} + c \qquad F3 = \frac{g}{(m-2)^h} + i$$

$$F2 = \frac{d}{(m-2)^e} + f \qquad F4 = \frac{-0.2688}{m-2} - 1.282$$

The estimate of the 90% confidence LCL factor $H_{0.10}$ is

$$H(s_y, m, 0.10) = s_y[F1 + F2 \exp(F3 \cdot s_y^{0.6793})] + F4 \qquad (19.10)$$

To compute the 90% UCL factor ($H_{0.90}$) for $m \geq 4$, let

$$a = 1.246697 \qquad f = -0.77479436$$
$$b = 0.22601945 \qquad g = -1.7337773$$
$$c = 2.1015161 \qquad h = 0.95301657$$
$$d = 0.38340342 \qquad i = 0.36349884$$
$$e = 1.5035618$$

Next compute the intermediate equations:

$$F1 = \frac{a+b}{(m-2)^{0.47}} \qquad F3 = s_y^{1.12}\left[\frac{1.0}{(m-2)^h} + i\right]$$

$$F2 = \frac{c}{(m-2)^d} + e \qquad F4 = 1 + f\exp(g \cdot F3)$$

The estimate of the 90% confidence UCL factor $H_{0.90}$ is

$$H(s_y, m, 0.90) = F1 + (F2 \cdot F3 \cdot F4) \tag{19.11}$$

To compute the 90% UCL factor ($H_{0.90}$) for $m = 3$, let

$$a = 1.5315769 \qquad f = -0.98701113$$
$$b = -1.4929561 \qquad g = 11.259201$$
$$c = -0.65067128 \qquad h = 0.10729354$$
$$d = 3.2239885 \qquad i = -2.6836421$$
$$e = 1.5201619 \qquad j = -0.0083461165$$

The estimate of the 90% confidence UCL factor $H_{0.90}$ is

$$H(s_y, m, 0.90) = \frac{a + cs_y + e\sqrt{s_y} + gs^3 + is^4}{1 + bs_y + d\sqrt{s_y} + fs^3 + hs^4 + js^5} \tag{19.12}$$

These approximations should provide accuracy to within ±2.5% of the tabulated values in Appendix A (i.e., Land, 1975) as long as $0.01 \leq s_y \leq 4$ and $3 < m < 1001$. Hewett and Ganser (1997) note that if the geometric standard deviation is nearly 1.00, it can be set conservatively to 1.01 [i.e., $s_y = \log_e(1.01) = 0.01$]. Similarly, if $s_y > 4$, then $s_y = 4$ will yield a reasonable approximation as well.

19.5 LOGNORMAL CONFIDENCE LIMITS FOR THE MEAN

95% Confidence A good approximation to the tabulated values in Appendix A for 95% confidence LCLs and UCLs is given by Hewett and Ganser (1997). To compute the 95% LCL factor ($H_{0.05}$), let

$$a = -0.85033767 \qquad f = 1.3213281$$
$$b = -0.5258052 \qquad g = 0.8155562$$
$$c = 0.92416176 \qquad h = -1.018148$$
$$d = 3.3298209 \qquad i = 0.25248895$$
$$e = 0.94348568$$

Next compute the intermediate equations:

$$F1 = \frac{a}{(m-2)^c} + b \qquad F3 = \frac{g}{(m-2)^i + h}$$

$$F2 = \frac{d}{(m-2)^f} + e \qquad F4 = \frac{-0.6226}{(m-2)^{0.2426}} + 1.1470$$

The estimate of the 95% confidence LCL factor $H_{0.05}$ is

$$H(s_y, m, 0.05) = \frac{-0.74295}{m-2} - 1.64765 + s_y[F3 + F2 \exp(F1 \cdot s_y^{F4})] \qquad (19.13)$$

To compute the 95% UCL factor ($H_{0.95}$), let

$$a = 0.76766658 \qquad f = 2.012669$$
$$b = 3.8716869 \qquad g = 0.21978875$$
$$c = 0.80598919 \qquad h = 0.41575588$$
$$d = 6.0321019 \qquad i = 0.29258276$$
$$e = 0.89998154$$

Next compute the intermediate equations:

$$F1 = s_y\left[i + \frac{1}{(m-2)^c}\right] \qquad F4 = [1 + g \cdot \exp(-h \cdot F1)]$$

$$F2 = b + \frac{d}{(m-2)^c} \qquad F5 = F2 \cdot \frac{F3}{F4}$$

$$F3 = F1 \cdot [1 - e \cdot \exp(-f \cdot F1)]$$

The estimate of the 95% confidence UCL factor $H_{0.95}$ is

$$H(s_y, m, 0.95) = 1.645 + \frac{a}{m-2} + F5 \tag{19.14}$$

These two approximations should provide accuracy to within ±5% of the tabulated values in Appendix A (i.e., Land, 1975) as long as $0.01 \leq s_y \leq 4$ and $3 < m < 1001$. Hewett and Ganser (1997) note that if the geometric standard deviation is nearly 1.00, it can be set conservatively to 1.01 [i.e., $s_y = \log_e(1.01) = 0.01$]. Similarly, if $s_y > 4$, then $s_y = 4$ will yield a reasonable approximation as well.

99% Confidence A good approximation to the tabulated values in Appendix A for 99% confidence LCLs and UCLs is given by Hewett and Ganser (personal communication, 1999). To compute the 99% LCL factor ($H_{0.01}$) for $m \geq 4$, let

$a = 1.8471501$ $f = 1.6515708$

$b = 0.33170599$ $g = -1.4260318$

$c = -1.5185346$ $h = 1.2726156$

$d = 15.03355$ $i = -0.61207013$

$e = 1.8247982$ $j = -3.1271057$

Next compute the intermediate equations:

$$F1 = \frac{a}{(m-2)^b} + c \quad\quad F4 = \frac{j}{(m-2)^{5.22979}} + 0.76322$$

$$F2 = \frac{d}{(m-2)^e} + f \quad\quad F5 = \frac{-3.3761}{(m-2)^{1.1285}} - 2.3491$$

$$F3 = \frac{g}{(m-2)^h} + i$$

The estimate of the 99% confidence LCL factor $H_{0.01}$ is

$$H(s_y, m, 0.01) = s_y[F1 + F2 \exp(F3 \cdot s_y^{F4})] + F5 \tag{19.15}$$

To compute the 99% LCL factor ($H_{0.01}$) for $m = 3$, let

$a = -5.6866837$ $f = 0.45985218$

$b = 4.0285791$ $g = -1.3999427$

$c = -5.7229139$ $h = -0.077249205$

19.5 LOGNORMAL CONFIDENCE LIMITS FOR THE MEAN

$$d = 1.2054977 \qquad i = 0.025431893$$
$$e = -1.6813656 \qquad j = 0.0041220243$$

The estimate of the 99% confidence LCL factor $H_{0.01}$ is

$$H(s_y, m, 0.01) = \frac{a + cs_y + e\sqrt{s_y} + gs^3 + is^4}{1 + bs_y + d\sqrt{s_y} + fs^3 + hs^4 + js^5} \qquad (19.16)$$

To compute the 99% UCL factor ($H_{0.99}$) for $m \geq 4$, let

$$a = 2.2938169 \qquad e = -3.8271316$$
$$b = 2.9455577 \qquad f = 1.0530915$$
$$c = 1.2703828 \qquad g = -0.96454397$$
$$d = 0.79332091$$

Next compute the intermediate equations:

$$F1 = a + \frac{b}{(m-2)^{0.8894}} \qquad F3 = s_y^d \left[\frac{e}{(m-2)^{0.9829}} - 0.7064 \right]$$

$$F2 = c + \frac{8.435}{(m-2)^{0.6585}} + \frac{54.1235}{(m-2)^{2.0879}} \qquad F4 = s_y^f [1 + g \exp(F3)]$$

The estimate of the 99% confidence UCL factor $H_{0.99}$ is

$$H(s_y, m, 0.99) = F1 + (F2 \cdot F4) \qquad (19.17)$$

To compute the 99% UCL factor ($H_{0.99}$) for $m = 3$, let

$$a = 5.2785049 \qquad f = -0.48280497$$
$$b = -7.9296781 \qquad g = 7355.5184$$
$$c = 22.112828 \qquad h = 0.35922551$$
$$d = 108.10625 \qquad i = 64.969622$$
$$e = -412.91153 \qquad j = -0.033149652$$

The estimate of the 99% confidence UCL factor $H_{0.99}$ is

$$H(s_y, m, 0.99) = \frac{a + cs_y + e\sqrt{s_y} + gs^3 + is^4}{1 + bs_y + d\sqrt{s_y} + fs^3 + hs^4 + js^5} \qquad (19.18)$$

These approximations should provide accuracy to within ±4% of the tabulated values in Appendix A (i.e., Land, 1975) as long as $0.01 \leq s_y \leq 4$ and $3 < m < 1001$. Hewett and Ganser (1997) note that if the geometric standard deviation is nearly 1.00, it can be set conservatively to 1.01 [i.e., $s_y = \log_e(1.01) = 0.01$]. Similarly, if $s_y > 4$, then $s_y = 4$ will yield a reasonable approximation as well.

19.5.3 Approximate Lognormal Confidence Limit Methods

Several approximations to lognormal confidence limits for the mean have also been proposed. These have conveniently been classified as either transformation methods or direct methods (Land, 1972). A *transformation method* is one in which the confidence limit is obtained for the expected value of some function of x and then transformed by some appropriate function to give an approximate limit for the expectation of x, which in the lognormal case is $E(x) = \mu + \frac{1}{2}\sigma^2$. A *direct method* is based on the estimate of $E(x)$ or some function of $E(x)$. This estimate is assumed to be normally distributed, and approximate confidence limits are computed accordingly.

The simplest transformation method is the *naive transformation*, which involves simply taking a log transformation of the data, computing the confidence limit on a log scale, and then exponentiating the limit. As noted previously, this is, in fact, a confidence limit for the median and not the mean. The method provides somewhat reasonable results as a confidence limit for the mean when σ_y is very small but deteriorates quickly as σ_y increases (Land, 1972).

Patterson (1966) proposed use of the transformation

$$\hat{\mu}_x = \exp(\bar{y} + \tfrac{1}{2}\sigma_y^2) \tag{19.19}$$

to remove the obvious bias of the naive method. *Patterson's transformation* would be exact if σ_x^2 were known; however, as a consequence, it too behaves poorly when σ_y increases (Land, 1970). More complicated alternatives described by Finney (1941) and Hoyle (1968) provide results similar to Patterson's transformation and are therefore not presented.

Direct methods offer an advantage over transformation methods in that they obtain confidence intervals directly for $E(x)$ or some function of $E(x)$. In light of this, these methods do not suffer from the bias introduced by failing to take into account the dependence of $E(x)$ on both μ and σ^2. However, by applying normality assumptions to $E(x)$, direct estimates can produce inadmissible confidence limits for $E(x)$. To this end, Aitchison and Brown (1957) have suggested computing the usual normal confidence limit, which under the central limit theorem should converge to exact limits as n becomes large. Hoyle (1968) suggested replacing $\bar{x}$ and s_x^2/m by their minimum variance unbiased estimates (MVUEs). Finney (1941) derived the MVUE of $E(x)$ as

$$\hat{\theta} = \exp(\bar{y})\psi[(1 - m^{-1})s_y^2] \tag{19.20}$$

19.5 LOGNORMAL CONFIDENCE LIMITS FOR THE MEAN

and Hoyle (1968) derived the MVUE for the variance of $E(x)$ as

$$\hat{\phi} = \exp(2\bar{y})\{\psi^2[(1 - m^{-1})s_y^2] - \psi[(2 - 4m^{-1})s_y^2]\} \quad (19.21)$$

where

$$\psi(g) = 1 + \frac{m-1}{m}g + \frac{(m-1)^3}{m^2 2!}\frac{g^2}{m+1} + \frac{(m-1)^5}{m^3 3!}\frac{g^3}{(m+1)(m+3)} + \cdots \quad (19.22)$$

is a Bessel function with argument g. In this method, the normal quantile z_α replaces $t_{m-1,\alpha}$, since there is no reason to believe that $\hat{\phi}$ is chi-square and independent of $\hat{\theta}$. Unfortunately, Land (1972) has shown that these methods are useful only for large m (i.e., $m > 100$) and even there, only for small values of s_y.

The final direct method, which is attributed to D. R. Cox (see Land, 1972), has been shown to give the best overall results of any of the approximate methods (Land, 1972). The MVUE of $\beta = \log E(x)$ is $\hat{\beta} = \bar{y} + \frac{1}{2}s_y^2$, and the MVUE of the variance γ^2 of $\hat{\beta}$ is

$$\hat{\gamma}^2 = \frac{s_y^2}{m} + \frac{\frac{1}{2}s_y^4}{m+1} \quad (19.23)$$

Assuming approximate normality for $\hat{\beta}$, we may obtain approximate confidence limits for $E(x)$ of the form

$$\text{LCL} = \exp(\hat{\beta} - z_\alpha \hat{\gamma}) \quad (19.24)$$

$$\text{UCL} = \exp(\hat{\beta} + z_\alpha \hat{\gamma}) \quad (19.25)$$

Illustration Consider the following data for hexavalent chromium (mg/L) obtained from a potentially affected downgradient monitoring well.

1996		1997		1998	
Date	Result	Date	Result	Date	Result
Aug. 2	29	Jan. 6	9	Feb. 2	57
Sept. 9	14	Feb. 10	<1		
Sept. 30	13	May 8	33		
Oct. 30	14	Aug. 5	150		
Dec. 4	19	Nov. 11	60		

The adjusted mean concentration is 36.136 mg/L with standard deviation 42.515 mg/L. The corresponding 95% normal LCL is 12.903 mg/L (see Table 19.1). Note that inspection of the raw data clearly does not provide evidence for normality of these data. Indeed, distributional testing revealed $W_{\text{raw}} = 0.716$, $W_{\text{log}} = 0.936$, and a critical

Table 19.1 Worksheet: comparison to standard, hex chrome (mg/L) at HA1-ROW, normal confidence limit

Step	Equation	Description
1	Percentile = mean of the on-site/downgradient distribution	
2	$\bar{x}_1 = \dfrac{\sum x_1}{m_1}$ $= \dfrac{397.5}{10}$ $= 39.75$	Compute the mean of the detected measurements.
3	$s_1 = \left[\dfrac{\sum x_1^2 - \sum (x_1)^2/m_1}{m_1 - 1}\right]^{1/2}$ $= \left(\dfrac{32{,}273.25 - 158{,}006.25/10}{10 - 1}\right)^{1/2}$ $= 42.782$	Compute the standard deviation of the detected measurements.
4	$\bar{x} = \left(1 - \dfrac{m_0}{m}\right)\bar{x}_1$ $= \left(1 - \dfrac{1}{11}\right)39.75$ $= 36.136$	Compute the adjusted mean.
5	$s = \left[\left(1 - \dfrac{m_0}{m}\right)s_1^2 + \dfrac{m_0}{m}\left(1 - \dfrac{m_0 - 1}{m - 1}\right)\bar{x}_1^2\right]^{1/2}$ $= \left[\left(1 - \dfrac{1}{11}\right)42.782^2 + \dfrac{1}{11}\left(1 - \dfrac{1 - 1}{11 - 1}\right)39.75^2\right]^{1/2}$ $= 42.515$	Compute the adjusted standard deviation.
6	$\text{LCL} = \bar{x} - \dfrac{ts}{m^{1/2}}$ $= 36.136 - \dfrac{1.812 \times 42.515}{11^{1/2}}$ $= 12.903$	Compute the lower confidence limit for the mean of the m on-site/downgradient measurements.
7	Confidence = 0.95	Report the confidence level for this location and constituent.

value of $W_{\text{crit}} = 0.842$, indicating that the data are consistent with a lognormal distribution. Taking natural logarithms and then exponentiating yields an LCL of 9.280 mg/L, which as discussed previously is for the median concentration and not the mean. Using Land's method, the lognormal 95% LCL for the mean is 22.469 mg/L, which is actually higher than the geometric mean of 20.863 mg/L (see Table 19.2).

19.5 LOGNORMAL CONFIDENCE LIMITS FOR THE MEAN

Table 19.2 Worksheet: comparison to standard, hex chrome (mg/L) at HA1-ROW, lognormal confidence limit

Step	Equation	Description
1	Percentile = mean of the on-site/downgradient distribution	
2	$y = \log_e(x + 1)$	Transform to natural logarithmic scale.
3	$\bar{y}_1 = \dfrac{\sum y_1}{m_1}$ $= \dfrac{33.418}{10}$ $= 3.342$	Compute the mean of the detected measurements.
4	$s_{y_1} = \left[\dfrac{\sum y_1^2 - \sum(y_1)^2/m_1}{m_1 - 1} \right]^{1/2}$ $= \left(\dfrac{118.237 - 1116.785/10}{10 - 1} \right)^{1/2}$ $= 0.854$	Compute the standard deviation of the detected measurements.
5	$\bar{y} = \left(1 - \dfrac{m_0}{m}\right) \bar{y}_1$ $= \left(1 - \dfrac{1}{11}\right) 3.342$ $= 3.038$	Compute the adjusted mean (log scale).
6	$GM = \exp(\bar{y}) - 1$ $= 20.864$	Compute the geometric mean.
7	$s_y = \left[\left(1 - \dfrac{m_0}{m}\right) s_{y_1}^2 + \dfrac{m_0}{m}\left(1 - \dfrac{m_0 - 1}{m - 1}\right) \bar{y}_1^2 \right]^{1/2}$ $= \left[\left(1 - \dfrac{1}{11}\right) 0.854^2 + \dfrac{1}{11}\left(1 - \dfrac{1 - 1}{11 - 1}\right) 3.342^2 \right]^{1/2}$ $= 1.295$	Compute the adjusted standard deviation (log scale).
8	$LCL = \exp\left[\bar{y} + 0.5 s_y^2 + \dfrac{H s_y}{(m - 1)^{1/2}} \right] - 1$ $= \exp\left[3.038 + 0.5 \times 1.295^2 - 1.761 \right.$ $\left. \times \dfrac{1.295}{(11 - 1)^{1/2}} \right] - 1$ $= 22.469$	Compute the lower confidence limit for the mean of the m on-site/downgradient measurements using Land's method.
9	Confidence = 0.95	Report the confidence level for this location and constituent.

19.6 NONPARAMETRIC CONFIDENCE LIMITS FOR THE MEDIAN

When data are neither normally or lognormally distributed, or the detection frequency is too low for a meaningful distributional analysis (e.g., < 50%), nonparametric confidence limits become the method of choice. The nonparametric confidence limit is defined by an order statistic (i.e., ranked observation) of the m on-site/downgradient measurements. Note that in the nonparametric case, we are restricted to computing confidence limits on percentiles of the distribution, for example, the 50th percentile or median of the on-site/downgradient distribution. Unless the distribution is symmetric (i.e., the mean and median are equivalent), there is no direct nonparametric way of constructing a confidence limit for the mean concentration.

To construct a confidence limit for the median concentration, we use the fact that the number of samples falling below the $p(100)$th percentile of the distribution (e.g., $p = 0.5$, where p is between 0 and 1) out of a set on m samples will follow a binomial distribution with parameters n and success probability p, where success is defined as the event that a sample measurement is below the $p(100)$th percentile. The cumulative binomial distribution $(\text{Bin}(x; m, p))$ represents the probability of getting x or fewer successes in m trials with success probability p, and can be evaluated as

$$\text{Bin}(x; m, p) = \sum_{i=1}^{x} \binom{m}{i} p^i (1-p)^{m-i} \tag{19.26}$$

The notation $\binom{m}{i}$ denotes the number of combinations of m things taken i at a time, where

$$\binom{m}{i} = \frac{m!}{i!(m-i)!}$$

and $k! = 1 \cdot 2 \cdot 3 \ldots k$ for any counting number, k. For example, the number of ways in which two things can be selected from three things is

$$\binom{3}{2} = \frac{3!}{2!(1)!} = \frac{1 \cdot 2 \cdot 3}{(1 \cdot 2)(1)} = \frac{6}{2} = 3$$

To compute a nonparametric confidence limit for the median, we begin by rank ordering the m measurements from smallest to largest as $x_{(1)}, x_{(2)}, \ldots, x_{(m)}$. Denote the candidate endpoints selected to bracket the 50th percentile [i.e., $(m + 1) \times 0.5$] as L^* and U^* for lower and upper bound, respectively. For the LCL, compute the probability

$$1 - \text{Bin}(L^* - 1; m, 0.5)$$

If the probability is less than the desired confidence level, $1 - \alpha$, select a new value of $L^* = L^* - 1$, and repeat the process until the desired confidence level is achieved. For the UCL, compute the probability

$$1 - \text{Bin}(U^* - 1; m, 0.5)$$

If the probability is less than the desired confidence level $1 - \alpha$, select a new value of $U^* = U^* + 1$ and repeat the process until the desired confidence level is achieved. If the desired confidence level cannot be achieved, set the LCL to the smallest value or the UCL to the largest value and report the confidence level achieved.

Illustration Returning to the hexavalent chromium data of Section 19.5, we have $m = 11$, and the third ordered observation (13 mg/L) is the corresponding LCL. The corresponding confidence level is 0.967. Note that this is a confidence bound on the median and not the mean, hence the similarity to the normal limit and lognormal limit for the median of Section 19.5.

19.7 CONFIDENCE LIMITS FOR OTHER PERCENTILES OF THE DISTRIBUTION

For some applications, there may be interest in a LCL or UCL for a specific percentile of the distribution (e.g., 90th, 95th, or 99th percentiles of the concentration distribution). For a normal distribution, which is symmetric, the mean, median, and 50th percentile are identical. However, this is not the case for the lognormal distribution, where the mean is larger than the median or 50th percentile. Of course, in the nonparametric case, only confidence limits for percentiles are available, such as the 50th percentile of the distribution (i.e., the concept of confidence limits for the mean does not exist without a specific parametric form of the distribution).

For those constituents with short-term exposure risks or in those cases in which one may wish to show added environmental protection, confidence limits for upper percentiles of the distribution may be used (e.g., 90th, 95th, or 99th percentiles). The interpretation here is that, for example, there is 99% confidence that 95% of the distribution is beneath the estimated confidence limit. Both LCLs and UCLs for upper percentiles can be computed and normal, lognormal, and nonparametric approaches have been described in general by Hahn and Meeker (1991) and are closely related to the statistical tolerance limits.

19.7.1 Normal Confidence Limits for a Percentile

To compute a normal confidence limit for a percentile of the distribution, we use the same factors that are used in computing one-sided normal tolerance limit. For example, a 99% UCL for the 95th percentile of the distribution implies that 95% of the distribution will lie below the UCL with 99% confidence. A 99% confidence, 95% coverage β-content tolerance limit implies that 95% of the distribution will be below the tolerance limit with 99% confidence. As such, upper tolerance limits (UTLs) and UCLs for percentiles are identical as long as the confidence levels are the same and the percentile of the UCL equals the coverage of the UTL. To compute the UCL for a percentile, we therefore use a good table of tolerance limit factors and obtain

$$\mathrm{UCL}_{1-\alpha,p} = \bar{x} + K_{1-\alpha,p} s \tag{19.27}$$

where $K_{1-\alpha,p}$ is the one-sided normal tolerance limit factor for $(1-\alpha)100\%$ confidence and $p(100)\%$ coverage. Table 19.3 presents values of K useful for computing UCLs with 80%, 90%, 95%, 97.5%, and 99% confidence for the 90th, 95th, and 99th percentile of the distribution.

For an LCL of a percentile, the situation is slightly different. Here we seek the $(1-\alpha)100\%$ lower bound on the $p(100)$th percentile of the distribution, which is computed as

$$\mathrm{LCL}_{1-\alpha,p} = \bar{x} + K_{\alpha,p} s \tag{19.28}$$

where $K_{\alpha,p}$ is the one-sided normal tolerance limit factor for $(\alpha)100\%$ confidence and $p(100)\%$ coverage. Table 19.4 presents values of K useful for computing LCLs with 80%, 90%, 95%, 97.5%, and 99% confidence for the 90th, 95th, and 99th percentile of the distribution. Note that the confidence values given in Table 19.4 actually represent tolerance limit factors with confidence levels of 20% 10%, 5%, 2.5%, and 1%, respectively.

19.7.2 Lognormal Confidence Limits for a Percentile

In the lognormal case, confidence limits for percentiles are obtained by computing the UCLs and LCLs of the preceding section on the natural logarithms of the measured values and exponentiating the resulting limits. Since the limits are for percentiles of the distribution and not the mean, the simple transformation estimator applies directly. For example,

$$\mathrm{UCL}_{1-\alpha,p} = \exp(\bar{y} + K_{1-\alpha,p} s_y) \tag{19.29}$$

$$\mathrm{LCL}_{1-\alpha,p} = \exp(\bar{y} + K_{\alpha,p} s_y) \tag{19.30}$$

where $\bar{y}$ and s_y are the mean and standard deviation of the natural log transformed data $y = \log_e(x)$. The factors used for computing these limits are given in Tables 19.3 and 19.4.

19.7.3 Nonparametric Confidence Limits for a Percentile

The nonparametric approach to computing confidence limits for the median concentration extend directly to computation of confidence limits for any percentile of

Table 19.3 One-sided factors for UCLs for percentiles of the distribution: 80%, 90%, 95%, 97.5%, and 99% confidence for the 90th, 95th, and 99th percentiles, $n = 4$ to 1000

n	90th Percentile					95th Percentile					99th Percentile				
	80%	90%	95%	97.5%	99%	80%	90%	95%	97.5%	99%	80%	90%	95%	97.5%	99%
4	2.372	3.188	4.162	5.354	7.380	2.968	3.957	5.144	6.602	9.083	4.110	5.438	7.042	9.018	12.387
5	2.145	2.742	3.407	4.166	5.362	2.683	3.400	4.203	5.124	6.578	3.711	4.666	5.741	6.980	8.939
6	2.012	2.494	3.006	3.568	4.411	2.517	3.092	3.708	4.385	5.406	3.482	4.243	5.062	5.967	7.335
7	1.923	2.333	2.755	3.206	3.859	2.407	2.894	3.399	3.940	4.728	3.331	3.972	4.642	5.361	6.412
8	1.859	2.219	2.582	2.960	3.497	2.328	2.754	3.187	3.640	4.285	3.224	3.783	4.354	4.954	5.812
9	1.809	2.133	2.454	2.783	3.240	2.268	2.650	3.031	3.424	3.972	3.142	3.641	4.143	4.662	5.389
10	1.770	2.066	2.355	2.647	3.048	2.220	2.568	2.911	3.259	3.738	3.078	3.532	3.981	4.440	5.074
11	1.738	2.011	2.275	2.540	2.898	2.182	2.503	2.815	3.129	3.556	3.026	3.443	3.852	4.265	4.829
12	1.711	1.966	2.210	2.452	2.777	2.149	2.448	2.736	3.023	3.410	2.982	3.371	3.747	4.124	4.633
13	1.689	1.928	2.155	2.379	2.677	2.122	2.402	2.671	2.936	3.290	2.946	3.309	3.659	4.006	4.472
14	1.669	1.895	2.109	2.317	2.593	2.098	2.363	2.614	2.861	3.189	2.914	3.257	3.585	3.907	4.337
15	1.652	1.867	2.068	2.264	2.521	2.078	2.329	2.566	2.797	3.102	2.887	3.212	3.520	3.822	4.222
16	1.637	1.842	2.033	2.218	2.459	2.059	2.299	2.524	2.742	3.028	2.863	3.172	3.464	3.749	4.123
17	1.623	1.819	2.002	2.177	2.405	2.043	2.272	2.486	2.693	2.963	2.841	3.137	3.414	3.684	4.037
18	1.611	1.800	1.974	2.141	2.357	2.029	2.249	2.453	2.650	2.905	2.822	3.105	3.370	3.627	3.960
19	1.600	1.782	1.949	2.108	2.314	2.016	2.227	2.423	2.611	2.854	2.804	3.077	3.331	3.575	3.892
20	1.590	1.765	1.926	2.079	2.276	2.004	2.208	2.396	2.576	2.808	2.789	3.052	3.295	3.529	3.832
21	1.581	1.750	1.905	2.053	2.241	1.993	2.190	2.371	2.544	2.766	2.774	3.028	3.263	3.487	3.777
22	1.572	1.737	1.886	2.028	2.209	1.983	2.174	2.349	2.515	2.729	2.761	3.007	3.233	3.449	3.727
23	1.564	1.724	1.869	2.006	2.180	1.973	2.159	2.328	2.489	2.694	2.749	2.987	3.206	3.414	3.681
24	1.557	1.712	1.853	1.985	2.154	1.965	2.145	2.309	2.465	2.662	2.738	2.969	3.181	3.382	3.640
25	1.550	1.702	1.838	1.966	2.129	1.957	2.132	2.292	2.442	2.633	2.727	2.952	3.158	3.353	3.601
26	1.544	1.691	1.824	1.949	2.106	1.949	2.120	2.275	2.421	2.606	2.718	2.937	3.136	3.325	3.566
27	1.538	1.682	1.811	1.932	2.085	1.943	2.109	2.260	2.402	2.581	2.708	2.922	3.116	3.300	3.533

(*Continued*)

Table 19.3 (Continued)

	90th Percentile					95th Percentile					99th Percentile				
n	80%	90%	95%	97.5%	99%	80%	90%	95%	97.5%	99%	80%	90%	95%	97.5%	99%
28	1.533	1.673	1.799	1.917	2.065	1.936	2.099	2.246	2.384	2.558	2.700	2.909	3.098	3.276	3.502
29	1.528	1.665	1.788	1.903	2.047	1.930	2.089	2.232	2.367	2.536	2.692	2.896	3.080	3.254	3.473
30	1.523	1.657	1.777	1.889	2.030	1.924	2.080	2.220	2.351	2.515	2.684	2.884	3.064	3.233	3.447
25	1.502	1.624	1.732	1.833	1.957	1.900	2.041	2.167	2.284	2.430	2.652	2.833	2.995	3.145	3.334
40	1.486	1.598	1.697	1.789	1.902	1.880	2.010	2.125	2.232	2.364	2.627	2.793	2.941	3.078	3.249
50	1.461	1.559	1.646	1.724	1.821	1.852	1.965	2.065	2.156	2.269	2.590	2.735	2.862	2.980	3.125
60	1.444	1.532	1.609	1.679	1.764	1.832	1.933	2.022	2.103	2.202	2.564	2.694	2.807	2.911	3.038
120	1.393	1.452	1.503	1.549	1.604	1.772	1.841	1.899	1.952	2.015	2.488	2.574	2.649	2.716	2.797
240	1.358	1.399	1.434	1.465	1.501	1.733	1.780	1.819	1.854	1.896	2.437	2.497	2.547	2.591	2.645
480	1.335	1.363	1.387	1.408	1.433	1.706	1.738	1.766	1.790	1.818	2.403	2.444	2.479	2.509	2.545
1000	1.282	1.282	1.282	1.282	1.282	1.645	1.645	1.645	1.645	1.645	2.326	2.326	2.326	2.326	2.326

Table 19.4 One-sided factors for LCLs for percentiles of the distribution: 80%, 90%, 95%, 97.5%, and 99% confidence for the 90th, 95th, and 99th percentiles, $n = 4$ to 1000

	90th Percentile					95th Percentile					99th Percentile				
n	80%	90%	95%	97.5%	99%	80%	90%	95%	97.5%	99%	80%	90%	95%	97.5%	99%
4	0.123	0.298	0.444	0.617	0.847	0.443	0.601	0.743	0.922	1.172	0.924	1.088	1.246	1.455	1.760
5	0.238	0.389	0.519	0.675	0.883	0.543	0.687	0.818	0.982	1.209	1.027	1.182	1.331	1.525	1.801
6	0.319	0.455	0.575	0.719	0.911	0.618	0.752	0.875	1.028	1.238	1.108	1.256	1.396	1.578	1.834
7	0.381	0.507	0.619	0.755	0.933	0.678	0.804	0.920	1.065	1.261	1.173	1.315	1.449	1.622	1.862
8	0.431	0.550	0.655	0.783	0.952	0.727	0.847	0.958	1.096	1.281	1.227	1.364	1.493	1.658	1.885
9	0.472	0.585	0.686	0.808	0.968	0.768	0.884	0.990	1.122	1.298	1.273	1.406	1.530	1.688	1.904
10	0.508	0.615	0.712	0.828	0.981	0.804	0.915	1.017	1.144	1.313	1.314	1.442	1.563	1.715	1.922
11	0.538	0.642	0.734	0.847	0.993	0.835	0.943	1.041	1.163	1.325	1.349	1.474	1.591	1.738	1.937
12	0.565	0.665	0.754	0.863	1.004	0.862	0.967	1.062	1.180	1.337	1.381	1.502	1.616	1.758	1.950
13	0.589	0.685	0.772	0.877	1.013	0.887	0.989	1.081	1.196	1.347	1.409	1.528	1.638	1.776	1.962
14	0.610	0.704	0.788	0.890	1.022	0.909	1.008	1.098	1.210	1.356	1.434	1.551	1.658	1.793	1.973
15	0.629	0.721	0.802	0.901	1.029	0.929	1.026	1.114	1.222	1.364	1.458	1.572	1.677	1.808	1.983
16	0.647	0.736	0.815	0.912	1.036	0.948	1.042	1.128	1.234	1.372	1.479	1.591	1.694	1.822	1.992
17	0.663	0.750	0.827	0.921	1.043	0.965	1.057	1.141	1.244	1.379	1.499	1.608	1.709	1.834	2.000
18	0.678	0.763	0.839	0.930	1.049	0.980	1.071	1.153	1.254	1.385	1.517	1.625	1.724	1.846	2.008
19	0.692	0.775	0.849	0.939	1.054	0.995	1.084	1.164	1.263	1.391	1.534	1.640	1.737	1.857	2.015
20	0.705	0.786	0.858	0.946	1.059	1.008	1.095	1.175	1.271	1.397	1.550	1.654	1.749	1.867	2.022
21	0.716	0.796	0.867	0.953	1.064	1.021	1.107	1.184	1.279	1.402	1.565	1.667	1.761	1.876	2.028
22	0.728	0.806	0.876	0.960	1.068	1.033	1.117	1.193	1.286	1.407	1.579	1.680	1.772	1.885	2.034
23	0.738	0.815	0.884	0.966	1.073	1.044	1.127	1.202	1.293	1.412	1.592	1.691	1.782	1.893	2.039
24	0.748	0.823	0.891	0.972	1.076	1.054	1.136	1.210	1.300	1.416	1.605	1.702	1.791	1.901	2.045
25	0.757	0.831	0.898	0.978	1.080	1.064	1.145	1.217	1.306	1.420	1.616	1.713	1.801	1.908	2.049
26	0.766	0.839	0.904	0.983	1.084	1.074	1.153	1.225	1.311	1.424	1.628	1.723	1.809	1.915	2.054
27	0.774	0.846	0.911	0.988	1.087	1.083	1.161	1.231	1.317	1.427	1.638	1.732	1.817	1.922	2.058

(*Continued*)

Table 19.4 (Continued)

	90th Percentile					95th Percentile					99th Percentile				
n	80%	90%	95%	97.5%	99%	80%	90%	95%	97.5%	99%	80%	90%	95%	97.5%	99%
28	0.782	0.853	0.917	0.993	1.090	1.091	1.168	1.238	1.322	1.431	1.648	1.741	1.825	1.928	2.063
29	0.790	0.860	0.922	0.997	1.093	1.099	1.175	1.244	1.327	1.434	1.658	1.749	1.833	1.934	2.067
30	0.797	0.866	0.928	1.002	1.096	1.107	1.182	1.250	1.332	1.437	1.667	1.757	1.840	1.940	2.070
35	0.828	0.893	0.951	1.020	1.108	1.141	1.212	1.276	1.352	1.451	1.708	1.793	1.871	1.965	2.087
40	0.854	0.916	0.970	1.036	1.119	1.169	1.236	1.297	1.369	1.462	1.741	1.823	1.896	1.986	2.101
50	0.894	0.950	1.000	1.059	1.134	1.212	1.274	1.329	1.396	1.480	1.793	1.869	1.936	2.018	2.122
60	0.924	0.976	1.022	1.077	1.146	1.245	1.303	1.354	1.415	1.493	1.833	1.903	1.966	2.042	2.138
120	1.020	1.059	1.093	1.134	1.184	1.352	1.395	1.433	1.478	1.535	1.963	2.016	2.063	2.119	2.189
240	1.092	1.121	1.146	1.175	1.211	1.431	1.463	1.492	1.525	1.565	2.061	2.100	2.135	2.176	2.227
480	1.145	1.166	1.184	1.205	1.231	1.491	1.514	1.535	1.558	1.588	2.134	2.163	2.189	2.218	2.255
1000	1.282	1.282	1.282	1.282	1.282	1.645	1.645	1.645	1.645	1.645	2.326	2.326	2.326	2.326	2.326

19.7 CONFIDENCE LIMITS FOR OTHER PERCENTILES OF THE DISTRIBUTION

Table 19.5 Worksheet: comparison to standard, hex chrome (mg/L) at HA1-ROW, normal confidence limit

Step	Equation	Description
1	Percentile = 90th percentile of the on-site/downgradient distribution	
2	$\bar{x}_1 = \dfrac{\sum x_1}{m_1}$ $= \dfrac{397.5}{10}$ $= 39.75$	Compute the mean of the detected measurements.
3	$s_1 = \left[\dfrac{\sum x_1^2 - \sum(x_1)^2/m_1}{m_1 - 1}\right]^{1/2}$ $= \left(\dfrac{32{,}273.25 - 158{,}006.25/10}{10 - 1}\right)^{1/2}$ $= 42.782$	Compute the standard deviation of the detected measurements.
4	$\bar{x} = \left(1 - \dfrac{m_0}{m}\right)\bar{x}_1$ $= \left(1 - \dfrac{1}{11}\right)39.75$ $= 36.136$	Compute the adjusted mean.
5	$s = \left[\left(1 - \dfrac{m_0}{m}\right)s_1^2 + \dfrac{m_0}{m}\left(1 - \dfrac{m_0 - 1}{m - 1}\right)\bar{x}_1^2\right]^{1/2}$ $= \left[\left(1 - \dfrac{1}{11}\right)42.782^2 + \dfrac{1}{11}\left(1 - \dfrac{1-1}{11-1}\right)39.75^2\right]^{1/2}$ $= 42.515$	Compute the adjusted standard deviation.
6	LCL = $\bar{x} + T_L s$ $= 36.136 + 0.734 \times 42.515$ $= 67.343$	Compute the lower confidence limit for the selected percentile of the distribution.
7	Confidence = 0.95	Report the confidence level for this location and constituent.

the distribution. For the LCL of the $p(100)$th percentile, we iteratively compute the probability

$$1 - \text{Bin}(L^* - 1; m, p)$$

until the desired confidence level $1 - \alpha$ is achieved or we reach the lowest value and report the achieved confidence level. For the UCL, we iteratively compute the

Table 19.6 Worksheet: comparison to standard, hex chrome (mg/L) at HA1-ROW, lognormal confidence limit

Step	Equation	Description
1	Percentile = 90th percentile of the on-site/downgradient distribution	
2	$y = \log_e(x + 1)$	Transform to natural logarithmic scale.
3	$\bar{y}_1 = \dfrac{\Sigma y_1}{m_1}$ $= \dfrac{33.418}{10}$ $= 3.342$	Compute the mean of the detected measurements.
4	$s_{y_1} = \left[\dfrac{\Sigma y_1^2 - \Sigma(y_1)^2/m_1}{m_1 - 1} \right]^{1/2}$ $= \left(\dfrac{118.237 - 1116.785/10}{10 - 1} \right)^{1/2}$ $= 0.854$	Compute the standard deviation of the detected measurements.
5	$\bar{y} = \left(1 - \dfrac{m_0}{m}\right) \bar{y}_1$ $= \left(1 - \dfrac{1}{11}\right) 3.342$ $= 3.038$	Compute the adjusted mean (log scale).
6	$\mathrm{GM} = \exp(\bar{y}) - 1$ $= 20.864$	Compute the geometric mean.
7	$s_y = \left[\left(1 - \dfrac{m_0}{m}\right) s_{y_1}^2 + \dfrac{m_0}{m}\left(1 - \dfrac{m_0 - 1}{m - 1}\right) \bar{y}_1^2 \right]^{1/2}$ $= \left[\left(1 - \dfrac{1}{11}\right) 0.854^2 + \dfrac{1}{11}\left(1 - \dfrac{1 - 1}{11 - 1}\right) 3.342^2 \right]^{1/2}$ $= 1.295$	Compute the adjusted standard deviation (log scale).
8	$\mathrm{LCL} = \exp(\bar{y} + t_L s_y) - 1$ $= \exp(3.038 + 0.734 \times 1.295) - 1$ $= 52.989$	Compute the lower confidence limit for the selected percentile of the distribution.
9	Confidence = 0.95	Report the confidence level for this location and constituent.

19.7 CONFIDENCE LIMITS FOR OTHER PERCENTILES OF THE DISTRIBUTION

probability

$$1 - \text{Bin}(U^* - 1; m, p)$$

until the desired confidence level $1 - \alpha$ is achieved or we reach the highest value and report the achieved confidence level. Note that for higher percentiles, much larger numbers of measurements (i.e., m) must be available in order to achieve reasonable levels of confidence (e.g., 95% confidence).

Illustration Returning to the hexavalent chromium data of the preceding sections, we can compute the 95% LCL for the upper 90th percentile of the distribution based on the normal, lognormal, and nonparametric methods described in this section. The normal 95% LCL for the upper 90th percentile is 67.343 mg/L (see Table 19.5). The lognormal 95% LCL for the upper 90th percentile is 52.989 mg/L (see Table 19.6). The nonparametric LCL for the upper 90th percentile is 33.000 mg/L, which corresponds to the eighth ordered observation, and provides a confidence of 0.981.

20

ASSESSMENT AND CORRECTIVE ACTION MONITORING: COMPARISON TO BACKGROUND

20.1 INTRODUCTION

In some cases, regulatory standards are unavailable and the relevant comparison is of the on-site or downgradient data to an off-site or upgradient background distribution. Even when a regulatory standard is available, the background concentrations of the constituent may already exceed the regulatory standard and the only relevant comparison is to background. Two general approaches are available. First, we can simultaneously compare each of the k individual on-site/downgradient measurements to a prediction limit for k new observations computed from n background (i.e., off-site/upgradient) measurements. Second, we can compare the mean or median of m on-site/downgradient measurements at each of k downgradient locations to the corresponding prediction limit (computed from the n off-site/upgradient measurements) for the mean or median of m future measurements. Of, course, there are normal, lognormal, and nonparametric forms of these prediction limits as there were in the case of confidence limits. As an example, if we pool the data across all $k = 5$ downgradient locations into a single site-wide comparison, we would have $m = 5$ and $k = 1$. By contrast, if we test the $k = 5$ downgradient locations individually, we would have $k = 5$ and $m = 1$.

20.2 NORMAL PREDICTION LIMITS FOR $m = 1$ FUTURE MEASUREMENTS AT EACH OF k LOCATIONS

In practice, it is rare to have an application in which only a single future measurement requires evaluation. Typically, we are interested in comparison of m future samples collected from each of k on-site or downgradient locations to a limit computed from n background measurements. A problem of particular interest is when there is $m = 1$ sample taken from k unique spatial coordinates within a defined and potentially affected area of a site. Here we may wish to compare concentrations simultaneously for a particular constituent individually to background, to rule out the possibility that only a subset is affected. The problem is, in fact, closely tied to detection monitoring, and many of the prediction limits discussed in Chapter 19 clearly apply. In assessment and/or corrective action monitoring, however, the notion of verification resampling may not be relevant. For example, in soil sampling, it is unclear how one would even obtain a verification resample.

The simplest approach is to assume independence. Under independence, if the probability of a false positive result for a single comparison is α, the probability of at least one of k comparisons being significant by chance alone is

$$\alpha^* = 1 - (1 - \alpha)^k \tag{20.1}$$

Here, α^* is the site-wide false positive rate since it considers simultaneously all k comparisons being performed on a given monitoring event. In detection monitoring, the value of k reflects the total number of statistical tests, which is the product of the number of monitoring wells and number of constituents. In assessment and/or corrective action monitoring, a case can be made to restrict the value of k to the number of locations monitored for a particular constituent. The reason is that in assessment and/or corrective action, we already have evidence of a potential site impact, and spreading the false positive rate over all locations and constituents may be unnecessarily conservative and lead to unrealistic false negative rates. In any event, one should use the smallest number of relevant constituents in all cases. The choice of k (i.e., wells or wells by constituents) is a site-specific question. For the purpose of this discussion, m is considered to represent a single measurement ($m = 1$) obtained from each of k locations.

For example, with 95% confidence for an individual comparison (i.e., $\alpha = 0.05$), $m = 1$, and $k = 10$ locations, the probability of at least one significant result by chance alone is

$$\alpha^* = 1 - (1 - 0.05)^{10} = 0.40$$

or a 40% chance of a statistically significant exceedance by chance alone. With 100 comparisons, $\alpha^* = 0.99$, or a 99% chance of a statistically significant exceedance by chance alone. Since it is not uncommon for assessment monitoring programs to have 20 or 30 sampled locations, the effect of these multiple comparisons on the site-wide false positive rate is considerable.

One solution to this problem is to compute a prediction limit that will provide $(1 - \alpha^*)100\%$ confidence of including all k future measurements. The simplest approach to this problem is through use of the Bonferroni inequality (see Chew, 1968; Miller, 1966), noting that

$$\alpha = \frac{\alpha^*}{k} \tag{20.2}$$

Application of this result reveals that to have a site-wide error rate at $\alpha^* = 0.05$ when $k = 10$ comparisons are made requires that we test each comparison at the $\alpha = 0.005\%$ level. The $(1 - \alpha^*)100\%$ prediction limit for the next k measurements from a normal distribution is therefore

$$\bar{x} + t_{[n-1, 1-\alpha^*/k]} s \sqrt{1 + \frac{1}{k}} \tag{20.3}$$

Table 20.1 displays one-sided values of $t_{[n-1, \alpha^*/k]}$ for $n = 4$ to 100 and $k = 5$ to 50.

Table 20.1 One-sided values of Student's *t*-statistic: 95% overall confidence for background $n = 4$ to 100 and $k = 4$ to 50 future measurements

					k					
n	5	10	15	20	25	30	35	40	45	50
4	4.54	5.84	6.74	7.45	8.05	8.57	9.04	9.46	9.85	10.21
8	3.00	3.50	3.81	4.03	4.21	4.35	4.48	4.59	4.69	4.78
12	2.71	3.10	3.32	3.48	3.60	3.71	3.79	3.87	3.93	3.99
16	2.60	2.94	3.14	3.28	3.39	3.48	3.55	3.61	3.67	3.72
20	2.54	2.86	3.04	3.17	3.27	3.35	3.42	3.48	3.53	3.57
24	2.50	2.81	2.98	3.10	3.20	3.27	3.34	3.39	3.44	3.48
28	2.47	2.77	2.94	3.06	3.15	3.22	3.28	3.33	3.38	3.42
32	2.45	2.74	2.91	3.02	3.11	3.18	3.24	3.29	3.33	3.37
36	2.44	2.72	2.88	3.00	3.08	3.15	3.21	3.26	3.30	3.34
40	2.43	2.71	2.87	2.98	3.06	3.13	3.18	3.23	3.27	3.31
44	2.42	2.69	2.85	2.96	3.04	3.11	3.16	3.21	3.25	3.29
48	2.41	2.68	2.84	2.95	3.03	3.09	3.15	3.19	3.24	3.27
52	2.40	2.68	2.83	2.93	3.01	3.08	3.13	3.18	3.22	3.26
56	2.40	2.67	2.82	2.92	3.00	3.07	3.12	3.17	3.21	3.24
60	2.39	2.66	2.81	2.92	3.00	3.06	3.11	3.16	3.20	3.23
64	2.39	2.66	2.81	2.91	2.99	3.05	3.10	3.15	3.19	3.22
68	2.38	2.65	2.80	2.90	2.98	3.04	3.10	3.14	3.18	3.22
72	2.38	2.65	2.80	2.90	2.98	3.04	3.09	3.13	3.17	3.21
76	2.38	2.64	2.79	2.89	2.97	3.03	3.08	3.13	3.17	3.20
80	2.37	2.64	2.79	2.89	2.97	3.03	3.08	3.12	3.16	3.20
84	2.37	2.64	2.78	2.88	2.96	3.02	3.07	3.12	3.16	3.19
88	2.37	2.63	2.78	2.88	2.96	3.02	3.07	3.11	3.15	3.19
92	2.37	2.63	2.78	2.88	2.95	3.01	3.07	3.11	3.15	3.18
96	2.37	2.63	2.77	2.87	2.95	3.01	3.06	3.11	3.14	3.18
100	2.36	2.63	2.77	2.87	2.95	3.01	3.06	3.10	3.14	3.17

20.2 NORMAL PREDICTION LIMITS

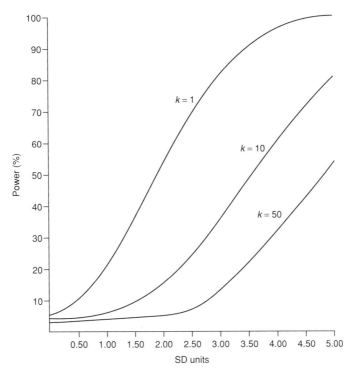

Figure 20.1 Power of 95% confidence Bonferroni normal prediction limits for 1, 10, and 50 future comparisons.

Although this prediction bound limits probability of any one of k future measurements exceeding the limit by chance alone to α^*, it does so at the expense of the false negative rate. To illustrate this point, Figure 20.1 displays statistical power curves for prediction limits for the next $k = 1$, 10, and 50 comparisons based on a background sample of $n = 8$ measurements and setting the individual comparison false positive rate to $\alpha = 0.05/k$. In Figure 20.1, contamination was introduced into a single location; hence only one of 1, 10, or 50 comparisons was contaminated. The power curves in Figure 20.1 therefore display the probability of detecting a very localized release that affects only one of k on-site/downgradient measurements. In practice, we would expect contamination to impact several locations; therefore, the probability estimates in Figure 20.1 represent a lower bound. Inspection of Figure 20.1 reveals that the false positive rates for $k = 1$, 10, and 50 future comparisons all approach the nominal level of 5%; however, false negative rates are affected dramatically by adjusting for larger numbers of future comparisons. For a difference of four standard deviation units, and eight background samples, the false negative rates are 4%, 39%, and 57% for $k = 1$, 10, and 50, respectively (i.e., 1 minus power (i.e., the probability of a significant result) at x-axis = 4 standard deviation units). These results indicate that a proliferation of sampling locations may not be in the best interest of environmental protection in that the likelihood of missing a site impact of given magnitude increases as the number of locations increases. As pointed out in Chapter 17, these effects can be reduced or eliminated completely through verification resampling (i.e., sequential testing). As noted previously, however, this option is not always available in assessment/corrective action monitoring and may rarely, if ever, be available in media other than groundwater.

The previous prediction limits for multiple future comparisons are simultaneous in the sense that they control the overall site-wide comparison rate (e.g., $\alpha^* = 0.05$), assuming that the multiple comparisons are independent. While this is true for a series of intralocation comparisons on a given sampling event (i.e., each location is compared to it's own background for a single constituent), it is not true in the context of comparisons to a common background. This type of comparison strategy introduces a correlation among the k comparisons for each constituent of magnitude $r = 1/(n + 1)$. An analogous situation occurs in the context of comparing multiple treatment group means to a single control (i.e., Dunnett's test; Dunnett and Sobel, 1955). The correlation among repeated comparisons makes use of the simple probability product for α^* too conservative. Obtaining the correct joint probability of failure on any one of the k comparisons requires evaluation of the equa-correlated multivariate normal distribution, and these probabilities must be integrated over the distribution of s, the sample standard deviation.

Suppose that there is interest in comparing k groups with a common background in terms of the means $\bar{x}_0, \bar{x}_1, \ldots, \bar{x}_k$ (and common standard deviation s) of $k + 1$ sets of observations which are assumed to be independently and normally distributed; $\bar{x}_0$ referring to the background and $\bar{x}_i$ to the ith comparison group ($i = 1, \ldots, k$) mean. In this case, Dunnett (1955) has provided a procedure for making confidence statements about expected values of k differences $\bar{x}_i - \bar{x}_0$, the procedure having the property that the probability of *all* k statements being correct simultaneously is equal to a specified probability level $1 - \alpha$. Dunnett's procedure and the associated tables were worked out for the case of equal sample sizes in all groups. Here we expand the procedure for the case when sample sizes are not equal.

Suppose that there are n background measurements, m_1 measurements on the first location, . . ., m_k measurements on the kth location, and denote these observations by x_{ij} ($i = 0, 1, \ldots, k; j = 1, 2, \ldots, m_i$) and the corresponding ith location mean as $\bar{x}_i$. We assume that the x_{ij} are independent and normally distributed with common variance σ^2 and means μ_i and that there is an estimate of σ^2 available (denoted s^2) based on ν degrees of freedom. Let

$$z_i = \frac{\bar{x}_i - \bar{x}_0 - (\mu_i - \mu_0)}{\sqrt{1/m_i + 1/n}} \tag{20.4}$$

and $t_i = z_i/s$ for $i = 1, 2, \ldots, k$. As Dunnett (1955) notes, the lower confidence limits with joint confidence coefficient $1 - \alpha$ for the k comparison group effects $\mu_i - \mu_0$ are given by

$$\bar{x}_i - \bar{x}_0 - d_i s \sqrt{\frac{1}{m_i} + \frac{1}{n}} \tag{20.5}$$

if the x constants d_i are chosen so that

$$\Pr(t_1 < d_1, t_2 < d_2, \ldots, t_k < d_k) = 1 - \alpha \qquad (20.6)$$

To find the k constants d_i that satisfy these equations, joint distribution of the t_i is required, which is the multivariate analog of Student's t-distribution defined by Dunnett and Sobel (1955). Dunnett (1955) has shown how the problem of tabulating the multivariate t-distribution can be reduced to the problem of tabulating the corresponding multivariate normal (MVN) distribution. For this, note that the joint distribution of the z_i is a MVN distribution with means 0 and variances σ^2. The correlation between z_i and z_j is given by

$$\rho_{ij} = 1 \bigg/ \sqrt{\left(\frac{n}{m_i} + 1\right)\left(\frac{n}{m_j} + 1\right)} \qquad (20.7)$$

Notice that the joint probability statement given above can be written in the following way:

$$\begin{aligned} 1 - \alpha &= \Pr(t_1 < d_1, t_2 < d_2, \ldots, t_k < d_k) \\ &= \Pr(z_1 < d_1 s, z_2 < d_2 s, \ldots, z_k < d_k s) \\ &= \int_{-\infty}^{+\infty} F(d_1 s, d_2 s, \ldots, d_k s) f(s) \, ds \end{aligned} \qquad (20.8)$$

where $F(d_1 s, d_2 s, \ldots, d_k s)$ is the MVN cumulative distribution function of the z_i and $f(s)$ is the probability density function of s. Thus, with probability values for $F(\cdot)$, equation (20.8) can be evaluated using numerical integration over the distribution of s. For this, note that the density function of s is given by Pearson and Hartley (1976) as

$$f(s) = \frac{\nu^{\nu/2}}{\Gamma(\nu/2) 2^{(\nu/2)-1}} \sigma^{-\nu} s^{\nu-1} \exp\left(-\frac{\nu s^2}{2\sigma^2}\right) \qquad (20.9)$$

Since $s^2/\sigma^2 = \chi^2/\nu$, we can rewrite the equation for $1 - \alpha$ in terms of integration over the distribution of $y = s/\sigma$ (which is defined on 0 to $+\infty$) as

$$\begin{aligned} 1 - \alpha &= \int_0^{+\infty} F(d_1 y, d_2 y, \ldots, d_k y) f(y) \frac{ds}{dy} \, dy \\ &= \int_0^{+\infty} F(d_1 y, d_2 y, \ldots, d_k y) \frac{\nu^{\nu/2}}{\Gamma(\nu/2) 2^{(\nu/2)-1}} y^{\nu-1} \exp\left(-\frac{\nu y^2}{2}\right) dy \end{aligned} \qquad (20.10)$$

Numerical integration over the distribution of y can then be performed to yield the associated probability $1 - \alpha$ for selected values of d, k, and ν.

In the present context, we are interested in comparing k new individual measurements (e.g., TOC levels in 10 downgradient monitoring wells) with collection of n background measurements, perhaps obtained from monitoring wells upgradient of the facility. In this case, $m_i = 1$, $i = 1, \ldots, k$, and the constant correlation is $\rho_{ij} = 1/(n + 1)$. As a result, the need for this correction decreases with increasing background sample size n since as n increases, ρ_{ij} goes to zero. In this special case the probability integral simplifies to

$$F_k(ds, ds, \ldots, ds; \rho) = \int_{-\infty}^{\infty} \left[F^k \left(\frac{ds + \rho^{1/2}y}{\sqrt{1 - \rho}} \right) \right] f(y) \, dy \qquad (20.11)$$

where $f(\tau) = \exp(-\frac{1}{2}\tau^2)/(2\pi)^{1/2}$ and $F(\tau) = \int_{-\infty}^{ds} f(\tau) \, d\tau$ (see Gupta, 1963).

To aid in application, Table 20.2 provides the constants d for background sample sizes n from 4 to 50 and number of future comparisons k from 2 to 50 for $1 - \alpha = 0.95$. These coefficients may be used in deriving prediction limits of the form

$$\bar{x} + d_{(n,k)} s \sqrt{1 + \frac{1}{n}} \qquad (20.12)$$

Illustration As an illustration, consider the hexavalent chromium data presented in Chapter 19 and assume, for the purpose of illustration, that they represent background conditions. For a single future sample, the 95% normal upper prediction limit (UPL) is

$$\begin{aligned} \text{UPL} &= \bar{x} + ts\sqrt{1 + \frac{1}{n}} \\ &= 36.136 + 1.812(42.515)\sqrt{1 + \tfrac{1}{11}} \\ &= 116.599 \text{ mg/L} \end{aligned}$$

Since there is a single measurement and a single future comparison (i.e., $m = 1$ and $k = 1$) the limit is based on Student's t-distribution. Note that with $k = 5$ locations, the multiplier for Student's t is 2.78 (Table 20.1) and 2.73 for multivariate t (Table 20.2), instead of 1.81 for $k = 1$. The corresponding prediction limits are 159.583 mg/L and 157.363 mg/L respectively.

20.3 NORMAL PREDICTION LIMITS FOR THE MEAN(S) OF $m > 1$ FUTURE MEASUREMENTS AT EACH OF k LOCATIONS

In certain cases, we may be interested in comparing an average concentration to background. The average concentration may be obtained as (1) the mean of a series

20.3 NORMAL PREDICTION LIMITS

Table 20.2 Dunnett multivariate t-statistics for $m = 1$: 95% overall confidence for background $n = 4$ to 100 and $k = 4$ to 50 future measurements

	k									
n	5	10	15	20	25	30	35	40	45	50
4	4.00	4.72	5.14	5.42	5.64	5.82	5.96	6.09	6.20	6.29
8	2.90	3.31	3.56	3.72	3.85	3.95	4.04	4.12	4.18	4.24
12	2.67	3.02	3.22	3.36	3.47	3.56	3.63	3.69	3.75	3.80
16	2.57	2.89	3.08	3.21	3.30	3.38	3.45	3.51	3.56	3.60
20	2.51	2.82	3.00	3.12	3.21	3.29	3.35	3.40	3.45	3.49
24	2.48	2.78	2.94	3.06	3.15	3.22	3.28	3.33	3.38	3.42
28	2.45	2.74	2.91	3.02	3.11	3.18	3.24	3.29	3.33	3.37
32	2.44	2.72	2.88	2.99	3.08	3.14	3.20	3.25	3.29	3.33
36	2.42	2.70	2.86	2.97	3.05	3.12	3.18	3.22	3.27	3.30
40	2.41	2.69	2.84	2.95	3.03	3.10	3.15	3.20	3.24	3.28
44	2.40	2.68	2.83	2.94	3.02	3.08	3.14	3.18	3.23	3.26
48	2.39	2.67	2.82	2.93	3.01	3.07	3.12	3.17	3.21	3.25
52	2.39	2.66	2.81	2.92	2.99	3.06	3.11	3.16	3.20	3.23
56	2.38	2.65	2.80	2.91	2.99	3.05	3.10	3.15	3.19	3.22
60	2.38	2.65	2.80	2.90	2.98	3.04	3.09	3.14	3.18	3.21
64	2.38	2.64	2.79	2.89	2.97	3.03	3.09	3.13	3.17	3.20
68	2.37	2.64	2.79	2.89	2.96	3.03	3.08	3.12	3.16	3.20
72	2.37	2.63	2.78	2.88	2.96	3.02	3.07	3.12	3.16	3.19
76	2.37	2.63	2.78	2.88	2.95	3.02	3.07	3.11	3.15	3.19
80	2.36	2.63	2.77	2.87	2.95	3.01	3.06	3.11	3.15	3.18
84	2.36	2.62	2.77	2.87	2.95	3.01	3.06	3.10	3.14	3.18
88	2.36	2.62	2.77	2.87	2.94	3.00	3.05	3.10	3.14	3.17
92	2.36	2.62	2.76	2.86	2.94	3.00	3.05	3.09	3.13	3.17
96	2.36	2.62	2.76	2.86	2.94	3.00	3.05	3.09	3.13	3.16
100	2.35	2.61	2.76	2.86	2.93	2.99	3.05	3.09	3.13	3.16

of different locations at a given time point, (2) the mean of a series of measurements collected over time at a single location, or (3) the mean of measurements collected over time at a number of locations. Although there are a number of statistical approaches for the comparison of mean values for normally distributed random variables [e.g., analysis of variance (ANOVA) or t-tests], these methods pool variance estimates from both background and on-site/downgradient measurements. Since contamination can affect both the absolute magnitude and variability of the concentration distribution, using potentially affected data to represent natural variability is not a judicious choice. If there is interest in comparing average on-site/downgradient concentrations to background, only the background data should be used in deriving the estimate of natural variability. Fortunately, this is exactly the case for prediction limits (i.e., β-expectation tolerance limits; Guttman, 1970) for future mean values. Since the variance of a single measurement σ^2 is much larger than the variance of the mean of m such measurements

σ^2/m, the corresponding $(1-\alpha)100\%$ prediction limit for the mean of m future values from a normal distribution

$$\bar{x} + t_{[n-1, 1-\alpha]} s \sqrt{\frac{1}{m} + \frac{1}{n}} \qquad (20.13)$$

is considerably smaller than the corresponding prediction limit for m future individual measurements. Also note that since the m measurements are now pooled into a single comparison for their mean, the total number of comparisons is reduced to a single comparison, further reducing the size of the statistical limit estimate. Of course, if we are interested in comparing the mean of m measurements at each of k locations, the Bonferroni adjusted prediction limit becomes

$$\bar{x} + t_{[n-1, 1-\alpha/k]} s \sqrt{\frac{1}{m} + \frac{1}{n}} \qquad (20.14)$$

Finally, if we are interested in comparing the grand mean averaging over both time and locations, the normal prediction limit becomes

$$\bar{x} + t_{[n-1, 1-\alpha]} s \sqrt{\frac{1}{km} + \frac{1}{n}} \qquad (20.15)$$

which once again reduces to a single comparison for the mean of the km measurements.

While the first and last cases involve a single comparison (per constituent), the second case (i.e., the mean of m samples at each of k locations) produces a nontrivial multiple comparison case in that the k locations are all compared to a common background. In this case, the correlation between the k comparisons is constant (assuming an equal number of measurements per location) with value

$$\rho = 1 \bigg/ \sqrt{\left(\frac{n}{m} + 1\right)\left(\frac{n}{m} + 1\right)} \qquad (20.16)$$

For example, with $n = 8$ background measurements and $m = 4$ on-site measurements in each of the k locations, the correlation between the comparisons is $\rho = 0.33$, as compared to $\rho = 0.11$ for a single ($m = 1$) measurement per well. As such, the coefficients obtained under the independence assumption will be more severely biased (i.e., too large) when comparing on-site/downgradient means as opposed to individual measurements to background. The variation on Dunnett's test described in Section 20.2 can be applied here as well. The tabulated values are different because the correlations change as a function of the number of measurements at each location. Table 20.3 presents the appropriate coefficients for $m = 4$ measurements at each of $k = 4 - 50$ locations. Table 20.4 presents the

20.3 NORMAL PREDICTION LIMITS

Table 20.3 Dunnett multivariate *t*-statistics for $m = 4$: 95% overall confidence for background $n = 4$ to 100 and $k = 4$ to 50 future measurements

n	\multicolumn{10}{c}{k}
	5	10	15	20	25	30	35	40	45	50
4	3.70	4.25	4.56	4.78	4.94	5.07	5.18	5.27	5.35	5.42
8	2.81	3.18	3.39	3.54	3.65	3.73	3.81	3.87	3.93	3.97
12	2.62	2.95	3.14	3.27	3.37	3.44	3.51	3.57	3.62	3.66
16	2.54	2.85	3.03	3.15	3.24	3.31	3.38	3.43	3.47	3.52
20	2.49	2.79	2.96	3.08	3.16	3.24	3.30	3.35	3.39	3.43
24	2.46	2.75	2.92	3.03	3.12	3.19	3.24	3.29	3.34	3.38
28	2.44	2.73	2.89	3.00	3.08	3.15	3.21	3.25	3.30	3.34
32	2.43	2.71	2.86	2.97	3.06	3.12	3.18	3.23	3.27	3.30
36	2.41	2.69	2.85	2.95	3.04	3.10	3.16	3.20	3.24	3.28
40	2.40	2.68	2.83	2.94	3.02	3.08	3.14	3.18	3.23	3.26
44	2.40	2.67	2.82	2.93	3.01	3.07	3.12	3.17	3.21	3.25
48	2.39	2.66	2.81	2.92	2.99	3.06	3.11	3.16	3.20	3.23
52	2.38	2.65	2.80	2.91	2.99	3.05	3.10	3.15	3.19	3.22
56	2.38	2.65	2.80	2.90	2.98	3.04	3.09	3.14	3.18	3.21
60	2.38	2.64	2.79	2.89	2.97	3.03	3.08	3.13	3.17	3.20
64	2.37	2.64	2.78	2.89	2.96	3.03	3.08	3.12	3.16	3.20
68	2.37	2.63	2.78	2.88	2.96	3.02	3.07	3.12	3.16	3.19
72	2.37	2.63	2.78	2.88	2.95	3.02	3.07	3.11	3.15	3.18
76	2.36	2.63	2.77	2.87	2.95	3.01	3.06	3.11	3.14	3.18
80	2.36	2.62	2.77	2.87	2.95	3.01	3.06	3.10	3.14	3.17
84	2.36	2.62	2.77	2.87	2.94	3.00	3.05	3.10	3.14	3.17
88	2.36	2.62	2.76	2.86	2.94	3.00	3.05	3.09	3.13	3.17
92	2.36	2.62	2.76	2.86	2.94	3.00	3.05	3.09	3.13	3.16
96	2.35	2.61	2.76	2.86	2.93	2.99	3.04	3.09	3.13	3.16
100	2.35	2.61	2.76	2.86	2.93	2.99	3.04	3.08	3.12	3.16

appropriate coefficients for $m = 8$ measurements at each of $k = 4 - 50$ locations. Table 20.5 presents the appropriate coefficients for $m = 12$ measurements at each of $k = 4 - 50$ locations.

Illustration Consider the hexavalent chromium data presented in Chapter 19 and assume for the purpose of illustration that they represent background conditions. For the mean of four future samples, the 95% normal upper prediction limit (UPL) is

$$\text{UPL} = \bar{x} + ts\sqrt{\frac{1}{m} + \frac{1}{n}}$$

$$= 36.136 + 1.812(42.515)\sqrt{\tfrac{1}{4} + \tfrac{1}{11}}$$

$$= 81.116 \text{ mg/L}$$

Table 20.4 Dunnett multivariate *t*-statistics for *m* = 8: 95% overall confidence for background *n* = 4 to 100 and *k* = 4 to 50 future measurements

					k				
n	5	10	15	20	25	30	35	40	45	50
4	3.48	3.92	4.16	4.33	4.46	4.56	4.65	4.72	4.78	4.84
8	2.73	3.05	3.23	3.36	3.45	3.53	3.59	3.64	3.69	3.73
12	2.57	2.88	3.04	3.16	3.25	3.32	3.38	3.43	3.47	3.51
16	2.51	2.80	2.96	3.07	3.16	3.23	3.28	3.33	3.37	3.41
20	2.47	2.75	2.91	3.02	3.10	3.17	3.23	3.28	3.32	3.35
24	2.44	2.72	2.88	2.99	3.07	3.14	3.19	3.24	3.28	3.32
28	2.43	2.70	2.86	2.96	3.04	3.11	3.16	3.21	3.25	3.29
32	2.41	2.69	2.84	2.95	3.03	3.09	3.14	3.19	3.23	3.27
36	2.40	2.67	2.83	2.93	3.01	3.07	3.13	3.17	3.21	3.25
40	2.39	2.66	2.81	2.92	3.00	3.06	3.11	3.16	3.20	3.23
44	2.39	2.66	2.81	2.91	2.99	3.05	3.10	3.15	3.19	3.22
48	2.38	2.65	2.80	2.90	2.98	3.04	3.09	3.14	3.18	3.21
52	2.38	2.64	2.79	2.89	2.97	3.03	3.08	3.13	3.17	3.20
56	2.37	2.64	2.79	2.89	2.96	3.03	3.08	3.12	3.16	3.20
60	2.37	2.63	2.78	2.88	2.96	3.02	3.07	3.12	3.15	3.19
64	2.37	2.63	2.78	2.88	2.95	3.01	3.07	3.11	3.15	3.18
68	2.36	2.63	2.77	2.87	2.95	3.01	3.06	3.11	3.14	3.18
72	2.36	2.62	2.77	2.87	2.94	3.01	3.06	3.10	3.14	3.17
76	2.36	2.62	2.77	2.87	2.94	3.00	3.05	3.10	3.13	3.17
80	2.36	2.62	2.76	2.86	2.94	3.00	3.05	3.09	3.13	3.16
84	2.36	2.62	2.76	2.86	2.93	3.00	3.05	3.09	3.13	3.16
88	2.35	2.61	2.76	2.86	2.93	2.99	3.04	3.09	3.12	3.16
92	2.35	2.61	2.76	2.85	2.93	2.99	3.04	3.08	3.12	3.15
96	2.35	2.61	2.75	2.85	2.93	2.99	3.04	3.08	3.12	3.15
100	2.35	2.61	2.75	2.85	2.92	2.98	3.04	3.08	3.12	3.15

Since there is a single future comparison (i.e., $k = 1$) the limit is based on Student's *t*-distribution. Now consider $k = 5$ locations. From Table 20.3 we obtain a multivariate *t* value of 2.67 and a corresponding UPL of 102.415 mg/L.

20.4 LOGNORMAL PREDICTION LIMITS FOR *m* = 1 FUTURE MEASUREMENTS AT EACH OF *k* LOCATIONS

When the distribution of the *n* background measurements is shown to be lognormal, the $(1 - \alpha^*)100\%$ lognormal prediction limit for the next *k* (i.e., $m = 1$) individual measurements is simply

$$\exp\left(\bar{y} + t_{[n-1, 1-\alpha/k]} s \sqrt{1 + \frac{1}{n}}\right) \qquad (20.17)$$

20.4 LOGNORMAL PREDICTION LIMITS

Table 20.5 Dunnett multivariate t-statistics for $m = 12$: 95% overall confidence for background $n = 4$ to 100 and $k = 4$ to 50 future measurements

n	\multicolumn{10}{c}{k}
	5	10	15	20	25	30	35	40	45	50
4	3.34	3.72	3.93	4.07	4.17	4.26	4.33	4.40	4.45	4.50
8	2.66	2.95	3.11	3.22	3.31	3.37	3.43	3.48	3.52	3.56
12	2.53	2.81	2.96	3.07	3.15	3.21	3.27	3.31	3.35	3.39
16	2.47	2.75	2.90	3.01	3.08	3.15	3.20	3.25	3.29	3.32
20	2.44	2.72	2.87	2.97	3.05	3.11	3.16	3.21	3.25	3.28
24	2.42	2.69	2.84	2.95	3.02	3.09	3.14	3.18	3.22	3.26
28	2.41	2.68	2.83	2.93	3.01	3.07	3.12	3.17	3.21	3.24
32	2.40	2.67	2.81	2.92	2.99	3.06	3.11	3.15	3.19	3.23
36	2.39	2.66	2.80	2.91	2.98	3.05	3.10	3.14	3.18	3.21
40	2.38	2.65	2.80	2.90	2.97	3.04	3.09	3.13	3.17	3.20
44	2.38	2.64	2.79	2.89	2.97	3.03	3.08	3.12	3.16	3.20
48	2.37	2.64	2.78	2.88	2.96	3.02	3.07	3.12	3.16	3.19
52	2.37	2.63	2.78	2.88	2.95	3.02	3.07	3.11	3.15	3.18
56	2.37	2.63	2.77	2.87	2.95	3.01	3.06	3.11	3.14	3.18
60	2.36	2.62	2.77	2.87	2.95	3.01	3.06	3.10	3.14	3.17
64	2.36	2.62	2.77	2.87	2.94	3.00	3.05	3.10	3.13	3.17
68	2.36	2.62	2.76	2.86	2.94	3.00	3.05	3.09	3.13	3.16
72	2.36	2.62	2.76	2.86	2.93	3.00	3.05	3.09	3.13	3.16
76	2.35	2.61	2.76	2.86	2.93	2.99	3.04	3.09	3.12	3.16
80	2.35	2.61	2.76	2.85	2.93	2.99	3.04	3.08	3.12	3.15
84	2.35	2.61	2.75	2.85	2.93	2.99	3.04	3.08	3.12	3.15
88	2.35	2.61	2.75	2.85	2.92	2.98	3.03	3.08	3.12	3.15
92	2.35	2.61	2.75	2.85	2.92	2.98	3.03	3.08	3.11	3.15
96	2.35	2.60	2.75	2.85	2.92	2.98	3.03	3.07	3.11	3.14
100	2.35	2.60	2.75	2.84	2.92	2.98	3.03	3.07	3.11	3.14

where $\bar{y}$ and s_y are the mean and standard deviation of the natural log transformed data. Unlike the problem encountered with confidence intervals for the mean, the prediction limit for individual future values can be computed directly on log-transformed data and exponentiated to obtain the appropriate limit. Alternatively, the multivariate t-statistics in Table 20.2 can be used in place of Student's t for the case of $k > 1$.

Illustration Consider the hexavalent chromium data presented in Chapter 19 and assume, for the purpose of illustration, that they represent background conditions. For a single future sample, the 95% lognormal upper prediction limit (UPL) is

$$\text{UPL} = \exp\left(\bar{y} + ts_y\sqrt{1 + \frac{1}{n}}\right)$$
$$= \exp\left[3.038 + 1.812(1.295)\sqrt{1 + \tfrac{1}{11}}\right] - 1$$
$$= 240.986 \text{ mg/L}$$

Since there is a single measurement and a single future comparison (i.e., $m = 1$ and $k = 1$) the limit is based on Student's t-distribution. Note that we subtract one from the limit because Aitchison's method was used to adjust for the one nondetect (see Table 19.2). Since Aitchison's method requires a positive random variable, one is added to each observation prior to logarithmic transformation and must be subtracted from the resulting limit estimate. Note that the lognormal limit is more than twice the size of the previously computed normal limit due to the positive skewness of the lognormal distribution. As discussed in Chapter 19, however, these data are better characterized by a lognormal distribution than a normal distribution.

20.5 LOGNORMAL PREDICTION LIMITS FOR THE MEDIAN OF $m > 1$ FUTURE MEASUREMENTS AT EACH OF k LOCATIONS

When the distribution of the n background measurements is shown to be lognormal, the $(1 - \alpha)100\%$ lognormal prediction limit for the median of the next m measurements is

$$\exp\left(\bar{y} + t_{[n-1, 1-\alpha]} s_y \sqrt{\frac{1}{m} + \frac{1}{n}}\right) \qquad (20.18)$$

where $\bar{y}$ and s_y are the mean and standard deviation of the natural log–transformed data. While in the normal case, the analogous prediction limit is for the mean; in the lognormal case, the exponentiated limit is for the median value.

When interest lies in comparing the median of m measurements simultaneously at each of k monitoring locations, the Bonferroni adjusted prediction limit becomes

$$\exp\left(\bar{y} + t_{[n-1, 1-\alpha/k]} s_y \sqrt{\frac{1}{m} + \frac{1}{n}}\right) \qquad (20.19)$$

and the more exact corresponding Dunnett-type prediction limit becomes

$$\exp\left(\bar{y} + d_{(n,m,k)} s_y \sqrt{\frac{1}{m} + \frac{1}{n}}\right) \qquad (20.20)$$

Illustration Again, returning to the hexavalent chromium data, the lognormal 95% prediction limit for the median of the $m = 4$ new measurements is

$$\begin{aligned}
\text{UPL} &= \exp\left(\bar{y} + t s_y \sqrt{\frac{1}{m} + \frac{1}{n}}\right) \\
&= \exp[3.038 + 1.812(1.295)\sqrt{\tfrac{1}{4} + \tfrac{1}{11}}] - 1 \\
&= 81.112 \text{ mg/L}
\end{aligned}$$

which is virtually identical to the normal limit for the mean of the $m = 4$ new measurements.

20.6 LOGNORMAL PREDICTION LIMITS FOR THE MEAN OF $m > 1$ FUTURE MEASUREMENTS AT EACH OF k LOCATIONS

When the data are lognormally distributed and the comparison of interest is in reference to the on-site/downgradient mean, we can use Land's coefficients to obtain an approximate $(1 - \alpha)100\%$ lognormal prediction limit for the mean of the m on-site/downgradient measurements, as

$$\exp\left(\bar{y} + 0.5s_y^2 + H_{1-\alpha} s_y \sqrt{\frac{1}{m} + \frac{1}{n}}\right) \quad (20.21)$$

When we are interested in the mean of m measurements at each of k separate locations, the Bonferroni adjusted prediction limit is

$$\exp\left(\bar{y} + 0.5s_y^2 + H_{1-\alpha/k} s_y \sqrt{\frac{1}{m} + \frac{1}{n}}\right) \quad (20.22)$$

Unfortunately, the Dunnett-type coefficients assume a symmetric multivariate normal distribution and therefore do not readily apply to the lognormal case considered here. Multivariate generalizations of Land's (1971) approach have not been considered in the statistical literature.

To examine the statistical properties of this Land-type prediction limit for a future lognormal mean, a limited simulation study was conducted. Letting $\bar{y}$ and s_y^2 denote the mean and variance of the natural log transformed n background measurements and $\hat{\mu}$ and $\hat{\sigma}^2$ denote the MVUE of the lognormal mean and variance (see Equations 19.20 and 19.21), we examined the actual confidence level obtained via simulation for a 95% confidence UPL for the mean of $m = 4$ new samples based on background sample sizes of $n = 8$, 16, and 32, $\bar{y} = 1.0$ and 5.0, and $s_y^2 = 0.1$, 0.5, and 1.0. The simulated confidence levels are displayed in Table 20.6. Inspection of Table 20.6 reveals that in all cases, the simulated confidence levels were in reasonably close agreement with the intended nominal rate of 0.95. The 95% UPLs tended to be somewhat conservative in that in all cases, they had confidence levels in excess of 95%. The actual confidence level appeared to be proportional to the magnitude of s_y^2, but even at $s_y^2 = 1$, the simulated confidence level was no more than 0.986, even with only $n = 8$ background samples. As such, the approximate method is not recommended for cases in which $s_y^2 > 1$ (which is illustrated in the following example). There was no effect of increasing the mean on the actual confidence level obtained. This is an important area for future statistical research.

Table 20.6 Simulated confidence levels for a land-type 95% confidence UPL for a future lognormal mean of $m = 4$ measurements as a function of the background mean $\bar{y}$, variance s_y^2, and number of background measurements n

	$s_y^2 = 0.1$			$s_y^2 = 0.5$			$s_y^2 = 1.0$		
	$n = 8$	$n = 16$	$n = 32$	$n = 8$	$n = 16$	$n = 32$	$n = 8$	$n = 16$	$n = 32$
$\bar{y} = 1$									
$\hat{\mu}$	2.390	2.848	2.853	3.366	3.425	3.457	4.130	4.291	4.382
$\hat{\sigma}$	0.902	0.913	0.920	2.432	2.588	2.688	4.297	4.878	5.293
95% UPL	4.278	3.989	3.866	11.489	8.630	7.682	36.525	20.320	16.127
Conf.	0.956	0.956	0.963	0.977	0.972	0.976	0.986	0.982	0.984
$\bar{y} = 5$									
$\hat{\mu}$	155.018	155.516	155.768	183.772	187.026	188.753	225.479	234.302	239.252
$\hat{\sigma}$	49.254	49.855	50.206	132.812	141.314	146.766	234.622	266.307	289.006
95% UPL	233.548	217.769	211.081	627.283	471.159	419.435	1994.179	1109.408	880.489
Conf.	0.956	0.956	0.963	0.977	0.972	0.976	0.986	0.982	0.984

Illustration Applying Land's method to the problem of predicting the mean of $m = 4$ new hexavalent chromium measurements based on the $n = 11$ background measurements used in the previous illustrations yields the 95% confidence

$$\begin{aligned}
\text{UPL} &= \exp\left(\bar{y} + 0.5 s_y^2 + H_{1-\alpha} s_y \sqrt{\frac{1}{m} + \frac{1}{n}}\right) \\
&= \exp\left[3.038 + 0.5(1.295^2) + 3.603(1.295)\sqrt{\tfrac{1}{4} + \tfrac{1}{11}}\,\right] - 1 \\
&= 734.095 \text{ mg/L}
\end{aligned}$$

which is almost an order of magnitude larger than the limit for the median of the four future measurements. Note that as the number of background measurements (n) gets small and the standard deviation (s_y) gets large (i.e., $s_y > 1.0$), the lognormal prediction limits for the mean can be so large as to be of little practical value. In these cases, it may be more useful to use a UPL for the median or one of the nonparametric alternatives discussed in the following sections.

20.7 NONPARAMETRIC PREDICTION LIMITS FOR $m = 1$ FUTURE MEASUREMENTS IN EACH OF k LOCATIONS

In the nonparametric case, prediction limits for k future samples based on a background of n measurements have now received considerable attention (Chou and Owen, 1986; Davis and McNichols, 1994, 1999; Gibbons, 1990, 1991, 1994; Hall et al., 1975). The statistical foundation for nonparametric prediction limits was presented in Chapter 19 for the case of passage of 1 of m samples in each of k monitoring locations. In the context of assessment/corrective action monitoring, we are more typically concerned with the comparison of each of k individual on-site/downgradient samples to the background prediction limit. In the nonparametric case, the background prediction limit is most typically the largest or second-largest background value. Table 20.7 presents the probability that all k on-site/downgradient samples will be below the maximum of n background measurements. Table 20.8 presents the probability that all k on-site/downgradient samples will be below the second largest of n background measurements.

Illustration For the hexavalent chromium data, the maximum value is 150 mg/L and with $n = 11$ and $m = k = 1$ (i.e., a single future sample), the associated confidence level is 0.917.

20.8 NONPARAMETRIC PREDICTION LIMITS FOR THE MEDIAN OF $m > 1$ FUTURE MEASUREMENTS AT EACH OF k LOCATIONS

In the nonparametric case, we can also construct a prediction bound for the median of m measurements at each of k on-site/downgradient locations based on a

Table 20.7 Probability that all k samples ($m = 1$) will be below the maximum of n background measurements

Number of Monitoring Wells, k

Previous n	1	2	3	4	5	6	7	8	9	10	11	12	13	14	15
4	.8000	.6667	.5714	.5000	.4444	.4000	.3636	.3333	.3077	.2857	.2667	.2500	.2353	.2222	.2105
5	.8333	.7143	.6250	.5556	.5000	.4545	.4167	.3846	.3571	.3333	.3125	.2941	.2778	.2632	.2500
6	.8571	.7500	.6667	.6000	.5455	.5000	.4615	.4286	.4000	.3750	.3529	.3333	.3158	.3000	.2857
7	.8750	.7778	.7000	.6364	.5833	.5385	.5000	.4667	.4375	.4118	.3889	.3684	.3500	.3333	.3182
8	.8889	.8000	.7273	.6667	.6154	.5714	.5333	.5000	.4706	.4444	.4211	.4000	.3810	.3636	.3478
9	.9000	.8182	.7500	.6923	.6429	.6000	.5625	.5294	.5000	.4737	.4500	.4286	.4091	.3913	.3750
10	.9091	.8333	.7692	.7143	.6667	.6250	.5882	.5556	.5263	.5000	.4762	.4545	.4348	.4167	.4000
11	.9167	.8462	.7857	.7333	.6875	.6471	.6111	.5789	.5500	.5238	.5000	.4783	.4583	.4400	.4231
12	.9231	.8571	.8000	.7500	.7059	.6667	.6316	.6000	.5714	.5455	.5217	.5000	.4800	.4615	.4444
13	.9286	.8667	.8125	.7647	.7222	.6842	.6500	.6190	.5909	.5652	.5417	.5200	.5000	.4815	.4643
14	.9333	.8750	.8235	.7778	.7368	.7000	.6667	.6364	.6087	.5833	.5600	.5385	.5185	.5000	.4828
15	.9375	.8824	.8333	.7895	.7500	.7143	.6818	.6522	.6250	.6000	.5769	.5556	.5357	.5172	.5000
16	.9412	.8889	.8421	.8000	.7619	.7273	.6957	.6667	.6400	.6154	.5926	.5714	.5517	.5333	.5161
17	.9444	.8947	.8500	.8095	.7727	.7391	.7083	.6800	.6538	.6296	.6071	.5862	.5667	.5484	.5313
18	.9474	.9000	.8571	.8182	.7826	.7500	.7200	.6923	.6667	.6429	.6207	.6000	.5806	.5625	.5455
19	.9500	.9048	.8636	.8261	.7917	.7600	.7308	.7037	.6786	.6552	.6333	.6129	.5938	.5758	.5588
20	.9524	.9091	.8696	.8333	.8000	.7692	.7407	.7143	.6897	.6667	.6452	.6250	.6061	.5882	.5714
25	.9615	.9259	.8929	.8621	.8333	.8065	.7813	.7576	.7353	.7143	.6944	.6757	.6579	.6410	.6250
30	.9677	.9375	.9091	.8824	.8571	.8333	.8108	.7895	.7692	.7500	.7317	.7143	.6977	.6818	.6667
35	.9722	.9459	.9211	.8974	.8750	.8537	.8333	.8140	.7955	.7778	.7609	.7447	.7292	.7143	.7000
40	.9756	.9524	.9302	.9091	.8889	.8696	.8511	.8333	.8163	.8000	.7843	.7692	.7547	.7407	.7273
45	.9783	.9574	.9375	.9184	.9000	.8824	.8654	.8491	.8333	.8182	.8036	.7895	.7759	.7627	.7500
50	.9804	.9615	.9434	.9259	.9091	.8929	.8772	.8621	.8475	.8333	.8197	.8065	.7937	.7813	.7692
60	.9836	.9677	.9524	.9375	.9231	.9091	.8955	.8824	.8696	.8571	.8451	.8333	.8219	.8108	.8000
70	.9859	.9722	.9589	.9459	.9333	.9211	.9091	.8974	.8861	.8750	.8642	.8537	.8434	.8333	.8235
80	.9877	.9756	.9639	.9524	.9412	.9302	.9195	.9091	.8989	.8889	.8791	.8696	.8602	.8511	.8421
90	.9890	.9783	.9677	.9574	.9474	.9375	.9278	.9184	.9091	.9000	.8911	.8824	.8738	.8654	.8571
100	.9901	.9804	.9709	.9615	.9524	.9434	.9346	.9259	.9174	.9091	.9009	.8929	.8850	.8772	.8696

(*Continued*)

Table 20.7 (Continued)

Previous n	\multicolumn{17}{c}{Number of Monitoring Wells, k}																
	20	25	30	35	40	45	50	55	60	65	70	75	80	90	100
4	.1667	.1379	.1176	.1026	.0909	.0816	.0741	.0678	.0625	.0580	.0541	.0506	.0476	.0426	.0385
5	.2000	.1667	.1429	.1250	.1111	.1000	.0909	.0833	.0769	.0714	.0667	.0625	.0588	.0526	.0476
6	.2308	.1935	.1667	.1463	.1304	.1176	.1071	.0984	.0909	.0845	.0789	.0741	.0698	.0625	.0566
7	.2593	.2188	.1892	.1667	.1489	.1346	.1228	.1129	.1045	.0972	.0909	.0854	.0805	.0722	.0654
8	.2857	.2424	.2105	.1860	.1667	.1509	.1379	.1270	.1176	.1096	.1026	.0964	.0909	.0816	.0741
9	.3103	.2647	.2308	.2045	.1837	.1667	.1525	.1406	.1304	.1216	.1139	.1071	.1011	.0909	.0826
10	.3333	.2857	.2500	.2222	.2000	.1818	.1667	.1538	.1429	.1333	.1250	.1176	.1111	.1000	.0909
11	.3548	.3056	.2683	.2391	.2157	.1964	.1803	.1667	.1549	.1447	.1358	.1279	.1209	.1089	.0991
12	.3750	.3243	.2857	.2553	.2308	.2105	.1935	.1791	.1667	.1558	.1463	.1379	.1304	.1176	.1071
13	.3939	.3421	.3023	.2708	.2453	.2241	.2063	.1912	.1781	.1667	.1566	.1477	.1398	.1262	.1150
14	.4118	.3590	.3182	.2857	.2593	.2373	.2188	.2029	.1892	.1772	.1667	.1573	.1489	.1346	.1228
15	.4286	.3750	.3333	.3000	.2727	.2500	.2308	.2143	.2000	.1875	.1765	.1667	.1579	.1429	.1304
16	.4444	.3902	.3478	.3137	.2857	.2623	.2424	.2254	.2105	.1975	.1860	.1758	.1667	.1509	.1379
17	.4595	.4048	.3617	.3269	.2982	.2742	.2537	.2361	.2208	.2073	.1954	.1848	.1753	.1589	.1453
18	.4737	.4186	.3750	.3396	.3103	.2857	.2647	.2466	.2308	.2169	.2045	.1935	.1837	.1667	.1525
19	.4872	.4318	.3878	.3519	.3220	.2969	.2754	.2568	.2405	.2262	.2135	.2021	.1919	.1743	.1597
20	.5000	.4444	.4000	.3636	.3333	.3077	.2857	.2667	.2500	.2353	.2222	.2105	.2000	.1818	.1667
25	.5556	.5000	.4545	.4167	.3846	.3571	.3333	.3125	.2941	.2778	.2632	.2500	.2381	.2174	.2000
30	.6000	.5455	.5000	.4615	.4286	.4000	.3750	.3529	.3333	.3158	.3000	.2857	.2727	.2500	.2308
35	.6364	.5833	.5385	.5000	.4667	.4375	.4118	.3889	.3684	.3500	.3333	.3182	.3043	.2800	.2593
40	.6667	.6154	.5714	.5333	.5000	.4706	.4444	.4211	.4000	.3810	.3636	.3478	.3333	.3077	.2857
45	.6923	.6429	.6000	.5625	.5294	.5000	.4737	.4500	.4286	.4091	.3913	.3750	.3600	.3333	.3103
50	.7143	.6667	.6250	.5882	.5556	.5263	.5000	.4762	.4545	.4348	.4167	.4000	.3846	.3571	.3333
60	.7500	.7059	.6667	.6316	.6000	.5714	.5455	.5217	.5000	.4800	.4615	.4444	.4286	.4000	.3750
70	.7778	.7368	.7000	.6667	.6364	.6087	.5833	.5600	.5385	.5185	.5000	.4828	.4667	.4375	.4118
80	.8000	.7619	.7273	.6957	.6667	.6400	.6154	.5926	.5714	.5517	.5333	.5161	.5000	.4706	.4444
90	.8182	.7826	.7500	.7200	.6923	.6667	.6429	.6207	.6000	.5806	.5625	.5455	.5294	.5000	.4737
100	.8333	.8000	.7692	.7407	.7143	.6897	.6667	.6452	.6250	.6061	.5882	.5714	.5556	.5263	.5000

Source: Reproduced with permission from R. D. Gibbons, 1994.
"Statistical Methods for Groundwater Monitoring," Wiley, New York. Originally prepared by Charles Davis based on results in Davis and McNichols (1999).

Table 20.8 Probability that all k samples ($m = 1$) will be below the second largest of n background measurements

Previous n	Number of Monitoring Wells, k														
	1	2	3	4	5	6	7	8	9	10	11	12	13	14	15
4	.9333	.8810	.8381	.8020	.7710	.7439	.7199	.6984	.6791	.6614	.6453	.6304	.6167	.6039	.5920
5	.9524	.9127	.8788	.8493	.8232	.7999	.7788	.7597	.7422	.7260	.7111	.6972	.6842	.6720	.6606
6	.9643	.9333	.9061	.8817	.8598	.8398	.8215	.8046	.7890	.7745	.7609	.7481	.7362	.7248	.7141
7	.9722	.9475	.9252	.9049	.8863	.8692	.8533	.8385	.8246	.8116	.7994	.7878	.7769	.7665	.7566
8	.9778	.9576	.9391	.9220	.9061	.8913	.8775	.8645	.8522	.8406	.8296	.8192	.8092	.7997	.7906
9	.9818	.9650	.9495	.9349	.9213	.9084	.8963	.8849	.8740	.8636	.8538	.8443	.8353	.8266	.8183
10	.9848	.9707	.9574	.9449	.9331	.9219	.9112	.9011	.8914	.8822	.8733	.8648	.8566	.8487	.8411
11	.9872	.9751	.9637	.9528	.9425	.9326	.9232	.9142	.9056	.8973	.8893	.8816	.8741	.8670	.8600
12	.9890	.9786	.9686	.9591	.9500	.9413	.9330	.9249	.9172	.9097	.9025	.8955	.8888	.8822	.8759
13	.9905	.9814	.9727	.9643	.9562	.9485	.9410	.9338	.9268	.9201	.9136	.9072	.9011	.8951	.8893
14	.9917	.9837	.9760	.9685	.9614	.9544	.9477	.9412	.9349	.9288	.9229	.9171	.9115	.9060	.9007
15	.9926	.9856	.9787	.9721	.9656	.9594	.9534	.9475	.9418	.9362	.9308	.9255	.9204	.9154	.9105
16	.9935	.9871	.9810	.9750	.9693	.9636	.9582	.9528	.9476	.9426	.9376	.9328	.9281	.9235	.9190
17	.9942	.9885	.9829	.9776	.9723	.9672	.9623	.9574	.9527	.9480	.9435	.9391	.9347	.9305	.9263
18	.9947	.9896	.9846	.9797	.9750	.9703	.9658	.9613	.9570	.9527	.9486	.9445	.9405	.9366	.9327
19	.9952	.9906	.9860	.9816	.9773	.9730	.9689	.9648	.9608	.9569	.9530	.9493	.9456	.9419	.9384
20	.9957	.9914	.9873	.9832	.9793	.9754	.9715	.9678	.9641	.9605	.9569	.9534	.9500	.9466	.9433
25	.9972	.9943	.9916	.9889	.9862	.9835	.9809	.9783	.9758	.9733	.9708	.9683	.9659	.9635	.9612
30	.9980	.9960	.9940	.9921	.9901	.9882	.9863	.9844	.9826	.9808	.9789	.9771	.9753	.9736	.9718
35	.9985	.9970	.9955	.9941	.9926	.9912	.9897	.9883	.9869	.9855	.9841	.9827	.9814	.9800	.9787
40	.9988	.9977	.9965	.9954	.9943	.9931	.9920	.9909	.9898	.9887	.9876	.9865	.9854	.9844	.9833
45	.9991	.9982	.9972	.9963	.9954	.9945	.9936	.9927	.9918	.9910	.9901	.9892	.9883	.9875	.9866
50	.9992	.9985	.9977	.9970	.9963	.9955	.9948	.9941	.9933	.9926	.9919	.9912	.9904	.9897	.9890
60	.9995	.9989	.9984	.9979	.9974	.9969	.9963	.9958	.9953	.9948	.9943	.9938	.9932	.9927	.9922
70	.9996	.9992	.9988	.9984	.9981	.9977	.9973	.9969	.9965	.9961	.9957	.9954	.9950	.9946	.9942
80	.9997	.9994	.9991	.9988	.9985	.9982	.9979	.9976	.9973	.9970	.9967	.9964	.9961	.9958	.9955
90	.9998	.9995	.9993	.9990	.9988	.9986	.9983	.9981	.9979	.9976	.9974	.9972	.9969	.9967	.9965
100	.9998	.9996	.9994	.9992	.9990	.9988	.9986	.9985	.9983	.9981	.9979	.9977	.9975	.9973	.9971

(*Continued*)

Table 20.8 (Continued)

Previous n	Number of Monitoring Wells, k															
	20	25	30	35	40	45	50	55	60	65	70	75	80	90	100
4	.5424	.5045	.4741	.4491	.4279	.4096	.3937	.3796	.3670	.3557	.3454	.3360	.3274	.3120	.2988
5	.6121	.5741	.5431	.5171	.4949	.4756	.4586	.4434	.4298	.4174	.4062	.3959	.3864	.3694	.3546
6	.6680	.6310	.6003	.5743	.5517	.5320	.5144	.4987	.4845	.4715	.4597	.4488	.4387	.4205	.4046
7	.7133	.6779	.6481	.6226	.6002	.5805	.5628	.5469	.5324	.5191	.5069	.4956	.4852	.4663	.4496
8	.7504	.7170	.6884	.6637	.6418	.6224	.6049	.5890	.5744	.5611	.5488	.5373	.5267	.5073	.4902
9	.7812	.7497	.7226	.6989	.6777	.6588	.6416	.6259	.6115	.5983	.5860	.5745	.5638	.5443	.5269
10	.8068	.7775	.7518	.7292	.7089	.6905	.6738	.6585	.6444	.6313	.6191	.6077	.5971	.5776	.5601
11	.8284	.8011	.7769	.7554	.7360	.7184	.7022	.6873	.6736	.6607	.6488	.6376	.6270	.6077	.5903
12	.8468	.8213	.7986	.7782	.7597	.7428	.7273	.7129	.6995	.6871	.6754	.6644	.6540	.6349	.6177
13	.8625	.8387	.8174	.7982	.7806	.7645	.7496	.7357	.7228	.7107	.6993	.6886	.6784	.6597	.6427
14	.8759	.8538	.8339	.8157	.7991	.7837	.7694	.7561	.7436	.7319	.7209	.7104	.7005	.6822	.6656
15	.8876	.8670	.8483	.8312	.8154	.8007	.7871	.7743	.7623	.7510	.7403	.7302	.7206	.7028	.6865
16	.8978	.8786	.8610	.8449	.8299	.8160	.8030	.7907	.7792	.7683	.7580	.7482	.7389	.7215	.7056
17	.9067	.8888	.8723	.8571	.8429	.8296	.8172	.8055	.7945	.7840	.7741	.7646	.7556	.7387	.7232
18	.9145	.8978	.8823	.8679	.8545	.8419	.8301	.8189	.8083	.7982	.7887	.7795	.7708	.7545	.7394
19	.9214	.9057	.8912	.8777	.8649	.8530	.8417	.8310	.8209	.8112	.8020	.7932	.7848	.7689	.7543
20	.9275	.9129	.8992	.8864	.8743	.8630	.8522	.8420	.8323	.8230	.8142	.8057	.7976	.7822	.7680
25	.9498	.9390	.9288	.9190	.9098	.9009	.8924	.8843	.8765	.8689	.8617	.8547	.8479	.8350	.8229
30	.9633	.9551	.9472	.9397	.9324	.9254	.9187	.9121	.9058	.8996	.8937	.8879	.8823	.8715	.8613
35	.9721	.9657	.9595	.9535	.9477	.9421	.9366	.9313	.9261	.9210	.9161	.9113	.9066	.8975	.8889
40	.9781	.9730	.9680	.9631	.9584	.9538	.9493	.9449	.9406	.9364	.9323	.9283	.9243	.9167	.9093
45	.9823	.9782	.9741	.9701	.9662	.9624	.9586	.9550	.9514	.9478	.9444	.9409	.9376	.9310	.9247
50	.9855	.9820	.9786	.9753	.9720	.9688	.9657	.9626	.9595	.9565	.9535	.9506	.9477	.9421	.9367
60	.9897	.9872	.9848	.9824	.9800	.9776	.9753	.9730	.9707	.9685	.9663	.9641	.9619	.9577	.9535
70	.9923	.9905	.9886	.9868	.9850	.9832	.9814	.9797	.9779	.9762	.9745	.9728	.9711	.9678	.9646
80	.9941	.9926	.9912	.9898	.9883	.9869	.9855	.9842	.9828	.9814	.9801	.9787	.9774	.9748	.9722
90	.9953	.9941	.9930	.9918	.9907	.9896	.9884	.9873	.9862	.9851	.9840	.9829	.9818	.9797	.9776
100	.9962	.9952	.9943	.9933	.9924	.9915	.9905	.9896	.9887	.9878	.9869	.9860	.9851	.9833	.9816

Source: Reproduced with permission from R. D. Gibbons, 1994.
"Statistical Methods for Groundwater Monitoring." Wiley, New York. Originally prepared by Charles Davis based on results in Davis and McNichols (1999).

background of n samples. The idea is to identify the order statistic of the n background measurements that will provide $(1 - \alpha^*)100\%$ confidence of not being exceeded by any more than 50% of the m on-site/downgradient measurements at each of the k locations. Fligner and Wolfe (1979), Guilbaud (1983), and Hahn and Meeker (1991) illustrate how the inverse hypergeometric distribution (Guenther, 1975) can be used to identify the appropriate order statistic of the n background measurements that will provide the desired level of confidence $1 - \alpha^*$, for given values of n and m. The inverse hypergeometric distribution is computed as the function

$$H^*(l, u, r, m, n) = \sum_{i=1}^{u} h^*(i, r + i, m, n) \tag{20.23}$$

where

$$h^*(i, r + i, m, n) = \frac{\binom{r-1}{i}\binom{n-r}{n-i}}{\binom{n}{m}} \tag{20.24}$$

and l is the lowest and u the highest order statistic in the current interval, r is the median rank of the m on-site/downgradient samples, and n is the number of background measurements. To obtain a one-sided upper prediction limit, we iteratively solve for

$$H^*(l, u - 1, r, m, n) \geq 1 - \alpha^* \tag{20.25}$$

setting $l = 0$ and beginning from $u = r + 1$.

Illustration For the hexavalent chromium data, the tenth largest value out of the $n = 11$ measurements provides a confidence level of 0.967 of including the median of the next $m = 4$ samples. The UPL is the corresponding concentration (i.e., 60 mg/L).

21

ASSESSMENT AND CORRECTIVE ACTION MONITORING: CASE STUDIES

21.1 CASE STUDY 1: LONG-TERM MONITORING

This example illustrates the use of statistical procedures at a site undergoing long-term monitoring. At the X Site (Site), manufacturing operations have ceased, all production equipment has been relocated, and the Site has been sold to a third party.

Hexavalent chromium was detected in soil and groundwater in the vicinity of a former chromium plating area located in the southeast quadrant of the facility. Site investigations defined the distribution of chemical constituents and physical characteristics of this area, and an interim groundwater remediation system was designed and installed as the primary remedial action component for groundwater. The system includes a blasted bedrock trench, a groundwater recovery system with electrical controls, a remote telemetry unit (RTU), and an ion-exchange groundwater treatment system that discharges treated groundwater to the sanitary sewer system.

A baseline sampling event occurred August 1 through 5, 1996, prior to remedial system startup and permanent operation. After initiation of permanent operation, selected wells were monitored biweekly for the first month of operation and quarterly thereafter. Other selected wells are monitored on an annual basis. The groundwater protection standard for total chromium is 50 parts per billion (ppb) in a groundwater protection area.

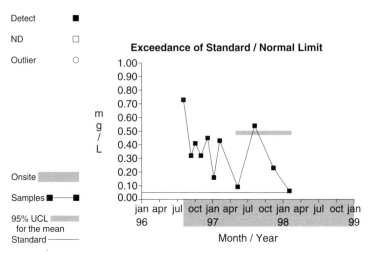

Figure 21.1 Case study 1: total chromium in well HA12A-ROW.

Following the previously defined monitoring strategy, we compare the 95% UCL for the mean concentration for the last four available measurements for each monitoring well and constituent to the relevant criterion. Additionally, we test for both increasing and decreasing trend in all the data for each well and constituent using Sen's test. To illustrate the method, data for total chromium in monitoring well HA12A-ROW were analyzed (see Figure 21.1 and Table 21.1).

Inspection of the graph and worksheet reveal that (1) there is a decreasing trend in the well, which demonstrates the beneficial effects of the remediation; (2) the most current measurement is now close to the cleanup criterion of 50 ppb; and (3) the UCL for the mean concentration is 488 ppb, which is still an order of magnitude above the criterion, indicating that remediation should continue.

21.2 CASE STUDY 2: SOIL PAOC AND SOIL PHASE III EVALUATION

At the Y Site, two areas exhibited elevated levels of soil lead concentrations: PAOC 2 and PAOC 18. This case study will illustrate how statistics are used to evaluate both a single PAOC to determine if a standard/criteria exceedance exists (see Figure 18.2), and to evaluate the phase III extent of contamination to determine the status of the entire "plume" of soil contamination (see Figure 18.3).

21.2.1. PAOC 2

A portion of the area may have been used as a plating waste sludge disposal area between 1942 and 1944. The capacity of the pond was approximately 5 million gallons. The entire area had been filled and graded by 1955. The area was subsequently developed and is now covered by the eastern portion of the assembly building, concrete

Table 21.1 Worksheet: Comparison to standard, total chrome at HA12A-ROW, normal confidence limit

Step	Equation	Description
1	Percentile = mean of the on-site/downgradient distribution	
2	$\bar{x} = \dfrac{\Sigma x}{m} = \dfrac{0.924}{4} = 0.231$	Compute the mean of the m on-site/downgradient measurements.
3	$s = \left[\dfrac{(\Sigma x^2 - \Sigma(x)^2/m)}{m-1} \right]^{1/2} = \left(\dfrac{0.357 - 0.854/4}{4-1} \right)^{1/2} = 0.219$	Compute the standard deviation of the m on-site/downgradient measurements.
4	$\text{UCL} = \bar{x} + \dfrac{ts}{m^{1/2}} = 0.231 + \dfrac{2.353 \times 0.219}{4^{1/2}} = 0.488$	Compute the upper confidence limit for the mean of the m on-site/downgradient measurements.
5	Confidence = 0.95	Report the confidence level for this location and constituent.

paved parking areas, the main electrical substation, and switch gear house. Investigations were conducted between June 1995 and April 1996 to determine whether regulated constituents were present in soil or groundwater at levels that could pose an unacceptable risk to human health or the environment at this location. Based on the results of the initial investigation, additional investigation was implemented to characterize the vertical and horizontal extent of lead impacts to unsaturated soil.

In April 1996, 26 soil borings were installed to delineate the extent of lead-impacted soil. Lead contamination in soil in excess of the direct contact criteria (DCC) was defined to the north, west, and south. The southeastern limit of contamination was not determined. Concentrations of lead in soil ranged from BDL to 40,000 mg/kg.

Two additional soil borings (SB-29 and SB-30) were installed in September 1996 in an unsuccessful attempt to define the southeastern extent of lead contamination. In December 1996, four additional borings (SB-31 to SB-34) were installed to define the southeastern extent of the lead-impacted soil. The area of lead-impacted soil merged with an area of lead-impacted soil on the O portion of the property. Borings to define the extent of lead-impacted soil at the O property were installed concurrently and definition of the affected soil was completed.

21.2.2 PAOC 18

"Paint sludge" was dumped in the area between the fence line and pond at the O property during the 1940s and 1950s. In February 1994, a response activity was implemented to remove an estimated 90 cubic yards of visible surface waste material, including approximately 30 drum remnants, from this area. Waste and soil were removed until visually clean underlying soils were encountered.

In May 1994, additional soil samples were collected from the excavated areas in an attempt to further assess the horizontal and vertical extent of lead contamination. Based on results of the initial investigation, additional studies were performed in December 1996 and March 1997 to define the horizontal and vertical extent of lead-impacted soil. Concentrations of lead in soil ranged form 1.1 to 170,000 mg/kg. The area of lead impacts above the generic industrial DCC has been defined to the south and east. Impacts to the west coalesce with lead-impacted soil in the former area and impacts to the north extend to the bank of the pond.

Relevant criteria includes the residential/commercial DCC of 400 mg/kg, the industrial DCC of 900 mg/kg, and the commercial subcategory III/IV DCC of 400 mg/kg (other criteria are less stringent).

Analysis of these data involves comparison of on-site mean concentrations in each PAOC to both background and relevant criteria, depending on which is larger. The LCL is used to compare to criteria because interest is in assessing whether or not these areas are affected by lead contamination.

21.2.3 Results

For comparison to background, the first step is to determine the best-fitting distribution (normal, lognormal, or nonparametric). Analysis of the background data ($n = 17$) revealed that a lognormal distribution fit the data (i.e., $G = 1.914$ critical value $= 2.326$), whereas the normal distribution did not fit the data ($G = 2.788$). The lognormal 95% confidence prediction limit for the geometric mean of the 28 samples from SB-2 is computed as 25.7 mg/kg and the prediction limit for the on-site arithmetic mean is 175.6 mg/kg. The actual geometric mean of the 28 on-site measurements is 202.1 mg/kg and the arithmetic mean is 1008.6 mg/kg, both of which exceed the background prediction limit. The lognormal 95% confidence prediction limit for the geometric mean of the 43 samples from SB-18 is computed as 24.1 mg/kg, and the prediction limit for the arithmetic mean is 154.9. The actual geometric mean of the 43 on-site measurements is 127.7 mg/kg, and the arithmetic mean is 1102.3 mg/kg, both of which exceed the background prediction limit. These results clearly reveal the potentially large difference in the median and mean of a lognormal distribution and corresponding statistical limit estimates.

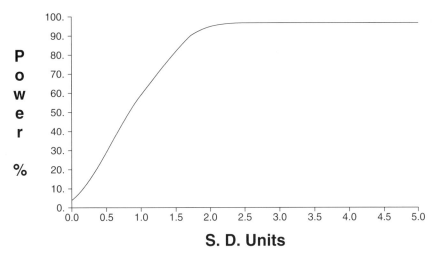

Figure 21.2 Case study 2: statistical power curve.

Statistical power computations reveal that the overall false positive rate is 5% and the test is sensitive to an on-site arithmetic mean increase of 1.5 to 2.0 standard deviation units over background (see Figure 21.2).

Since the lowest relevant criterion is 400 mg/kg and both background prediction limits are below this value, the appropriate comparison is to the criterion and not background for both SB-2 and SB-18. Analysis of the on-site data for SB-2 ($n = 28$) revealed that a lognormal ($W = 0.95$) distribution fit these data adequately (critical value $W = 0.92$), but a normal distribution did not ($W = 0.47$); therefore, a lognormal LCL was computed. Analysis of the on-site data for SB-18 ($n = 43$) revealed that neither normal ($W = 0.57$) nor lognormal ($W = 0.92$) distributions fit these data adequately (critical value $W = 0.94$); therefore, a nonparametric LCL was computed.

For SB-2 the lognormal LCL for the arithmetic mean is 758.0 mg/kg (arithmetic mean = 1008.6 mg/kg), which exceeds the residential and commercial DCC of 400 mg/kg but not the industrial DCC of 900 mg/kg. For SB-18 the nonparametric LCL for the 50th percentile (i.e., median) for a sample of size 43 is given by the 16th largest value, which in this case is 23 mg/kg, which is below the residential/commercial criterion of 400 mg/kg.

21.3 CASE STUDY 3: SITE-WIDE GROUNDWATER EVALUATION

The site includes a building approximately 25,000 square feet in size surrounded by asphalt pavement. An underground storage tank farm was formerly located near the northwestern corner of the building. The tanks and associated piping were removed in 1994. During the tank removal activities, soil and groundwater impacts were observed, possibly involving benzene. A phase I hydrogeologic investigation was performed in September 1994. The impacted soils were excavated and addressed under the Tier I levels for risk-based corrective action (RBCA) at leaking underground storage tank (LUST) sites. An additional investigation, including a geoprobe survey, was performed in April 1996 to determine the extent of affected groundwater. A final assessment report (FAR) was prepared for the site in October 1996. As part of the corrective action plan (CAP) described in the FAR, four quarters of monitoring were completed to determine if remediation by natural attenuation (RNA) is occurring at the site. Monitoring began in December 1996 and continued through November 1997.

One of the ways to evaluate groundwater data is to perform statistical analysis on all downgradient data to determine if criteria have been exceeded. In this case, data from the 1994 direct-push sampling event were evaluated to determine if groundwater concentrations exceeds relevant criteria (5 μg/L health-based residential and commercial, 53 μg/L GSI, and 9300 μg/L direct contact).

Analysis of these data involved computation of the 95% UCL for all 41 on-site measurements, which in turn can be used to determine if the relevant criteria have been exceeded. The UCL is used given that the site is being evaluated following corrective action and 95% confidence is required to ensure that the true concentration is below the relevant criterion. To begin, the Shapiro–Wilk test was used to test for normality and lognormality of the 36 detected benzene concentrations (i.e., there were five nondetects). The resulting W statistics were $W = 0.52$ for normal and $W = 0.90$ for lognormal with a 95% critical value of $W = 0.94$. Since both normal and lognormal W statistics are less than the critical value, both normality and lognormality are rejected at the 5% level. In light of this, a nonparametric UCL is used for this comparison (see Figure 21.3). The 95% confidence UCL for the 50th percentile is the 27th largest of the 41 on-site measurements, which is 26 μg/L, which exceeds the residential and commercial criterion but not the GSI or direct contact criteria.

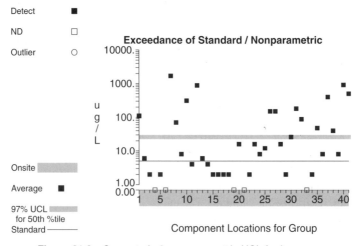

Figure 21.3 Case study 3: nonparametric UCL for benzene.

21.3 CASE STUDY 3: SITE-WIDE GROUNDWATER EVALUATION

For completeness, the computation is also illustrated using normal and lognormal UCLs. The 95% confidence normal UCL for the arithmetic mean is 225.4 μg/L and the 95% confidence lognormal UCL for the arithmetic mean is 606.3 μg/L. These results would have led to exceedance of all but the direct contact criterion.

In all cases, care must be taken to select the appropriate area for statistical evaluation. The investigator must be clear about the issues and exposure pathways involved in order to apply statistical principles correctly. Inspection of Figure 21.3 reveals considerable spatial variability in Benzene concentrations accross the 41 locations. This finding suggests that the site should be further subdivided into more homogenous PAOCS.

22

REVIEW OF AVAILABLE COMPUTER SOFTWARE

Surprisingly little computer software is available for environmental statistical applications in general and the estimation of detection and quantification limits in particular. For the latter, the programs AML and DETECT are available commercially from Discerning Systems in Vancouver, British Columbia, Canada (www.dsi-software.com). AML is a Windows program that computes the alternative minimum level as described by Gibbons et al. (1997a) and provides both tabular and graphical output of the recovery curve, variance function, and a worksheet detailing the intermediate computations. DETECT is similar to AML but provides the majority of detection and quantification limit estimators described in this book.

With respect to the statistical applications described in the book, there is somewhat more available software, but not much more. DUMPStat, which provides automated detection monitoring statistical analysis using the methods described in Chapter 17 and the ASTM Standard D6312, is distributed by Discerning Systems (www.dsi-software.com). Sanitas (www.IDT-LTD.com) provides some of the methods described in Chapter 17 for individual ground-water monitoring wells and constituents. CARStat, also distributed by Discerning Systems and Earthsoft (www.earthsoft.com), provides automated comparisons to standards and to background as described in Chapters 18 to 21. The program was originally developed for the automotive industry, hence the name.

A more general statistical package with environmental focus is S-Plus Spatial Statistics, which is distributed by Mathsoft (www.mathsoft.com). The focus of this package is on geostatistical applications and the analysis of spatial data.

23

SUMMARY

The purpose of environmental monitoring is to obtain the best possible estimate of the true concentration given available resources. The purpose of this book is to provide an overview of the tools necessary to enable this process. Even the most cursory reading of the first half of the book should make it exquisitely clear that a single measured concentration tells us very little about the true concentration that it is intended to measure. Characterizing the uncertainty in measured concentrations through the calibration function and related confidence regions is an important first step in this process. In many cases, the confidence region for true concentration may exceed a health-based standard by a sufficient degree to heighten concern and lead to further evaluation and potential regulatory action. Such decisions, however, are only a first step. The second half of this book provides a broader framework by which a series of representative environmental measurements can be used to derive bounds on true concentration, which can in turn be used to make an environmental impact decision. The true concentration (and corresponding interval estimate) in a particular sample may not be representative of the true concentration in the field. Indeed, confidence regions for the true concentration in a particular sample are invariably smaller than confidence regions for a series of such samples taken to be representative of a particular environment. Analytical variability and sampling variability (containing spatial and temporal components) are quite different things. Nevertheless, many environmental impact decisions are based on a few analytical measurements, sometimes from different laboratories. Understanding the uncertainty in these measurements is critical.

Throughout this book, concepts such as the critical level, detection limit, and quantification limit are discussed, operationalized, and perhaps promoted in one way or another. Of course, these are nothing more than point estimates along a continuum of analytical uncertainty that are useful, but not essential for practice. As an example, the critical level merely tells us if the confidence region for the underlying true concentration of a particular sample includes zero. Reporting a measured concentration and its confidence interval would achieve the same purpose and provide even more information. In the absence of such practice, it is important to make sure that these critical points along the continuum of analytical uncertainty are properly understood and well defined.

Finally, with respect to the discussion of statistical methods for assessment and corrective action monitoring, we have attempted to provide a detailed treatment of this topic, in some cases going well beyond the existing literature. It seems remarkable that so little rigorous statistical work has been done in this area of such great importance to human health and the environment. It is our hope that the material presented here provides a foundation for future practice and further statistical research.

APPENDIX

LAND'S TABLES*

*Tables are reproduced with permission from C. E. Land, *Selected Tables in Mathematical Statistics*, Vol. III, American Mathematical Society, Providence, R.I., 1975.

2 Degrees of freedom

$S(y)$	.0025	.005	.01	.025	.05	.10	Confidence .90	.95	.975	.99	.995	.9975
.10	−7.418	−5.803	−4.435	−2.988	−2.130	−1.431	1.686	2.750	4.367	8.328	14.66	28.24
.20	−5.712	−4.665	−3.720	−2.639	−1.949	−1.350	1.885	3.295	5.849	13.94	28.06	56.35
.30	−4.759	−3.983	−3.260	−2.396	−1.816	−1.289	2.156	4.109	8.166	20.88	42.10	84.53
.40	−4.149	−3.531	−2.943	−2.220	−1.717	−1.245	2.521	5.220	10.86	27.65	56.14	112.7
.50	−3.727	−3.211	−2.714	−2.090	−1.644	−1.213	2.990	6.495	13.59	34.82	70.18	140.9
.60	−3.421	−2.976	−2.544	−1.992	−1.589	−1.190	3.542	7.807	16.31	41.78	84.21	169.1
.70	−3.191	−2.800	−2.415	−1.919	−1.549	−1.176	4.136	9.120	19.04	48.75	98.25	197.2
.80	−3.015	−2.664	−2.317	−1.864	−1.521	−1.168	4.742	10.43	21.76	55.71	112.3	225.4
.90	−2.879	−2.560	−2.242	−1.823	−1.502	−1.165	5.349	11.74	24.49	62.68	126.3	253.6
1.00	−2.773	−2.479	−2.185	−1.794	−1.490	−1.166	5.955	13.05	27.21	69.65	140.4	281.8
1.25	−2.597	−2.350	−2.099	−1.759	−1.486	−1.184	7.466	16.33	34.02	87.06	175.4	352.2
1.50	−2.508	−2.291	−2.069	−1.761	−1.508	−1.217	8.973	19.60	40.83	104.5	210.5	422.7
1.75	−2.473	−2.277	−2.075	−1.789	−1.547	−1.260	10.48	22.87	47.63	121.9	245.6	493.1
2.00	−2.475	−2.294	−2.106	−1.834	−1.598	−1.310	11.98	26.14	54.44	139.3	280.7	563.6
2.50	−2.550	−2.389	−2.217	−1.960	−1.727	−1.426	14.99	32.69	68.05	174.1	350.9	704.5
3.00	−2.686	−2.535	−2.371	−2.118	−1.880	−1.560	18.00	39.23	81.66	208.9	421.1	845.3
3.50	−2.859	−2.714	−2.553	−2.299	−2.051	−1.710	21.00	45.71	95.27	243.8	491.3	986.2
4.00	−3.059	−2.917	−2.756	−2.496	−2.237	−1.871	24.00	52.31	108.9	278.6	561.4	1127
4.50	−3.277	−3.136	−2.973	−2.706	−2.434	−2.041	27.01	58.85	122.5	313.4	631.6	1268
5.00	−3.510	−3.368	−3.202	−2.925	−2.638	−2.217	30.01	65.39	136.1	348.2	701.8	1409
6.00	−4.005	−3.858	−3.683	−3.382	−3.062	−2.581	36.02	78.47	163.3	417.9	842.2	1691
7.00	−4.527	−4.372	−4.185	−3.856	−3.499	−2.955	42.02	91.55	190.6	487.5	982.5	1972
8.00	−5.066	−4.901	−4.700	−4.341	−3.945	−3.336	48.03	104.6	217.8	557.2	1123	2254
9.00	−5.616	−5.440	−5.223	−4.832	−4.397	−3.721	54.03	117.7	245.0	626.9	1263	2536
10.00	−6.173	−5.986	−5.753	−5.328	−4.852	−4.109	60.04	130.8	272.2	696.5	1404	2818

3 Degrees of freedom

$S(y)$	.0025	.005	.01	.025	.05	.10	.90	.95	.975	.99	.995	.9975
.10	−5.221	−4.267	−3.437	−2.504	−1.898	−1.351	1.506	2.222	3.100	4.665	6.334	8.677
.20	−4.464	−3.743	−3.089	−2.316	−1.791	−1.299	1.620	2.463	3.571	5.768	8.428	12.64
.30	−3.957	−3.377	−2.836	−2.176	−1.710	−1.260	1.763	2.777	4.210	7.336	11.37	17.82
.40	−3.598	−3.112	−2.649	−2.070	−1.650	−1.233	1.942	3.175	5.031	9.244	14.68	23.32
.50	−3.335	−2.915	−2.508	−1.989	−1.605	−1.214	2.160	3.658	5.989	11.29	18.10	28.90
.60	−3.138	−2.766	−2.402	−1.929	−1.572	−1.202	2.417	4.209	7.019	13.39	21.56	34.52
.70	−2.987	−2.652	−2.321	−1.885	−1.550	−1.197	2.708	4.801	8.083	15.52	25.04	40.16
.80	−2.872	−2.566	−2.260	−1.854	−1.537	−1.197	3.023	5.414	9.164	17.65	28.54	45.82
.90	−2.783	−2.500	−2.216	−1.833	−1.530	−1.201	3.353	6.038	10.25	19.80	32.05	51.48
1.00	−2.715	−2.451	−2.184	−1.820	−1.530	−1.208	3.691	6.669	11.35	21.95	35.56	57.15
1.25	−2.611	−2.382	−2.147	−1.819	−1.549	−1.240	4.558	8.265	14.11	27.35	44.36	71.35
1.50	−2.574	−2.368	−2.153	−1.849	−1.590	−1.285	5.436	9.874	16.88	32.77	53.18	85.56
1.75	−2.581	−2.390	−2.190	−1.899	−1.647	−1.341	6.319	11.49	19.65	38.19	61.99	99.77
2.00	−2.619	−2.439	−2.247	−1.965	−1.714	−1.403	7.206	13.11	22.43	43.61	70.82	114.0
2.50	−2.756	−2.589	−2.408	−2.132	−1.877	−1.547	8.986	16.35	28.00	54.47	88.48	142.4
3.00	−2.948	−2.788	−2.610	−2.331	−2.065	−1.712	10.77	19.60	33.58	65.34	106.1	170.9
3.50	−3.178	−3.019	−2.839	−2.552	−2.272	−1.889	12.56	22.85	39.16	76.21	123.8	199.4
4.00	−3.430	−3.271	−3.087	−2.789	−2.491	−2.078	14.34	26.11	44.74	87.08	141.5	227.8
4.50	−3.700	−3.538	−3.349	−3.037	−2.720	−2.274	16.13	29.36	50.32	97.96	159.2	256.3
5.00	−3.983	−3.817	−3.622	−3.294	−2.957	−2.475	17.92	32.62	55.90	108.8	176.8	284.8
6.00	−4.578	−4.401	−4.189	−3.826	−3.444	−2.889	21.49	39.13	67.07	130.6	212.2	341.7
7.00	−5.199	−5.007	−4.775	−4.372	−3.943	−3.314	25.07	45.65	78.24	152.3	247.5	398.6
8.00	−5.833	−5.627	−5.374	−4.929	−4.451	−3.744	28.65	52.16	89.41	174.1	282.9	455.6
9.00	−6.478	−6.254	−5.980	−5.492	−4.965	−4.180	32.23	58.66	100.6	195.9	318.2	512.5
10.00	−7.132	−6.891	−6.593	−6.061	−5.483	−4.618	35.81	65.20	111.8	217.6	353.6	569.5

Confidence

4 Degrees of freedom

$S(y)$	.0025	.005	.01	.025	.05	.10	Confidence .90	.95	.975	.99	.995	.9975
.10	−4.354	−3.667	−3.047	−2.314	−1.806	−1.320	1.438	2.035	2.703	3.760	4.750	5.968
.20	−3.896	−3.338	−2.819	−2.183	−1.729	−1.281	1.522	2.198	2.987	4.310	5.637	7.389
.30	−3.565	−3.094	−2.646	−2.083	−1.669	−1.252	1.627	2.402	3.348	5.035	6.833	9.329
.40	−3.319	−2.910	−2.514	−2.007	−1.625	−1.233	1.755	2.651	3.794	5.934	8.296	11.62
.50	−3.133	−2.770	−2.414	−1.950	−1.594	−1.221	1.907	2.947	4.322	6.966	9.916	14.07
.60	−2.991	−2.663	−2.338	−1.908	−1.573	−1.215	2.084	3.287	4.914	8.077	11.61	16.60
.70	−2.883	−2.582	−2.282	−1.879	−1.560	−1.215	2.284	3.662	5.548	9.231	13.35	19.17
.80	−2.800	−2.522	−2.242	−1.860	−1.555	−1.219	2.503	4.062	6.208	10.41	15.12	21.76
.90	−2.738	−2.478	−2.214	−1.850	−1.556	−1.227	2.736	4.478	6.885	11.60	16.90	24.37
1.00	−2.693	−2.448	−2.196	−1.848	−1.562	−1.239	2.980	4.905	7.572	12.81	18.69	26.99
1.25	−2.635	−2.416	−2.189	−1.867	−1.596	−1.280	3.617	6.001	9.320	15.85	23.20	33.56
1.50	−2.632	−2.431	−2.220	−1.914	−1.650	−1.334	4.276	7.120	11.09	18.92	27.73	40.17
1.75	−2.668	−2.479	−2.277	−1.981	−1.719	−1.398	4.944	8.250	12.88	22.01	32.27	46.78
2.00	−2.731	−2.550	−2.355	−2.062	−1.799	−1.470	5.619	9.387	14.67	25.10	36.83	53.41
2.50	−2.914	−2.742	−2.552	−2.259	−1.986	−1.634	6.979	11.67	18.27	31.29	45.96	66.68
3.00	−3.148	−2.979	−2.788	−2.487	−2.199	−1.817	8.346	13.97	21.87	37.50	55.09	79.96
3.50	−3.417	−3.246	−3.050	−2.736	−2.429	−2.014	9.717	16.27	25.49	43.72	64.24	93.25
4.00	−3.708	−3.533	−3.331	−3.001	−2.672	−2.221	11.09	18.58	29.11	49.94	73.39	106.5
4.50	−4.017	−3.836	−3.626	−3.276	−2.924	−2.435	12.47	20.88	32.73	56.16	82.54	119.8
5.00	−4.338	−4.151	−3.930	−3.560	−3.183	−2.654	13.84	23.19	36.35	62.38	91.70	133.1
6.00	−5.007	−4.803	−4.559	−4.145	−3.715	−3.104	16.60	27.81	43.59	74.84	110.0	159.7
7.00	−5.699	−5.476	−5.208	−4.744	−4.260	−3.564	19.35	32.43	50.84	87.29	128.3	186.3
8.00	−6.407	−6.163	−5.868	−5.354	−4.812	−4.030	22.11	37.06	58.10	99.75	146.6	212.9
9.00	−7.125	−6.860	−6.536	−5.971	−5.371	−4.500	24.87	41.68	65.35	112.2	165.0	239.5
10.00	−7.850	−7.563	−7.211	−6.592	−5.933	−4.973	27.63	46.31	72.60	124.7	183.3	266.2

5 Degrees of freedom

							Confidence						
$S(y)$	.0025	.005	.01	.025	.05	.10	.90	.95	.975	.99	.995	.9975
.10	−3.918	−3.364	−2.849	−2.215	−1.759	−1.305	1.403	1.942	2.513	3.360	4.100	4.955
.20	−3.592	−3.122	−2.677	−2.113	−1.697	−1.273	1.472	2.069	2.723	3.731	4.652	5.766
.30	−3.346	−2.939	−2.544	−2.034	−1.650	−1.251	1.558	2.226	2.982	4.199	5.363	6.827
.40	−3.160	−2.798	−2.442	−1.975	−1.615	−1.236	1.662	2.415	3.296	4.771	6.232	8.116
.50	−3.017	−2.690	−2.364	−1.932	−1.592	−1.228	1.785	2.638	3.664	5.434	7.225	9.558
.60	−2.909	−2.609	−2.307	−1.901	−1.578	−1.226	1.926	2.892	4.081	6.167	8.298	11.09
.70	−2.826	−2.548	−2.266	−1.882	−1.572	−1.229	2.085	3.173	4.534	6.947	9.423	12.67
.80	−2.764	−2.504	−2.238	−1.871	−1.572	−1.237	2.260	3.477	5.014	7.757	10.58	14.28
.90	−2.720	−2.474	−2.221	−1.869	−1.577	−1.248	2.447	3.796	5.512	8.856	11.76	15.92
1.00	−2.690	−2.456	−2.214	−1.873	−1.588	−1.262	2.644	4.127	6.024	9.430	12.95	17.57
1.25	−2.663	−2.450	−2.227	−1.907	−1.632	−1.310	3.167	4.990	7.339	11.58	15.96	21.73
1.50	−2.684	−2.486	−2.275	−1.966	−1.696	−1.371	3.713	5.880	8.684	13.76	19.01	25.93
1.75	−2.741	−2.551	−2.348	−2.045	−1.774	−1.442	4.273	6.786	10.05	15.95	22.08	30.15
2.00	−2.823	−2.639	−2.440	−2.138	−1.864	−1.521	4.842	7.701	11.42	18.16	25.16	34.38
2.50	−3.040	−2.862	−2.665	−2.357	−2.070	−1.700	5.990	9.546	14.18	22.60	31.34	42.86
3.00	−3.308	−3.129	−2.927	−2.607	−2.301	−1.897	7.147	11.40	16.96	27.05	37.54	51.36
3.50	−3.608	−3.425	−3.216	−2.879	−2.550	−2.108	8.312	13.27	19.74	31.52	43.74	59.86
4.00	−3.930	−3.742	−3.523	−3.164	−2.810	−2.329	9.480	15.14	22.53	35.98	49.96	68.38
4.50	−4.269	−4.072	−3.842	−3.461	−3.080	−2.557	10.65	17.01	25.32	40.45	56.17	76.89
5.00	−4.621	−4.413	−4.171	−3.766	−3.356	−2.789	11.82	18.88	28.12	44.93	62.39	85.41
6.00	−5.347	−5.120	−4.850	−4.393	−3.923	−3.267	14.17	22.63	33.71	53.88	74.83	102.5
7.00	−6.098	−5.848	−5.548	−5.033	−4.502	−3.753	16.51	26.39	39.31	62.84	87.28	119.5
8.00	−6.865	−6.590	−6.258	−5.685	−5.090	−4.246	18.86	30.14	44.91	71.79	99.73	136.6
9.00	−7.640	−7.340	−6.975	−6.343	−5.684	−4.742	21.21	33.90	50.51	80.75	112.2	153.6
10.00	−8.423	−8.096	−7.698	−7.006	−6.280	−5.243	23.56	37.66	56.11	89.72	124.6	170.7

6 Degrees of freedom

S(y)						Confidence						
	.0025	.005	.01	.025	.05	.10	.90	.95	.975	.99	.995	.9975
.10	−3.663	−3.184	−2.730	−2.157	−1.731	−1.296	1.381	1.886	2.403	3.137	3.751	4.433
.20	−3.408	−2.992	−2.590	−2.071	−1.678	−1.268	1.442	1.992	2.573	3.422	4.157	5.000
.30	−3.212	−2.844	−2.482	−2.006	−1.639	−1.250	1.517	2.125	2.781	3.775	4.665	5.719
.40	−3.062	−2.729	−2.399	−1.958	−1.611	−1.239	1.607	2.282	3.030	4.199	5.279	6.588
.50	−2.946	−2.642	−2.337	−1.923	−1.594	−1.234	1.712	2.465	3.319	4.691	5.986	7.579
.60	−2.859	−2.577	−2.292	−1.900	−1.584	−1.235	1.834	2.673	3.647	5.240	6.764	8.654
.70	−2.794	−2.530	−2.261	−1.887	−1.582	−1.241	1.970	2.904	4.005	5.831	7.591	9.783
.80	−2.747	−2.498	−2.241	−1.883	−1.586	−1.251	2.119	3.155	4.389	6.452	8.451	10.95
.90	−2.715	−2.478	−2.232	−1.885	−1.595	−1.264	2.280	3.420	4.791	7.095	9.335	12.13
1.00	−2.696	−2.469	−2.232	−1.894	−1.610	−1.281	2.450	3.698	5.206	7.753	10.23	13.34
1.25	−2.690	−2.481	−2.260	−1.939	−1.662	−1.334	2.904	4.426	6.285	9.442	12.53	16.40
1.50	−2.731	−2.533	−2.322	−2.009	−1.733	−1.400	3.383	5.184	7.397	11.17	14.86	19.50
1.75	−2.804	−2.613	−2.407	−2.097	−1.819	−1.477	3.877	5.960	8.528	12.92	17.22	22.63
2.00	−2.901	−2.715	−2.511	−2.200	−1.917	−1.562	4.380	6.747	9.671	14.68	19.59	25.77
2.50	−3.147	−2.963	−2.758	−2.438	−2.138	−1.751	5.401	8.339	11.98	18.22	24.36	32.07
3.00	−3.441	−3.254	−3.042	−2.706	−2.384	−1.960	6.434	9.945	14.30	21.79	29.14	38.39
3.50	−3.767	−3.574	−3.352	−2.994	−2.647	−2.183	7.473	11.56	16.64	25.36	33.94	44.73
4.00	−4.115	−3.913	−3.680	−3.298	−2.922	−2.415	8.516	13.18	18.98	28.94	38.75	51.07
4.50	−4.478	−4.267	−4.020	−3.612	−3.206	−2.653	9.562	14.80	21.32	32.53	43.56	57.42
5.00	−4.854	−4.632	−4.370	−3.934	−3.497	−2.897	10.61	16.43	23.67	36.12	48.37	63.78
6.00	−5.631	−5.383	−5.089	−4.594	−4.092	−3.396	12.71	19.68	28.37	43.30	58.00	76.49
7.00	−6.430	−6.156	−5.827	−5.269	−4.699	−3.904	14.81	22.94	33.07	50.49	67.63	89.20
8.00	−7.244	−6.941	−6.577	−5.955	−5.315	−4.418	16.91	26.20	37.77	57.68	77.27	101.9
9.00	−8.068	−7.736	−7.334	−6.646	−5.936	−4.937	19.02	29.46	42.48	64.87	86.91	114.6
10.00	−8.899	−8.537	−8.098	−7.343	−6.560	−5.459	21.12	32.73	47.19	72.07	95.56	127.4

7 Degrees of freedom

$S(y)$	.0025	.005	.01	.025	.05	.10	.90	.95	.975	.99	.995	.9975
.10	−3.498	−3.068	−2.653	−2.117	−1.712	−1.291	1.367	1.849	2.330	2.994	3.534	4.117
.20	−3.286	−2.906	−2.534	−2.044	−1.667	−1.267	1.422	1.943	2.476	3.231	3.861	4.558
.30	−3.122	−2.781	−2.441	−1.988	−1.633	−1.251	1.489	2.058	2.653	3.519	4.261	5.105
.40	−2.996	−2.684	−2.371	−1.948	−1.610	−1.243	1.569	2.195	2.864	3.862	4.741	5.761
.50	−2.900	−2.612	−2.320	−1.919	−1.596	−1.240	1.664	2.354	3.107	4.258	5.292	6.513
.60	−2.828	−2.558	−2.283	−1.902	−1.591	−1.243	1.773	2.534	3.382	4.702	5.904	7.338
.70	−2.775	−2.521	−2.260	−1.894	−1.592	−1.251	1.894	2.735	3.684	5.183	6.563	8.216
.80	−2.739	−2.498	−2.247	−1.894	−1.599	−1.262	2.027	2.952	4.009	5.693	7.253	9.130
.90	−2.716	−2.486	−2.244	−1.901	−1.611	−1.277	2.171	3.184	4.351	6.225	7.966	10.07
1.00	−2.705	−2.483	−2.249	−1.913	−1.628	−1.296	2.324	3.426	4.707	6.772	8.699	11.03
1.25	−2.718	−2.510	−2.290	−1.967	−1.687	−1.353	2.732	4.068	5.636	8.186	10.58	13.47
1.50	−2.772	−2.575	−2.362	−2.045	−1.764	−1.424	3.166	4.741	6.602	9.641	12.50	15.96
1.75	−2.859	−2.667	−2.457	−2.141	−1.857	−1.505	3.615	5.432	7.588	11.12	14.45	18.48
2.00	−2.969	−2.779	−2.571	−2.252	−1.960	−1.595	4.075	6.135	8.588	12.61	16.41	21.01
2.50	−3.239	−3.049	−2.836	−2.505	−2.193	−1.794	5.010	7.563	10.61	15.63	20.36	26.11
3.00	−3.555	−3.361	−3.140	−2.788	−2.452	−2.013	5.958	9.006	12.65	18.66	24.33	31.23
3.50	−3.903	−3.700	−3.467	−3.091	−2.727	−2.244	6.913	10.46	14.71	21.71	28.32	36.36
4.00	−4.272	−4.059	−3.812	−3.409	−3.015	−2.485	7.873	11.92	16.77	24.76	32.32	41.50
4.50	−4.656	−4.432	−4.170	−3.738	−3.310	−2.733	8.836	13.38	18.83	27.82	36.32	46.65
5.00	−5.053	−4.816	−4.537	−4.074	−3.613	−2.986	9.800	14.84	20.89	30.88	40.32	51.80
6.00	−5.872	−5.606	−5.291	−4.763	−4.231	−3.503	11.74	17.78	25.03	37.01	48.34	61.11
7.00	−6.713	−6.416	−6.064	−5.467	−4.862	−4.029	13.67	20.72	29.18	43.14	56.36	72.43
8.00	−7.567	−7.240	−6.847	−6.181	−5.502	−4.561	15.61	23.66	33.33	49.26	64.39	82.75
9.00	−8.433	−8.073	−7.639	−6.901	−6.146	−5.098	17.55	26.60	37.47	55.43	72.41	93.07
10.00	−9.305	−8.912	−8.437	−7.626	−6.795	−5.638	19.49	29.54	41.62	61.57	80.45	103.4

Confidence

8 Degrees of freedom

						Confidence						
$S(y)$	.0025	.005	.01	.025	.05	.10	.90	.95	.975	.99	.995	.9975
.10	−3.382	−2.986	−2.598	−2.090	−1.699	−1.287	1.356	1.822	2.879	2.897	3.386	3.905
.20	−3.201	−2.846	−2.494	−2.025	−1.658	−1.266	1.407	1.908	2.409	3.101	3.662	4.270
.30	−3.059	−2.737	−2.413	−1.976	−1.629	−1.253	1.469	2.011	2.565	3.348	3.998	4.717
.40	−2.950	−2.654	−2.353	−1.941	−1.610	−1.246	1.543	2.134	2.750	3.640	4.396	5.248
.50	−2.868	−2.592	−2.309	−1.918	−1.599	−1.245	1.630	2.277	2.963	3.976	4.854	5.857
.60	−2.807	−2.547	−2.279	−1.905	−1.597	−1.250	1.729	2.439	3.204	4.353	5.363	6.530
.70	−2.764	−2.518	−2.262	−1.901	−1.600	−1.259	1.840	2.618	3.469	4.764	5.914	7.252
.80	−2.736	−2.501	−2.255	−1.904	−1.610	−1.272	1.962	2.813	3.754	5.201	6.496	8.009
.90	−2.721	−2.494	−2.256	−1.915	−1.625	−1.289	2.094	3.021	4.056	5.659	7.101	8.791
1.00	−2.716	−2.497	−2.265	−1.931	−1.644	−1.309	2.234	3.239	4.371	6.133	7.724	9.592
1.25	−2.742	−2.536	−2.316	−1.992	−1.708	−1.370	2.610	3.820	5.199	7.365	9.333	11.65
1.50	−2.810	−2.612	−2.397	−2.076	−1.791	−1.444	3.012	4.433	6.064	8.640	10.99	13.76
1.75	−2.907	−2.713	−2.501	−2.179	−1.889	−1.530	3.429	5.065	6.951	9.940	12.67	15.89
2.00	−3.028	−2.835	−2.623	−2.296	−1.998	−1.623	3.857	5.710	7.853	11.26	14.37	18.05
2.50	−3.318	−3.123	−2.904	−2.562	−2.241	−1.830	4.730	7.021	9.681	13.92	17.80	22.39
3.00	−3.654	−3.452	−3.223	−2.858	−2.510	−2.057	5.617	8.350	11.53	16.60	21.25	26.75
3.50	−4.020	−3.809	−3.566	−3.174	−2.795	−2.296	6.511	9.688	13.39	19.30	24.71	31.13
4.00	−4.408	−4.185	−3.926	−3.505	−3.093	−2.545	7.411	11.03	15.26	22.00	28.19	35.52
4.50	−4.811	−4.575	−4.299	−3.846	−3.399	−2.801	8.314	12.38	17.13	24.71	31.67	39.91
5.00	−5.227	−4.976	−4.681	−4.194	−3.712	−3.061	9.219	13.73	19.00	27.42	35.15	44.31
6.00	−6.081	−5.799	−5.465	−4.908	−4.351	−3.593	11.03	16.44	22.76	32.86	42.14	53.13
7.00	−6.958	−6.643	−6.267	−5.637	−5.002	−4.135	12.85	19.16	26.52	38.30	49.11	61.92
8.00	−7.850	−7.499	−7.081	−6.375	−5.661	−4.683	14.67	21.87	30.28	43.74	56.11	70.74
9.00	−8.751	−8.365	−7.902	−7.120	−6.326	−5.234	16.50	24.59	34.05	49.19	63.09	79.57
10.00	−9.659	−9.237	−8.730	−7.869	−6.994	−5.789	18.32	27.31	37.82	54.64	70.08	88.39

9 Degrees of freedom

$S(y)$	.0025	.005	.01	.025	.05	.10	.90	.95	.975	.99	.995	.9975
							Confidence					
.10	−3.298	−2.926	−2.558	−2.070	−1.690	−1.285	1.349	1.802	2.242	2.825	3.280	3.755
.20	−3.138	−2.801	−2.465	−2.012	−1.653	−1.266	1.396	1.881	2.359	3.006	3.522	4.069
.30	−3.013	−2.704	−2.393	−1.968	−1.627	−1.254	1.453	1.977	2.501	3.225	3.813	4.449
.40	−2.917	−2.632	−2.340	−1.938	−1.611	−1.249	1.523	2.089	2.667	3.482	4.157	4.900
.50	−2.845	−2.578	−2.302	−1.918	−1.603	−1.250	1.604	2.220	2.859	3.778	4.552	5.416
.60	−2.793	−2.540	−2.278	−1.908	−1.602	−1.256	1.696	2.368	3.076	4.109	4.992	5.988
.70	−2.758	−2.517	−2.265	−1.908	−1.608	−1.266	1.800	2.532	3.314	4.471	5.470	6.605
.80	−2.737	−2.505	−2.262	−1.914	−1.620	−1.280	1.914	2.710	3.572	4.858	5.978	7.256
.90	−2.727	−2.503	−2.268	−1.927	−1.637	−1.298	2.036	2.902	3.844	5.264	6.507	7.931
1.00	−2.728	−2.511	−2.280	−1.946	−1.658	−1.320	2.167	3.103	4.130	5.686	7.055	8.626
1.25	−2.767	−2.561	−2.339	−2.013	−1.727	−1.384	2.518	3.639	4.884	6.789	8.476	10.42
1.50	−2.844	−2.645	−2.428	−2.104	−1.814	−1.462	2.896	4.207	5.676	7.936	9.944	12.26
1.75	−2.952	−2.755	−2.540	−2.212	−1.916	−1.551	3.289	4.795	6.490	9.109	11.44	14.14
2.00	−3.080	−2.885	−2.668	−2.335	−2.029	−1.647	3.693	5.396	7.320	10.30	12.96	16.03
2.50	−3.389	−3.189	−2.964	−2.612	−2.283	−1.862	4.518	6.621	9.006	12.71	16.02	19.86
3.00	−3.741	−3.533	−3.296	−2.919	−2.560	−2.095	5.359	7.864	10.71	15.14	19.11	23.70
3.50	−4.125	−3.905	−3.652	−3.246	−2.855	−2.341	6.208	9.118	12.43	17.59	22.21	27.57
4.00	−4.529	−4.296	−4.026	−3.588	−3.161	−2.596	7.062	10.38	14.16	20.05	25.32	31.44
4.50	−4.949	−4.701	−4.412	−3.940	−3.476	−2.858	7.919	11.64	15.89	22.51	28.44	35.33
5.00	−5.380	−5.116	−4.808	−4.300	−3.798	−3.126	8.779	12.91	17.63	24.98	31.56	39.21
6.00	−6.266	−5.969	−5.618	−5.035	−4.455	−3.671	10.50	15.45	21.10	29.92	37.82	47.00
7.00	−7.175	−6.842	−6.446	−5.785	−5.123	−4.226	12.23	18.00	24.59	34.87	44.08	54.78
8.00	−8.099	−7.728	−7.286	−6.545	−5.800	−4.787	13.96	20.55	28.08	39.82	50.35	62.58
9.00	−9.031	−8.622	−8.133	−7.311	−6.482	−5.352	15.70	23.10	31.57	44.77	56.62	70.38
10.00	−9.972	−9.523	−8.987	−8.082	−7.168	−5.920	17.43	25.66	35.06	49.73	62.89	78.18

10 Degrees of freedom

						Confidence					
S(y)	.0025	.005	.01	.025	.05	.10	.90	.95	.975	.99	.995	.9975
.10	−3.233	−2.880	−2.527	−2.055	−1.683	−1.283	1.343	1.787	2.212	2.770	3.199	3.641
.20	−3.090	−2.767	−2.442	−2.001	−1.649	−1.266	1.387	1.860	2.321	2.935	3.416	3.919
.30	−2.978	−2.680	−2.378	−1.962	−1.626	−1.255	1.441	1.949	2.451	3.132	3.676	4.254
.40	−2.892	−2.615	−2.330	−1.935	−1.612	−1.252	1.507	2.054	2.604	3.364	3.982	4.649
.50	−2.829	−2.568	−2.298	−1.919	−1.606	−1.254	1.583	2.176	2.780	3.631	4.332	5.100
.60	−2.784	−2.536	−2.278	−1.913	−1.608	−1.261	1.671	2.314	2.979	3.929	4.723	5.601
.70	−2.755	−2.518	−2.269	−1.914	−1.615	−1.273	1.768	2.466	3.198	4.255	5.148	6.144
.80	−2.739	−2.511	−2.270	−1.923	−1.629	−1.288	1.876	2.632	3.434	4.604	5.601	6.718
.90	−2.734	−2.513	−2.279	−1.939	−1.647	−1.307	1.992	2.810	3.685	4.973	6.076	7.317
1.00	−2.740	−2.524	−2.295	−1.959	−1.670	−1.329	2.115	2.998	3.949	5.357	6.568	7.935
1.25	−2.788	−2.582	−2.361	−2.032	−1.743	−1.396	2.448	3.500	4.647	6.363	7.851	9.537
1.50	−2.875	−2.675	−2.456	−2.128	−1.834	−1.477	2.806	4.033	5.383	7.414	9.183	11.19
1.75	−2.991	−2.792	−2.574	−2.242	−1.940	−1.569	3.180	4.587	6.142	8.492	10.54	12.88
2.00	−3.128	−2.929	−2.709	−2.369	−2.058	−1.669	3.564	5.154	6.916	9.587	11.92	14.58
2.50	−3.452	−3.247	−3.017	−2.656	−2.319	−1.889	4.353	6.312	8.493	11.81	14.72	18.03
3.00	−3.819	−3.605	−3.361	−2.973	−2.604	−2.128	5.157	7.489	10.09	14.06	17.54	21.51
3.50	−4.217	−3.990	−3.729	−3.310	−2.907	−2.380	5.970	8.677	11.70	16.32	20.37	25.00
4.00	−4.636	−4.394	−4.115	−3.661	−3.221	−2.641	6.788	9.872	13.32	18.59	23.22	28.51
4.50	−5.070	−4.813	−4.513	−4.023	−3.544	−2.910	7.610	11.07	14.95	20.87	26.08	32.02
5.00	−5.515	−5.241	−4.920	−4.392	−3.873	−3.183	8.434	12.27	16.58	23.15	28.93	35.53
6.00	−6.432	−6.120	−5.754	−5.147	−4.546	−3.740	10.09	14.69	19.85	27.73	34.66	42.57
7.00	−7.369	−7.019	−6.605	−5.916	−5.230	−4.306	11.75	17.10	23.12	32.31	40.39	49.62
8.00	−8.321	−7.931	−7.468	−6.695	−5.922	−4.879	13.41	19.53	26.39	36.89	46.13	56.68
9.00	−9.283	−8.852	−8.339	−7.480	−6.620	−5.455	15.07	21.95	29.67	41.48	51.87	63.74
10.00	−10.25	−9.778	−9.215	−8.270	−7.321	−6.035	16.73	24.38	32.95	46.07	57.61	70.80

11 Degrees of freedom

S(y)	.0025	.005	.01	.025	.05	.10	Confidence .90	.95	.975	.99	.995	.9975
.10	−3.182	−2.844	−2.503	−2.042	−1.677	−1.281	1.338	1.775	2.190	2.727	3.136	3.553
.20	−3.052	−2.741	−2.425	−1.994	−1.646	−1.266	1.380	1.843	2.291	2.878	3.333	3.804
.30	−2.951	−2.662	−2.366	−1.958	−1.625	−1.257	1.432	1.927	2.411	3.060	3.570	4.106
.40	−2.873	−2.603	−2.324	−1.934	−1.613	−1.254	1.494	2.026	2.554	3.273	3.847	4.459
.50	−2.817	−2.561	−2.295	−1.920	−1.609	−1.257	1.567	2.141	2.718	3.517	4.164	4.863
.60	−2.778	−2.534	−2.279	−1.917	−1.612	−1.266	1.650	2.271	2.903	3.790	4.519	5.312
.70	−2.754	−2.520	−2.274	−1.921	−1.622	−1.278	1.743	2.414	3.106	4.089	4.905	5.799
.80	−2.742	−2.517	−2.277	−1.932	−1.636	−1.294	1.845	2.570	3.327	4.110	5.316	6.316
.90	−2.742	−2.523	−2.289	−1.949	−1.656	−1.314	1.955	2.738	3.561	4.750	5.750	6.857
1.00	−2.752	−2.537	−2.308	−1.972	−1.681	−1.337	2.073	2.915	3.807	5.103	6.200	7.417
1.25	−2.809	−2.603	−2.380	−2.049	−1.758	−1.407	2.391	3.389	4.461	6.036	7.376	8.875
1.50	−2.903	−2.702	−2.481	−2.150	−1.853	−1.491	2.733	3.896	5.153	7.012	8.603	10.39
1.75	−3.027	−2.826	−2.605	−2.268	−1.962	−1.585	3.092	4.422	5.869	8.016	9.860	11.93
2.00	−3.172	−2.970	−2.746	−2.400	−2.083	−1.688	3.461	4.962	6.599	9.039	11.14	13.49
2.50	−3.509	−3.300	−3.064	−2.696	−2.351	−1.913	4.220	6.067	8.091	11.12	13.72	16.66
3.00	−3.890	−3.670	−3.419	−3.022	−2.644	−2.157	4.994	7.191	9.605	13.22	16.34	19.85
3.50	−4.301	−4.068	−3.799	−3.367	−2.953	−2.415	5.778	8.326	11.13	15.34	18.97	23.07
4.00	−4.734	−4.484	−4.195	−3.727	−3.275	−2.681	6.566	9.469	12.67	17.47	21.62	26.29
4.50	−5.181	−4.914	−4.603	−4.097	−3.605	−2.955	7.360	10.62	14.21	19.60	24.27	29.52
5.00	−5.640	−5.354	−5.021	−4.475	−3.941	−3.233	8.155	11.77	15.75	21.74	26.92	32.76
6.00	−6.581	−6.256	−5.875	−5.247	−4.627	−3.800	9.751	14.08	18.85	26.03	32.24	39.24
7.00	−7.544	−7.179	−6.748	−6.033	−5.325	−4.377	11.35	16.39	21.96	30.33	37.57	45.73
8.00	−8.522	−8.114	−7.632	−6.829	−6.031	−4.960	12.96	18.71	25.07	34.63	42.90	52.23
9.00	−9.509	−9.058	−8.523	−7.631	−6.742	−5.547	14.56	21.03	28.18	38.93	48.24	58.73
10.00	−10.50	−10.01	−9.420	−8.438	−7.458	−6.137	16.17	23.35	31.29	43.24	53.58	65.23

12 Degrees of freedom

					Confidence						
S(y)	.0025	.005	.01	.025	.05	.10	.90	.95	.975	.99	.995	.9975
.10	−3.143	−2.814	−2.484	−2.032	−1.673	−1.281	1.334	1.763	2.169	2.691	3.084	3.482
.20	−3.021	−2.719	−2.411	−1.987	−1.644	−1.266	1.374	1.830	2.265	2.833	3.267	3.713
.30	−2.928	−2.647	−2.357	−1.954	−1.625	−1.258	1.424	1.909	2.380	3.002	3.486	3.988
.40	−2.858	−2.593	−2.319	−1.933	−1.614	−1.257	1.483	2.003	2.514	3.200	3.741	4.310
.50	−2.807	−2.556	−2.294	−1.922	−1.612	−1.261	1.553	2.112	2.668	3.426	4.033	4.678
.60	−2.773	−2.533	−2.281	−1.920	−1.617	−1.270	1.633	2.235	2.842	3.680	4.358	5.087
.70	−2.754	−2.522	−2.278	−1.926	−1.628	−1.283	1.722	2.371	3.033	3.958	4.713	5.531
.80	−2.747	−2.522	−2.284	−1.939	−1.644	−1.301	1.820	2.520	3.240	4.256	5.093	6.004
.90	−2.750	−2.532	−2.298	−1.958	−1.665	−1.321	1.926	2.679	3.461	4.572	5.493	6.501
1.00	−2.763	−2.549	−2.320	−1.983	−1.690	−1.345	2.038	2.848	3.693	4.903	5.910	7.015
1.25	−2.827	−2.621	−2.398	−2.064	−1.770	−1.417	2.344	3.300	4.312	5.775	7.004	8.360
1.50	−2.929	−2.726	−2.504	−2.169	−1.869	−1.503	2.674	3.784	4.968	6.693	8.148	9.757
1.75	−3.059	−2.857	−2.633	−2.291	−1.981	−1.599	3.019	4.288	5.648	7.638	9.322	11.19
2.00	−3.211	−3.006	−2.778	−2.428	−2.106	−1.704	3.376	4.805	6.344	8.602	10.52	12.64
2.50	−3.560	−3.347	−3.107	−2.731	−2.380	−1.934	4.110	5.866	7.765	10.56	12.94	15.59
3.00	−3.954	−3.729	−3.472	−3.065	−2.679	−2.183	4.860	6.947	9.210	12.54	15.39	18.56
3.50	−4.377	−4.137	−3.861	−3.418	−2.995	−2.446	5.619	8.039	10.67	14.56	17.87	21.55
4.00	−4.822	−4.564	−4.267	−3.786	−3.323	−2.717	6.384	9.140	12.14	16.57	20.35	24.56
4.50	−5.281	−5.005	−4.685	−4.164	−3.659	−2.995	7.154	10.24	13.61	18.59	22.84	27.57
5.00	−5.752	−5.456	−5.112	−4.550	−4.001	−3.278	7.924	11.35	15.09	20.62	25.33	30.59
6.00	−6.717	−6.380	−5.986	−5.337	−4.700	−3.855	9.473	13.58	18.05	24.68	30.33	36.63
7.00	−7.704	−7.324	−6.877	−6.139	−5.411	−4.441	11.03	15.81	21.02	28.75	35.34	42.69
8.00	−8.705	−8.281	−7.780	−6.950	−6.129	−5.033	12.58	18.04	23.99	32.82	40.35	48.75
9.00	−9.715	−9.246	−8.690	−7.768	−6.853	−5.629	14.14	20.28	26.97	36.90	45.37	54.81
10.00	−10.73	−10.22	−9.607	−8.590	−7.581	−6.228	15.70	22.51	29.95	40.98	50.39	60.88

13 Degrees of freedom

							Confidence						
S(y)	.0025	.005	.01	.025	.05	.10	.90	.95	.975	.99	.995	.9975
.10	−3.108	−2.790	−2.467	−2.025	−1.669	−1.280	1.330	1.756	2.155	2.663	3.043	3.426
.20	−2.996	−2.702	−2.400	−1.982	−1.642	−1.266	1.369	1.818	2.245	2.796	3.214	3.639
.30	−2.910	−2.634	−2.350	−1.952	−1.624	−1.259	1.417	1.894	2.353	2.955	3.417	3.893
.40	−2.846	−2.586	−2.315	−1.933	−1.615	−1.258	1.474	1.984	2.480	3.140	3.654	4.191
.50	−2.800	−2.553	−2.293	−1.924	−1.615	−1.264	1.542	2.088	2.626	3.353	3.926	4.530
.60	−2.771	−2.533	−2.283	−1.924	−1.621	−1.274	1.619	2.206	2.791	3.590	4.229	4.907
.70	−2.755	−2.526	−2.283	−1.932	−1.633	−1.288	1.705	2.336	2.973	3.851	4.559	5.318
.80	−2.752	−2.529	−2.291	−1.946	−1.651	−1.306	1.799	2.479	3.169	4.131	4.914	5.755
.90	−2.758	−2.541	−2.308	−1.967	−1.673	−1.327	1.901	2.631	3.379	4.428	5.287	6.216
1.00	−2.774	−2.560	−2.331	−1.993	−1.699	−1.353	2.010	2.792	3.599	4.740	5.677	6.695
1.25	−2.845	−2.639	−2.414	−2.079	−1.782	−1.426	2.305	3.226	4.189	5.564	6.704	7.948
1.50	−2.954	−2.749	−2.525	−2.187	−1.883	−1.514	2.623	3.691	4.815	6.432	7.780	9.265
1.75	−3.090	−2.885	−2.659	−2.313	−1.998	−1.612	2.959	4.176	5.466	7.330	8.888	10.60
2.00	−3.247	−3.040	−2.809	−2.452	−2.126	−1.719	3.305	4.675	6.133	8.245	10.01	11.96
2.50	−3.608	−3.391	−3.147	−2.764	−2.406	−1.953	4.017	5.698	7.497	10.11	12.31	14.73
3.00	−4.013	−3.782	−3.521	−3.105	−2.711	−2.207	4.746	6.743	8.884	12.01	14.63	17.52
3.50	−4.448	−4.201	−3.918	−3.465	−3.033	−2.473	5.486	7.799	10.29	13.91	16.97	20.34
4.00	−4.903	−4.638	−4.333	−3.840	−3.366	−2.749	6.229	8.864	11.70	15.84	19.32	23.17
4.50	−5.373	−5.019	−4.760	−4.226	−3.708	−3.031	6.978	9.933	13.11	17.76	21.68	26.00
5.00	−5.855	−5.550	−5.195	−4.618	−4.056	−3.319	7.729	11.01	14.54	19.70	24.05	28.85
6.00	−6.841	−6.494	−6.087	−5.419	−4.766	−3.904	9.238	13.16	17.39	23.57	28.79	34.54
7.00	−7.850	−7.457	−6.995	−6.235	−5.488	−4.498	10.75	15.32	20.25	27.45	33.53	40.24
8.00	−8.872	−8.433	−7.916	−7.060	−6.218	−5.099	12.27	17.48	23.11	31.34	38.29	45.95
9.00	−9.904	−9.417	−8.843	−7.892	−6.954	−5.703	13.79	19.65	25.97	35.23	43.05	51.67
10.00	−10.94	−10.41	−9.776	−8.728	−7.692	−6.311	15.31	21.82	28.84	39.13	47.81	57.38

14 Degrees of freedom

							Confidence						
S(y)	.0025	.005	.01	.025	.05	.10	.90	.95	.975	.99	.995	.9975
.10	−3.081	−2.770	−2.454	−2.018	−1.666	−1.279	1.328	1.749	2.141	2.638	3.008	3.377
.20	−2.976	−2.687	−2.390	−1.978	−1.640	−1.266	1.365	1.809	2.227	2.764	3.168	3.577
.30	−2.896	−2.625	−2.344	−1.950	−1.625	−1.260	1.411	1.882	2.331	2.914	3.360	3.815
.40	−2.836	−2.580	−2.312	−1.933	−1.617	−1.261	1.467	1.968	2.452	3.090	3.583	4.092
.50	−2.795	−2.550	−2.293	−1.926	−1.618	−1.266	1.532	2.068	2.592	3.291	3.838	4.408
.60	−2.769	−2.534	−2.285	−1.928	−1.625	−1.277	1.606	2.181	2.749	3.515	4.122	4.760
.70	−2.757	−2.529	−2.287	−1.937	−1.638	−1.292	1.690	2.306	2.922	3.762	4.432	5.143
.80	−2.756	−2.534	−2.298	−1.953	−1.656	−1.311	1.781	2.443	3.109	4.027	4.766	5.553
.90	−2.766	−2.549	−2.316	−1.975	−1.680	−1.333	1.880	2.589	3.310	4.309	5.118	5.984
1.00	−2.784	−2.571	−2.341	−2.003	−1.707	−1.358	1.985	2.744	3.521	4.605	5.486	6.433
1.25	−2.862	−2.654	−2.429	−2.091	−1.793	−1.434	2.271	3.163	4.086	5.388	6.457	7.612
1.50	−2.976	−2.770	−2.545	−2.203	−1.896	−1.523	2.581	3.612	4.688	6.217	7.477	8.845
1.75	−3.118	−2.911	−2.682	−2.332	−2.015	−1.624	2.907	4.081	5.314	7.074	8.530	10.11
2.00	−3.280	−3.070	−2.836	−2.476	−2.144	−1.733	3.244	4.564	5.956	7.949	9.601	11.40
2.50	−3.652	−3.431	−3.183	−2.793	−2.430	−1.971	3.938	5.557	7.271	9.735	11.79	14.02
3.00	−4.067	−3.832	−3.564	−3.141	−2.740	−2.229	4.650	6.570	8.610	11.55	14.00	16.67
3.50	−4.512	−4.259	−3.970	−3.508	−3.067	−2.499	5.370	7.596	9.964	13.38	16.23	19.34
4.00	−4.977	−4.706	−4.393	−3.889	−3.406	−2.778	6.097	8.630	11.33	15.23	18.47	22.03
4.50	−5.458	−5.166	−4.828	−4.281	−3.753	−3.064	6.829	9.669	12.70	17.07	20.73	24.72
5.00	−5.949	−5.636	−5.272	−4.680	−4.107	−3.356	7.563	10.71	14.07	18.93	22.98	27.42
6.00	−6.956	−6.598	−6.179	−5.494	−4.827	−3.949	9.037	12.81	16.83	22.65	27.51	32.82
7.00	−7.984	−7.579	−7.104	−6.324	−5.559	−4.549	10.52	14.90	19.59	26.38	32.04	38.24
8.00	−9.027	−8.573	−8.040	−7.161	−6.300	−5.159	12.00	17.01	22.36	30.11	36.58	43.66
9.00	−10.08	−9.575	−8.983	−8.006	−7.045	−5.771	13.48	19.11	25.13	33.84	41.12	49.09
10.00	−11.14	−10.58	−9.932	−8.855	−7.794	−6.386	14.97	21.22	27.90	37.58	45.67	54.52

15 Degrees of freedom

					Confidence					
S(y)	.0025	.005	.01	.025	.05	.10	.90	.95	.975	.99	.995	.9975
.10	−3.057	−2.753	−2.442	−2.012	−1.663	−1.278	1.325	1.743	2.130	2.618	2.978	3.337
.20	−2.959	−2.675	−2.383	−1.974	−1.639	−1.267	1.361	1.800	2.212	2.737	3.130	3.525
.30	−2.883	−2.616	−2.339	−1.949	−1.625	−1.261	1.406	1.871	2.311	2.880	3.312	3.749
.40	−2.828	−2.575	−2.310	−1.934	−1.618	−1.262	1.460	1.954	2.428	3.047	3.523	4.010
.50	−2.791	−2.548	−2.293	−1.928	−1.620	−1.269	1.524	2.050	2.562	3.239	3.763	4.307
.60	−2.769	−2.535	−2.288	−1.931	−1.629	−1.280	1.596	2.160	2.712	3.453	4.032	4.638
.70	−2.759	−2.533	−2.292	−1.942	−1.643	−1.296	1.677	2.280	2.879	3.687	4.326	4.998
.80	−2.761	−2.540	−2.304	−1.959	−1.662	−1.315	1.765	2.412	3.059	3.940	4.642	5.384
.90	−2.774	−2.557	−2.324	−1.983	−1.686	−1.338	1.861	2.554	3.251	4.209	4.976	5.791
1.00	−2.794	−2.581	−2.351	−2.012	−1.715	−1.364	1.963	2.704	3.454	4.491	5.325	6.215
1.25	−2.878	−2.670	−2.443	−2.104	−1.803	−1.441	2.242	3.109	3.998	5.240	6.249	7.332
1.50	−2.997	−2.790	−2.563	−2.218	−1.909	−1.533	2.544	3.544	4.579	6.034	7.223	8.502
1.75	−3.144	−2.935	−2.704	−2.351	−2.029	−1.634	2.862	4.000	5.183	6.857	8.228	9.707
2.00	−3.311	−3.099	−2.862	−2.496	−2.162	−1.746	3.191	4.470	5.804	7.699	9.254	10.93
2.50	−3.693	−3.468	−3.216	−2.821	−2.452	−1.987	3.870	5.435	7.078	9.415	11.34	13.43
3.00	−4.118	−3.878	−3.605	−3.174	−2.767	−2.248	4.565	6.422	8.376	11.17	13.47	15.96
3.50	−4.572	−4.314	−4.019	−3.547	−3.099	−2.522	5.271	7.422	9.689	12.93	15.61	18.51
4.00	−5.047	−4.769	−4.449	−3.935	−3.443	−2.805	5.983	8.429	11.01	14.71	17.76	21.07
4.50	−5.536	−5.237	−4.891	−4.332	−3.794	−3.095	6.699	9.442	12.34	16.49	19.92	23.64
5.00	−6.037	−5.716	−5.343	−4.738	−4.153	−3.390	7.418	10.46	13.67	18.28	22.09	26.22
6.00	−7.062	−6.694	−6.264	−5.564	−4.882	−3.989	8.862	12.50	16.35	21.87	26.43	31.39
7.00	−8.109	−7.692	−7.204	−6.404	−5.624	−4.599	10.31	14.55	19.03	25.46	30.78	36.56
8.00	−9.170	−8.702	−8.154	−7.254	−6.374	−5.213	11.77	16.60	21.72	29.06	35.14	41.74
9.00	−10.24	−9.721	−9.113	−8.111	−7.129	−5.833	13.22	18.65	24.41	32.67	39.51	46.93
10.00	−11.32	−10.75	−10.08	−8.972	−7.888	−6.455	14.68	20.71	27.10	36.28	43.87	52.12

16 Degrees of freedom

$S(y)$	.0025	.005	.01	.025	.05	.10	Confidence .90	.95	.975	.99	.995	.9975
.10	−3.036	−2.738	−2.432	−2.008	−1.661	−1.278	1.323	1.738	2.120	2.600	2.953	3.302
.20	−2.944	−2.665	−2.376	−1.972	−1.638	−1.267	1.358	1.793	2.199	2.714	3.097	3.481
.30	−2.873	−2.609	−2.335	−1.947	−1.625	−1.262	1.402	1.861	2.295	2.851	3.270	3.693
.40	−2.822	−2.571	−2.308	−1.934	−1.620	−1.264	1.455	1.942	2.407	3.011	3.471	3.940
.50	−2.788	−2.547	−2.294	−1.930	−1.622	−1.271	1.516	2.035	2.536	3.194	3.700	4.221
.60	−2.769	−2.536	−2.290	−1.934	−1.632	−1.283	1.586	2.141	2.681	3.398	3.956	4.534
.70	−2.762	−2.536	−2.296	−1.946	−1.647	−1.299	1.666	2.258	2.841	3.623	4.236	4.875
.80	−2.766	−2.546	−2.310	−1.965	−1.667	−1.319	1.752	2.386	3.015	3.865	4.537	5.241
.90	−2.781	−2.565	−2.332	−1.990	−1.692	−1.342	1.845	2.523	3.200	4.123	4.855	5.628
1.00	−2.804	−2.590	−2.360	−2.019	−1.722	−1.369	1.945	2.669	3.397	4.394	5.189	6.031
1.25	−2.892	−2.684	−2.456	−2.114	−1.812	−1.448	2.217	3.062	3.922	5.114	6.073	7.095
1.50	−3.017	−2.808	−2.579	−2.232	−1.920	−1.541	2.512	3.485	4.485	5.878	7.007	8.213
1.75	−3.168	−2.957	−2.724	−2.367	−2.043	−1.645	2.823	3.929	5.070	6.671	7.971	9.364
2.00	−3.340	−3.125	−2.886	−2.516	−2.177	−1.757	3.145	4.387	5.674	7.483	8.958	10.54
2.50	−3.730	−3.503	−3.247	−2.845	−2.472	−2.002	3.810	5.328	6.911	9.145	10.97	12.93
3.00	−4.164	−3.920	−3.643	−3.205	−2.792	−2.266	4.492	6.293	8.174	10.84	13.01	15.36
3.50	−4.627	−4.364	−4.063	−3.583	−3.128	−2.544	5.184	7.269	9.451	12.54	15.08	17.81
4.00	−5.110	−4.827	−4.500	−3.976	−3.476	−2.830	5.883	8.254	10.74	14.26	17.15	20.26
4.50	−5.609	−5.303	−4.950	−4.380	−3.833	−3.123	6.586	9.244	12.03	15.99	19.23	22.73
5.00	−6.118	−5.790	−5.408	−4.790	−4.195	−3.421	7.292	10.24	13.33	17.72	21.32	25.21
6.00	−7.161	−6.784	−6.343	−5.628	−4.934	−4.027	8.710	12.23	15.93	21.19	25.51	30.17
7.00	−8.225	−7.797	−7.297	−6.480	−5.685	−4.642	10.13	14.24	18.55	24.68	29.71	35.14
8.00	−9.303	−8.823	−8.261	−7.340	−6.443	−5.264	11.56	16.24	21.17	28.17	33.91	40.12
9.00	−10.39	−9.857	−9.232	−8.208	−7.207	−5.890	12.99	18.25	23.79	31.66	38.12	45.10
10.00	−11.48	−10.90	−10.21	−9.079	−7.974	−6.518	14.42	20.26	26.41	35.15	42.34	50.08

17 Degrees of freedom

						Confidence					
$S(y)$	.0025	.005	.01	.025	.05	.10	.90	.95	.975	.99	.995	.9975
.10	−3.019	−2.726	−2.424	−2.003	−1.659	−1.278	1.322	1.733	2.112	2.584	2.930	3.272
.20	−2.931	−2.656	−2.370	−1.969	−1.638	−1.267	1.355	1.787	2.188	2.694	3.069	3.442
.30	−2.864	−2.604	−2.332	−1.946	−1.626	−1.263	1.398	1.853	2.280	2.826	3.234	3.644
.40	−2.817	−2.568	−2.307	−1.935	−1.622	−1.266	1.449	1.931	2.388	2.979	3.426	3.880
.50	−2.786	−2.547	−2.294	−1.932	−1.625	−1.273	1.509	2.021	2.513	3.155	3.646	4.147
.60	−2.769	−2.538	−2.292	−1.938	−1.635	−1.286	1.578	2.124	2.653	3.351	3.890	4.445
.70	−2.765	−2.540	−2.300	−1.951	−1.651	−1.302	1.655	2.238	2.808	3.567	4.157	4.770
.80	−2.771	−2.552	−2.315	−1.971	−1.672	−1.323	1.739	2.362	2.976	3.800	4.446	5.119
.90	−2.788	−2.572	−2.339	−1.996	−1.698	−1.346	1.831	2.496	3.157	4.049	4.751	5.488
1.00	−2.813	−2.600	−2.369	−2.027	−1.728	−1.374	1.929	2.638	3.347	4.309	5.072	5.873
1.25	−2.907	−2.697	−2.468	−2.125	−1.820	−1.455	2.195	3.021	3.856	5.004	5.921	6.892
1.50	−3.035	−2.825	−2.595	−2.245	−1.930	−1.548	2.483	3.434	4.402	5.743	6.819	7.964
1.75	−3.191	−2.978	−2.743	−2.383	−2.055	−1.654	2.788	3.867	4.972	6.510	7.750	9.071
2.00	−3.368	−3.150	−2.908	−2.534	−2.192	−1.767	3.104	4.314	5.559	7.297	8.702	10.20
2.50	−3.766	−3.535	−3.275	−2.869	−2.491	−2.016	3.757	5.236	6.765	8.907	10.65	12.50
3.00	−4.208	−3.960	−3.679	−3.233	−2.815	−2.283	4.427	6.179	7.996	10.55	12.62	14.84
3.50	−4.680	−4.412	−4.105	−3.617	−3.155	−2.563	5.107	7.136	9.242	12.21	14.62	17.20
4.00	−5.171	−4.881	−4.549	−4.015	−3.507	−2.853	5.794	8.100	10.50	13.88	16.62	19.57
4.50	−5.677	−5.365	−5.005	−4.424	−3.868	−3.149	6.485	9.070	11.76	15.55	18.64	21.95
5.00	−6.195	−5.858	−5.469	−4.840	−4.235	−3.450	7.179	10.04	13.03	17.24	20.66	24.34
6.00	−7.254	−6.867	−6.418	−5.687	−4.981	−4.062	8.575	12.00	15.57	20.61	24.72	29.12
7.00	−8.333	−7.895	−7.383	−6.549	−5.741	−4.683	9.975	13.96	18.13	24.00	28.78	33.91
8.00	−9.427	−8.935	−8.360	−7.420	−6.507	−5.311	11.38	15.93	20.68	27.39	32.85	38.71
9.00	−10.53	−9.984	−9.344	−8.298	−7.278	−5.942	12.78	17.90	23.24	30.78	36.93	43.52
10.00	−11.64	−11.04	−10.33	−9.179	−8.054	−6.578	14.19	19.87	25.80	34.18	41.01	48.33

18 Degrees of freedom

S(y)	Confidence										
	.0025	.005	.01	.025	.05	.10	.90	.95	.975	.99	.995	.9975
.10	−3.004	−2.715	−2.416	−2.000	−1.658	−1.278	1.320	1.729	2.104	2.571	2.911	3.200
.20	−2.920	−2.648	−2.365	−1.967	−1.637	−1.268	1.353	1.781	2.178	2.676	3.044	3.408
.30	−2.857	−2.599	−2.329	−1.946	−1.626	−1.265	1.394	1.845	2.267	2.803	3.202	3.602
.40	−2.812	−2.566	−2.306	−1.935	−1.622	−1.267	1.444	1.921	2.372	2.951	3.387	3.827
.50	−2.784	−2.546	−2.295	−1.933	−1.627	−1.275	1.503	2.009	2.493	3.121	3.598	4.083
.60	−2.769	−2.539	−2.295	−1.940	−1.638	−1.288	1.570	2.110	2.630	3.311	3.833	4.368
.70	−2.768	−2.543	−2.304	−1.955	−1.654	−1.305	1.646	2.221	2.780	3.519	4.090	4.679
.80	−2.777	−2.557	−2.321	−1.976	−1.677	−1.326	1.728	2.342	2.943	3.744	4.367	5.014
.90	−2.795	−2.579	−2.346	−2.003	−1.703	−1.351	1.819	2.472	3.117	3.983	4.661	5.367
1.00	−2.822	−2.608	−2.377	−2.035	−1.734	−1.378	1.914	2.611	3.302	4.235	4.970	5.737
1.25	−2.920	−2.709	−2.479	−2.134	−1.828	−1.460	2.174	2.984	3.798	4.908	5.788	6.715
1.50	−3.052	−2.841	−2.609	−2.257	−1.940	−1.555	2.458	3.388	4.330	5.625	6.660	7.749
1.75	−3.211	−2.998	−2.760	−2.396	−2.067	−1.662	2.757	3.812	4.887	6.369	7.557	8.815
2.00	−3.392	−3.174	−2.929	−2.551	−2.205	−1.777	3.069	4.251	5.461	7.134	8.479	9.905
2.50	−3.799	−3.565	−3.302	−2.890	−2.508	−2.029	3.710	5.153	6.636	8.700	10.36	12.13
3.00	−4.249	−3.997	−3.711	−3.260	−2.836	−2.298	4.369	6.078	7.840	10.30	12.28	14.39
3.50	−4.728	−4.456	−4.144	−3.649	−3.180	−2.581	5.039	7.016	9.058	11.91	14.22	16.67
4.00	−5.227	−4.932	−4.593	−4.052	−3.536	−2.874	5.715	7.963	10.29	13.54	16.16	18.96
4.50	−5.742	−5.423	−5.055	−4.465	−3.901	−3.173	6.396	8.916	11.52	15.17	18.12	21.27
5.00	−6.267	−5.923	−5.526	−4.886	−4.272	−3.477	7.080	9.872	12.76	16.81	20.08	23.58
6.00	−7.340	−6.946	−6.486	−5.743	−5.026	−4.094	8.454	11.79	15.25	20.10	24.02	28.21
7.00	−8.435	−7.987	−7.465	−6.614	−5.793	−4.721	9.833	13.72	17.75	23.40	27.97	32.85
8.00	−9.543	−9.040	−8.453	−7.495	−6.566	−5.354	11.22	15.65	20.25	26.70	31.92	37.50
9.00	−10.66	−10.10	−9.449	−8.382	−7.346	−5.992	12.60	17.59	22.76	30.01	35.88	42.15
10.00	−11.79	−11.17	−10.45	−9.273	−8.129	−6.632	13.99	19.52	25.27	33.32	39.85	46.81

20 Degrees of freedom

$S(y)$	.0025	.005	.01	.025	.05	.10	Confidence .90	.95	.975	.99	.995	.9975
.10	−2.979	−2.696	−2.404	−1.993	−1.655	−1.277	1.317	1.722	2.091	2.548	2.878	3.201
.20	−2.902	−2.635	−2.357	−1.964	−1.636	−1.268	1.348	1.771	2.161	2.647	3.002	3.352
.30	−2.845	−2.591	−2.325	−1.945	−1.627	−1.266	1.388	1.833	2.246	2.767	3.150	3.532
.40	−2.805	−2.563	−2.306	−1.936	−1.625	−1.270	1.437	1.905	2.345	2.904	3.322	3.740
.50	−2.782	−2.547	−2.298	−1.937	−1.631	−1.279	1.494	1.989	2.460	3.064	3.518	3.976
.60	−2.772	−2.543	−2.300	−1.946	−1.643	−1.292	1.558	2.085	2.588	3.242	3.737	4.240
.70	−2.774	−2.551	−2.312	−1.962	−1.661	−1.310	1.631	2.191	2.731	3.438	3.977	4.529
.80	−2.786	−2.567	−2.331	−1.985	−1.685	−1.332	1.710	2.307	2.886	3.649	4.236	4.839
.90	−2.808	−2.592	−2.358	−2.014	−1.713	−1.358	1.797	2.432	3.052	3.875	4.510	5.168
1.00	−2.838	−2.625	−2.392	−2.047	−1.745	−1.387	1.889	2.564	3.227	4.112	4.800	5.511
1.25	−2.944	−2.732	−2.500	−2.151	−1.842	−1.470	2.141	2.923	3.700	4.749	5.571	6.424
1.50	−3.084	−2.870	−2.635	−2.278	−1.958	−1.568	2.415	3.311	4.209	5.426	6.387	7.392
1.75	−3.251	−3.034	−2.792	−2.423	−2.088	−1.677	2.705	3.719	4.740	6.134	7.237	8.394
2.00	−3.439	−3.215	−2.966	−2.581	−2.230	−1.795	3.005	4.141	5.289	6.861	8.108	9.419
2.50	−3.860	−3.621	−3.351	−2.930	−2.540	−2.051	3.629	5.013	6.419	8.353	9.893	11.51
3.00	−4.324	−4.065	−3.771	−3.308	−2.874	−2.326	4.270	5.907	7.576	9.875	11.71	13.65
3.50	−4.817	−4.536	−4.215	−3.706	−3.226	−2.615	4.921	6.815	8.748	11.42	13.55	15.80
4.00	−5.330	−5.025	−4.676	−4.118	−3.589	−2.913	5.580	7.731	9.930	12.97	15.40	17.96
4.50	−5.858	−5.528	−5.148	−4.539	−3.960	−3.217	6.243	8.652	11.12	14.53	17.26	20.14
5.00	−6.397	−6.041	−5.630	−4.969	−4.338	−3.525	6.909	9.579	12.31	16.10	19.12	22.32
6.00	−7.498	−7.088	−6.612	−5.844	−5.106	−4.153	8.248	11.44	14.71	19.24	22.87	26.70
7.00	−8.621	−8.154	−7.611	−6.732	−5.886	−4.790	9.592	13.31	17.12	22.39	26.62	31.08
8.00	−9.756	−9.232	−8.621	−7.630	−6.674	−5.433	10.94	15.18	19.53	25.55	30.38	35.48
9.00	−10.90	−10.32	−9.640	−8.535	−7.468	−6.080	12.29	17.05	21.94	28.72	34.14	39.88
10.00	−12.05	−11.41	−10.66	−9.443	−8.264	−6.730	13.64	18.93	24.36	31.88	37.91	44.28

22 Degrees of freedom

$S(y)$	.0025	.005	.01	.025	.05	.10	.90	.95	.975	.99	.995	.9975
.10	−2.958	−2.682	−2.395	−1.989	−1.653	−1.277	1.315	1.716	2.081	2.529	2.852	3.166
.20	−2.887	−2.625	−2.351	−1.961	−1.636	−1.270	1.345	1.763	2.147	2.623	2.969	3.307
.30	−2.835	−2.585	−2.322	−1.945	−1.628	−1.268	1.383	1.822	2.228	2.735	3.108	3.475
.40	−2.800	−2.560	−2.305	−1.938	−1.627	−1.272	1.430	1.892	2.323	2.867	3.270	3.670
.50	−2.781	−2.548	−2.300	−1.941	−1.634	−1.281	1.485	1.973	2.432	3.017	3.454	3.892
.60	−2.774	−2.547	−2.305	−1.951	−1.648	−1.296	1.548	2.065	2.555	3.186	3.661	4.139
.70	−2.780	−2.557	−2.319	−1.969	−1.667	−1.315	1.618	2.167	2.692	3.372	3.887	4.409
.80	−2.795	−2.577	−2.341	−1.993	−1.691	−1.338	1.695	2.279	2.840	3.573	4.131	4.700
.90	−2.820	−2.604	−2.370	−2.023	−1.721	−1.364	1.779	2.399	2.999	3.787	4.391	5.010
1.00	−2.853	−2.639	−2.406	−2.059	−1.755	−1.393	1.868	2.526	3.167	4.013	4.665	5.334
1.25	−2.966	−2.752	−2.519	−2.167	−1.854	−1.479	2.113	2.873	3.621	4.620	5.394	6.196
1.50	−3.114	−2.897	−2.659	−2.298	−1.973	−1.579	2.379	3.248	4.109	5.267	6.172	7.110
1.75	−3.286	−3.066	−2.821	−2.446	−2.107	−1.690	2.662	3.643	4.622	5.944	6.981	8.060
2.00	−3.481	−3.254	−3.000	−2.608	−2.251	−1.810	2.954	4.052	5.151	6.641	7.812	9.033
2.50	−3.914	−3.670	−3.394	−2.965	−2.568	−2.072	3.562	4.898	6.243	8.073	9.516	11.02
3.00	−4.391	−4.126	−3.825	−3.351	−2.908	−2.351	4.188	5.766	7.361	9.536	11.26	13.05
3.50	−4.897	−4.608	−4.279	−3.757	−3.266	−2.644	4.825	6.649	8.495	11.02	13.01	15.10
4.00	−5.422	−5.109	−4.749	−4.176	−3.635	−2.946	5.468	7.540	9.639	12.51	14.79	17.17
4.50	−5.963	−5.622	−5.231	−4.606	−4.013	−3.255	6.116	8.437	10.79	14.01	16.57	19.24
5.00	−6.514	−6.146	−5.723	−5.043	−4.397	−3.567	6.767	9.338	11.95	15.52	18.35	21.32
6.00	−7.640	−7.215	−6.724	−5.933	−5.177	−4.204	8.076	11.15	14.27	18.55	21.94	25.50
7.00	−8.787	−8.303	−7.742	−6.837	−5.970	−4.850	9.391	12.97	16.60	21.58	25.54	29.68
8.00	−9.947	−9.403	−8.772	−7.750	−6.770	−5.002	10.71	14.79	18.94	24.63	29.14	33.87
9.00	−11.12	−10.51	−9.809	−8.670	−7.575	−6.158	12.03	16.62	21.28	27.67	32.75	38.07
10.00	−12.29	−11.63	−10.85	−9.594	−8.385	−6.817	13.35	18.44	23.62	30.72	36.36	42.27

Confidence

24 Degrees of freedom

$S(y)$	.0025	.005	.01	.025	.05	.10	Confidence .90	.95	.975	.99	.995	.9975
.10	−2.942	−2.670	−2.386	−1.985	−1.651	−1.277	1.313	1.711	2.072	2.514	2.830	3.137
.20	−2.876	−2.617	−2.346	−1.959	−1.635	−1.270	1.342	1.756	2.135	2.602	2.941	3.271
.30	−2.828	−2.581	−2.320	−1.945	−1.629	−1.269	1.379	1.813	2.213	2.710	3.072	3.429
.40	−2.797	−2.558	−2.305	−1.940	−1.629	−1.274	1.425	1.881	2.305	2.836	3.227	3.614
.50	−2.781	−2.549	−2.302	−1.944	−1.638	−1.284	1.478	1.959	2.409	2.979	3.402	3.823
.60	−2.777	−2.552	−2.309	−1.956	−1.652	−1.299	1.539	2.048	2.528	3.141	3.598	4.057
.70	−2.786	−2.564	−2.325	−1.975	−1.672	−1.319	1.607	2.147	2.659	3.318	3.813	4.312
.80	−2.804	−2.585	−2.349	−2.001	−1.698	−1.342	1.682	2.255	2.802	3.510	4.046	4.588
.90	−2.832	−2.615	−2.380	−2.033	−1.728	−1.369	1.764	2.371	2.955	3.716	4.294	4.881
1.00	−2.867	−2.652	−2.418	−2.069	−1.763	−1.399	1.851	2.495	3.116	3.931	4.554	5.189
1.25	−2.986	−2.771	−2.535	−2.181	−1.866	−1.487	2.089	2.830	3.555	4.513	5.251	6.009
1.50	−3.139	−2.921	−2.680	−2.315	−1.987	−1.589	2.349	3.195	4.027	5.136	5.996	6.881
1.75	−3.318	−3.095	−2.847	−2.467	−2.123	−1.703	2.625	3.579	4.524	5.768	6.772	7.788
2.00	−3.518	−3.287	−3.030	−2.633	−2.271	−1.825	2.911	3.977	5.037	6.460	7.570	8.718
2.50	−3.964	−3.715	−3.434	−2.997	−2.593	−2.090	3.506	4.802	6.096	7.842	9.208	10.62
3.00	−4.452	−4.180	−3.873	−3.389	−2.939	−2.373	4.119	5.649	7.182	9.256	10.88	12.57
3.50	−4.968	−4.672	−4.335	−3.802	−3.302	−2.670	4.743	6.510	8.284	10.69	12.57	14.54
4.00	−5.505	−5.183	−4.814	−4.229	−3.677	−2.976	5.374	7.380	9.397	12.13	14.28	16.52
4.50	−6.057	−5.707	−5.305	−4.665	−4.060	−3.288	6.009	8.257	10.52	13.59	16.00	18.51
5.00	−6.619	−6.241	−5.805	−5.110	−4.449	−3.605	6.648	9.137	11.64	15.05	17.72	20.51
6.00	−7.768	−7.330	−6.824	−6.013	−5.241	−4.250	7.933	10.91	13.90	17.98	21.18	24.52
7.00	−8.936	−8.438	−7.860	−6.931	−6.045	−4.904	9.222	12.68	16.17	20.92	24.65	28.54
8.00	−10.12	−9.557	−8.906	−7.858	−6.855	−5.564	10.52	14.47	18.45	23.86	28.12	32.56
9.00	−11.31	−10.68	−9.961	−8.791	−7.672	−6.228	11.81	16.25	20.72	26.81	31.60	36.60
10.00	−12.51	−11.82	−11.02	−9.729	−8.491	−6.894	13.11	18.04	23.00	29.77	35.09	40.63

27 Degrees of freedom

$S(y)$	.0025	.005	.01	.025	.05	.10	.90	.95	.975	.99	.995	.9975
							Confidence					
.10	−2.892	−2.656	−2.377	−1.980	−1.649	−1.277	1.310	1.706	2.062	2.495	2.804	3.102
.20	−2.862	−2.608	−2.340	−1.957	−1.636	−1.271	1.338	1.749	2.121	2.579	2.907	3.226
.30	−2.820	−2.576	−2.317	−1.945	−1.630	−1.271	1.374	1.802	2.194	2.679	3.030	3.373
.40	−2.794	−2.558	−2.306	−1.942	−1.632	−1.277	1.417	1.867	2.281	2.798	3.174	3.545
.50	−2.782	−2.552	−2.306	−1.948	−1.642	−1.288	1.469	1.942	2.381	2.933	3.339	3.740
.60	−2.783	−2.557	−2.316	−1.962	−1.658	−1.304	1.528	2.027	2.494	3.085	3.523	3.958
.70	−2.795	−2.573	−2.334	−1.983	−1.679	−1.324	1.594	2.122	2.619	3.253	3.725	4.197
.80	−2.817	−2.598	−2.361	−2.011	−1.706	−1.349	1.667	2.225	2.755	3.434	3.943	4.454
.90	−2.848	−2.630	−2.394	−2.044	−1.738	−1.377	1.745	2.337	2.901	3.628	4.176	4.728
1.00	−2.887	−2.670	−2.434	−2.083	−1.774	−1.408	1.830	2.456	3.056	3.833	4.421	5.016
1.25	−3.014	−2.796	−2.558	−2.199	−1.880	−1.498	2.060	2.779	3.474	4.385	5.079	5.786
1.50	−3.174	−2.953	−2.709	−2.338	−2.005	−1.602	2.312	3.130	3.927	4.978	5.784	6.606
1.75	−3.361	−3.134	−2.881	−2.495	−2.145	−1.718	2.579	3.501	4.404	5.599	6.520	7.462
2.00	−3.569	−3.333	−3.070	−2.665	−2.296	−1.843	2.858	3.886	4.897	6.241	7.278	8.342
2.50	−4.029	−3.773	−3.486	−3.038	−2.625	−2.113	3.436	4.683	5.916	7.562	8.836	10.15
3.00	−4.532	−4.253	−3.936	−3.440	−2.979	−2.402	4.033	5.504	6.963	8.916	10.43	11.99
3.50	−5.064	−4.758	−4.410	−3.862	−3.349	−2.704	4.641	6.340	8.027	10.29	12.05	13.66
4.00	−5.616	−5.282	−4.901	−4.298	−3.731	−3.015	5.257	7.184	9.101	11.67	13.67	15.74
4.50	−6.183	−5.820	−5.404	−4.744	−4.122	−3.333	5.876	8.034	10.18	13.07	15.31	17.61
5.00	−6.760	−6.367	−5.916	−5.197	−4.518	−3.655	6.500	8.889	11.27	14.47	16.96	19.53
6.00	−7.938	−7.482	−6.958	−6.119	−5.325	−4.311	7.753	10.61	13.45	17.28	20.26	23.34
7.00	−9.136	−8.616	−8.017	−7.056	−6.142	−4.975	9.013	12.33	15.65	20.10	23.58	27.16
8.00	−10.35	−9.762	−9.086	−8.001	−6.968	−5.645	10.28	14.06	17.84	22.93	26.90	30.99
9.00	−11.57	−10.92	−10.16	−8.952	−7.798	−6.319	11.54	15.80	20.05	25.76	30.22	34.82
10.00	−12.80	−12.08	−11.25	−9.908	−8.632	−6.996	12.81	17.53	22.25	28.60	33.55	38.66

30 Degrees of freedom

$S(y)$	.0025	.005	.01	.025	.05	.10	Confidence .90	.95	.975	.99	.995	.9975
.10	−2.907	−2.645	−2.369	−1.977	−1.648	−1.277	1.308	1.701	2.053	2.480	2.784	3.075
.20	−2.852	−2.601	−2.336	−1.956	−1.636	−1.272	1.335	1.742	2.110	2.559	2.881	3.191
.30	−2.814	−2.572	−2.316	−1.945	−1.632	−1.272	1.370	1.793	2.180	2.655	2.997	3.330
.40	−2.792	−2.557	−2.308	−1.944	−1.635	−1.279	1.412	1.856	2.263	2.767	3.133	3.491
.50	−2.783	−2.555	−2.310	−1.952	−1.646	−1.291	1.462	1.928	2.359	2.896	3.289	3.675
.60	−2.787	−2.563	−2.322	−1.968	−1.662	−1.307	1.519	2.010	2.467	3.041	3.462	3.880
.70	−2.803	−2.581	−2.342	−1.991	−1.686	−1.329	1.583	2.102	2.587	3.200	3.655	4.105
.80	−2.828	−2.608	−2.371	−2.020	−1.714	−1.354	1.654	2.202	2.717	3.373	3.862	4.349
.90	−2.862	−2.643	−2.406	−2.055	−1.747	−1.383	1.731	2.310	2.858	3.559	4.083	4.608
1.00	−2.903	−2.686	−2.449	−2.095	−1.784	−1.414	1.812	2.423	3.007	3.755	4.317	4.880
1.25	−3.037	−2.818	−2.578	−2.215	−1.893	−1.507	2.036	2.737	3.410	4.283	4.943	5.610
1.50	−3.205	−2.981	−2.734	−2.358	−2.020	−1.613	2.282	3.077	3.847	4.852	5.616	6.391
1.75	−3.399	−3.168	−2.911	−2.518	−2.164	−1.732	2.543	3.437	4.307	5.449	6.320	7.206
2.00	−3.613	−3.373	−3.105	−2.693	−2.318	−1.859	2.814	3.812	4.784	6.066	7.046	8.046
2.50	−4.086	−3.825	−3.531	−3.074	−2.654	−2.133	3.380	4.588	5.772	7.339	8.542	9.771
3.00	−4.603	−4.316	−3.992	−3.484	−3.014	−2.427	3.964	5.388	6.787	8.645	10.07	11.54
3.50	−5.149	−4.834	−4.476	−3.914	−3.391	−2.733	4.559	6.201	7.820	9.970	11.63	13.32
4.00	−5.714	−5.369	−4.977	−4.358	−3.779	−3.050	5.161	7.024	8.868	11.31	13.19	15.13
4.50	−6.293	−5.919	−5.491	−4.812	−4.176	−3.372	5.769	7.854	9.913	12.66	14.77	16.94
5.00	−6.884	−6.477	−6.012	−5.273	−4.579	−3.698	6.379	8.686	10.97	14.01	16.35	18.76
6.00	−8.088	−7.616	−7.075	−6.212	−5.397	−4.363	7.607	10.36	13.09	16.73	19.53	22.41
7.00	−9.312	−8.773	−8.154	−7.164	−6.227	−5.037	8.842	12.05	15.22	19.45	22.73	26.08
8.00	−10.55	−9.943	−9.244	−8.125	−7.066	−5.715	10.08	13.74	17.36	22.19	25.92	29.75
9.00	−11.80	−11.12	−10.34	−9.092	−7.909	−6.399	11.32	15.43	19.50	24.93	29.12	33.43
10.00	−13.05	−12.30	−11.44	−10.06	−8.755	−7.085	12.56	17.13	21.64	27.67	32.33	37.11

35 Degrees of freedom

						Confidence					
$S(y)$	.0025	.005	.01	.025	.05	.10	.90	.95	.975	.99	.995	.9975
.10	−2.888	−2.631	−2.361	−1.972	−1.647	−1.277	1.306	1.695	2.043	2.462	2.757	3.040
.20	−2.840	−2.593	−2.331	−1.954	−1.636	−1.272	1.332	1.734	2.096	2.534	2.846	3.146
.30	−2.808	−2.569	−2.315	−1.946	−1.633	−1.275	1.364	1.783	2.161	2.623	2.954	3.274
.40	−2.790	−2.558	−2.310	−1.948	−1.639	−1.282	1.404	1.841	2.239	2.729	3.080	3.422
.50	−2.787	−2.560	−2.316	−1.958	−1.651	−1.295	1.452	1.910	2.329	2.849	3.225	3.592
.60	−2.796	−2.572	−2.330	−1.976	−1.669	−1.313	1.507	1.988	2.432	2.984	3.387	3.782
.70	−2.815	−2.594	−2.354	−2.001	−1.694	−1.336	1.568	2.075	2.545	3.134	3.565	3.990
.80	−2.845	−2.625	−2.386	−2.032	−1.724	−1.361	1.636	2.171	2.668	3.296	3.758	4.215
.90	−2.883	−2.663	−2.425	−2.069	−1.759	−1.391	1.710	2.273	2.801	3.471	3.965	4.455
1.00	−2.929	−2.709	−2.470	−2.112	−1.798	−1.424	1.789	2.383	2.943	3.655	4.183	4.708
1.25	−3.072	−2.850	−2.606	−2.237	−1.911	−1.519	2.005	2.682	3.327	4.153	4.770	5.389
1.50	−3.250	−3.021	−2.769	−2.386	−2.043	−1.629	2.242	3.008	3.743	4.691	5.403	6.119
1.75	−3.453	−3.216	−2.954	−2.552	−2.190	−1.750	2.494	3.355	4.183	5.256	6.066	6.882
2.00	−3.676	−3.430	−3.155	−2.733	−2.349	−1.881	2.758	3.715	4.639	5.842	6.752	7.670
2.50	−4.169	−3.899	−3.595	−3.125	−2.694	−2.161	3.305	4.463	5.585	7.052	8.167	9.294
3.00	−4.705	−4.407	−4.071	−3.547	−3.063	−2.461	3.872	5.234	6.559	8.296	9.617	10.96
3.50	−5.269	−4.941	−4.570	−3.988	−3.448	−2.775	4.450	6.020	7.551	9.560	11.09	12.64
4.00	−5.853	−5.494	−5.086	−4.444	−3.846	−3.097	5.036	6.816	8.554	10.84	12.58	14.35
4.50	−6.452	−6.060	−5.614	−4.910	−4.252	−3.426	5.626	7.618	9.564	12.12	14.08	16.06
5.00	−7.061	−6.635	−6.150	−5.382	−4.664	−3.759	6.219	8.424	10.58	13.42	15.58	17.78
6.00	−8.302	−7.807	−7.241	−6.343	−5.500	−4.436	7.415	10.05	12.62	16.01	18.60	21.23
7.00	−9.563	−8.997	−8.348	−7.318	−6.348	−5.122	8.616	11.68	14.67	18.62	21.63	24.70
8.00	−10.84	−10.20	−9.467	−8.301	−7.204	−5.815	9.821	13.31	16.73	21.23	24.68	28.17
9.00	−12.12	−11.41	−10.59	−9.292	−8.064	−6.510	11.03	14.95	18.79	23.85	27.72	31.65
10.00	−13.41	−12.63	−11.72	−10.29	−8.928	−7.208	12.24	16.59	20.85	26.47	30.77	35.14

40 Degrees of freedom

						Confidence						
S(y)	.0025	.005	.01	.025	.05	.10	.90	.95	.975	.99	.995	.9975
.10	−2.874	−2.622	−2.354	−1.969	−1.645	−1.277	1.304	1.690	2.034	2.447	2.738	3.014
.20	−2.831	−2.587	−2.328	−1.954	−1.636	−1.274	1.328	1.727	2.085	2.516	2.821	3.113
.30	−2.803	−2.567	−2.315	−1.948	−1.635	−1.277	1.360	1.773	2.146	2.599	2.922	3.232
.40	−2.790	−2.559	−2.312	−1.951	−1.642	−1.285	1.399	1.830	2.221	2.699	3.041	3.371
.50	−2.791	−2.564	−2.320	−1.964	−1.656	−1.299	1.445	1.896	2.307	2.813	3.177	3.530
.60	−2.804	−2.580	−2.338	−1.983	−1.676	−1.318	1.498	1.971	2.404	2.941	3.330	3.708
.70	−2.827	−2.605	−2.365	−2.010	−1.701	−1.341	1.558	2.055	2.513	3.083	3.498	3.904
.80	−2.860	−2.638	−2.399	−2.043	−1.733	−1.368	1.624	2.146	2.631	3.238	3.681	4.116
.90	−2.901	−2.680	−2.440	−2.082	−1.769	−1.399	1.695	2.246	2.759	3.404	3.876	4.342
1.00	−2.950	−2.729	−2.487	−2.126	−1.809	−1.432	1.771	2.352	2.894	3.579	4.083	4.581
1.25	−3.102	−2.877	−2.629	−2.256	−1.925	−1.530	1.981	2.641	3.263	4.054	4.641	5.224
1.50	−3.287	−3.055	−2.798	−2.409	−2.061	−1.642	2.211	2.956	3.664	4.569	5.243	5.916
1.75	−3.499	−3.257	−2.989	−2.581	−2.213	−1.766	2.457	3.292	4.087	5.111	5.876	6.642
2.00	−3.731	−3.478	−3.197	−2.766	−2.375	−1.899	2.713	3.640	4.528	5.673	6.531	7.391
2.50	−4.239	−3.961	−3.649	−3.168	−2.727	−2.185	3.248	4.367	5.443	6.836	7.884	8.938
3.00	−4.791	−4.484	−4.138	−3.600	−3.104	−2.491	3.801	5.117	6.386	8.033	9.275	10.52
3.50	−5.371	−5.032	−4.649	−4.051	−3.497	−2.809	4.366	5.881	7.346	9.250	10.69	12.14
4.00	−5.971	−5.599	−5.177	−4.516	−3.902	−3.137	4.938	6.656	8.317	10.48	12.12	13.76
4.50	−6.585	−6.178	−5.717	−4.991	−4.316	−3.471	5.515	7.437	9.297	11.72	13.55	15.40
5.00	−7.210	−6.768	−6.265	−5.473	−4.734	−3.810	6.096	8.222	10.28	12.97	15.00	17.05
6.00	−8.483	−7.968	−7.380	−6.453	−5.585	−4.497	7.266	9.803	12.26	15.47	17.90	20.35
7.00	−9.776	−9.186	−8.512	−7.446	−6.448	−5.194	8.442	11.39	14.25	17.99	20.82	23.67
8.00	−11.08	−10.42	−9.655	−8.449	−7.319	−5.897	9.622	12.99	16.25	20.51	23.74	27.00
9.00	−12.40	−11.65	−10.80	−9.458	−8.194	−6.603	10.81	14.58	18.25	23.04	26.67	30.33
10.00	−13.72	−12.90	−11.96	−10.47	−9.073	−7.312	11.99	16.18	20.25	25.57	29.60	33.67

45 Degrees of freedom

$S(y)$	.0025	.005	.01	.025	.05	.10	Confidence .90	.95	.975	.99	.995	.9975
.10	−2.864	−2.614	−2.349	−1.967	−1.644	−1.277	1.303	1.687	2.028	2.436	2.722	2.994
.20	−2.825	−2.583	−2.326	−1.953	−1.636	−1.275	1.326	1.722	2.076	2.501	2.801	3.087
.30	−2.801	−2.566	−2.315	−1.949	−1.637	−1.278	1.356	1.766	2.135	2.581	2.897	3.199
.40	−2.792	−2.561	−2.315	−1.954	−1.645	−1.288	1.394	1.821	2.206	2.675	3.010	3.331
.50	−2.795	−2.569	−2.325	−1.968	−1.659	−1.302	1.439	1.884	2.290	2.785	3.139	3.482
.60	−2.811	−2.587	−2.345	−1.989	−1.680	−1.321	1.490	1.957	2.383	2.908	3.285	3.651
.70	−2.837	−2.615	−2.374	−2.018	−1.708	−1.345	1.549	2.038	2.488	3.044	3.446	3.837
.80	−2.873	−2.651	−2.410	−2.052	−1.740	−1.373	1.613	2.127	2.602	3.192	3.621	4.039
.90	−2.917	−2.695	−2.453	−2.093	−1.777	−1.405	1.682	2.224	2.725	3.351	3.808	4.255
1.00	−2.969	−2.745	−2.502	−2.139	−1.818	−1.440	1.757	2.327	2.857	3.520	4.006	4.483
1.25	−3.127	−2.899	−2.650	−2.272	−1.938	−1.539	1.962	2.607	3.213	3.977	4.540	5.097
1.50	−3.320	−3.084	−2.824	−2.430	−2.077	−1.653	2.187	2.915	3.601	4.473	5.118	5.758
1.75	−3.538	−3.292	−3.020	−2.605	−2.231	−1.779	2.427	3.241	4.013	4.997	5.727	6.454
2.00	−3.777	−3.519	−3.233	−2.794	−2.397	−1.914	2.677	3.582	4.441	5.540	6.359	7.174
2.50	−4.299	−4.015	−3.696	−3.204	−2.755	−2.205	3.201	4.290	5.330	6.666	7.664	8.662
3.00	−4.864	−4.549	−4.194	−3.643	−3.138	−2.515	3.744	5.023	6.248	7.826	9.008	10.19
3.50	−5.458	−5.110	−4.717	−4.103	−3.538	−2.838	4.298	5.771	7.184	9.007	10.37	11.74
4.00	−6.072	−5.689	−5.255	−4.576	−3.949	−3.171	4.861	6.528	8.131	10.20	11.75	13.31
4.50	−6.700	−6.280	−5.805	−5.060	−4.369	−3.510	5.427	7.293	9.085	11.40	13.15	14.89
5.00	−7.338	−6.882	−6.364	−5.550	−4.795	−3.852	5.998	8.061	10.05	12.61	14.54	16.48
6.00	−8.637	−8.106	−7.500	−6.546	−5.658	−4.549	7.147	9.608	11.98	15.05	17.35	19.66
7.00	−9.958	−9.347	−8.652	−7.556	−6.534	−5.255	8.302	11.16	13.92	17.49	20.18	22.87
8.00	−11.29	−10.60	−9.815	−8.574	−7.416	−5.966	9.461	12.73	15.87	19.94	23.01	26.08
9.00	−12.63	−11.86	−10.98	−9.599	−8.305	−6.681	10.62	14.29	17.82	22.40	25.84	29.30
10.00	−13.98	−13.13	−12.16	−10.63	−9.196	−7.399	11.79	15.85	19.78	24.86	28.68	32.52

50 Degrees of freedom

					Confidence					
$S(y)$	.0025	.005	.01	.025	.05	.10	.90	.95	.975	.99	.995	.9975
.10	−2.856	−2.609	−2.346	−1.966	−1.644	−1.278	1.301	1.684	2.023	2.428	2.710	2.978
.20	−2.820	−2.580	−2.325	−1.953	−1.637	−1.275	1.324	1.718	2.068	2.489	2.785	3.066
.30	−2.799	−2.566	−2.315	−1.950	−1.638	−1.280	1.354	1.761	2.126	2.566	2.877	3.174
.40	−2.793	−2.564	−2.317	−1.957	−1.647	−1.289	1.390	1.813	2.194	2.657	2.984	3.299
.50	−2.800	−2.573	−2.330	−1.972	−1.663	−1.304	1.434	1.876	2.275	2.762	3.109	3.444
.60	−2.818	−2.593	−2.352	−1.995	−1.685	−1.324	1.485	1.946	2.366	2.880	3.249	3.606
.70	−2.846	−2.623	−2.382	−2.024	−1.713	−1.349	1.541	2.025	2.467	3.012	3.404	3.784
.80	−2.884	−2.661	−2.420	−2.060	−1.747	−1.377	1.604	2.112	2.578	3.155	3.572	3.977
.90	−2.930	−2.707	−2.465	−2.102	−1.785	−1.409	1.672	2.206	2.698	3.309	3.752	4.184
1.00	−2.985	−2.761	−2.516	−2.149	−1.827	−1.445	1.745	2.306	2.825	3.472	3.943	4.403
1.25	−3.150	−2.919	−2.667	−2.287	−1.949	−1.547	1.946	2.580	3.172	3.915	4.460	4.994
1.50	−3.348	−3.109	−2.846	−2.447	−2.091	−1.662	2.166	2.881	3.551	4.396	5.018	5.633
1.75	−3.572	−3.323	−3.047	−2.626	−2.247	−1.790	2.402	3.200	3.952	4.904	5.608	6.304
2.00	−3.817	−3.555	−3.264	−2.818	−2.416	−1.928	2.648	3.533	4.369	5.432	6.220	7.000
2.50	−4.352	−4.061	−3.737	−3.236	−2.780	−2.223	3.163	4.228	5.238	6.528	7.487	8.440
3.00	−4.929	−4.606	−4.244	−3.683	−3.169	−2.536	3.697	4.947	6.136	7.658	8.792	9.920
3.50	−5.535	−5.178	−4.775	−4.149	−3.574	−2.864	4.242	5.681	7.051	8.810	10.12	11.43
4.00	−6.161	−5.767	−5.322	−4.629	−3.990	−3.200	4.796	6.424	7.979	9.974	11.46	12.95
4.50	−6.800	−6.370	−5.881	−5.119	−4.416	−3.542	5.354	7.174	8.913	11.15	12.82	14.48
5.00	−7.451	−6.981	−6.450	−5.617	−4.847	−3.889	5.916	7.929	9.854	12.33	14.18	16.02
6.00	−8.774	−8.226	−7.604	−6.627	−5.721	−4.594	7.048	9.449	11.75	14.70	16.91	19.11
7.00	−10.12	−9.488	−8.774	−7.651	−6.608	−5.307	8.186	10.98	13.65	17.09	19.66	22.23
8.00	−11.47	−10.76	−9.955	−8.684	−7.502	−6.026	9.329	12.51	15.56	19.48	22.42	25.34
9.00	−12.84	−12.04	−11.14	−9.722	−8.401	−6.748	10.48	14.05	17.47	21.88	25.17	28.47
10.00	−14.21	−13.33	−12.34	−10.77	−9.302	−7.474	11.62	15.59	19.39	24.28	27.94	31.59

60 Degrees of freedom

						Confidence						
S(y)	.0025	.005	.01	.025	.05	.10	.90	.95	.975	.99	.995	.9975
.10	−2.844	−2.600	−2.340	−1.963	−1.643	−1.278	1.299	1.679	2.015	2.414	2.692	2.954
.20	−2.814	−2.576	−2.322	−1.953	−1.638	−1.277	1.321	1.711	2.058	2.472	2.761	3.035
.30	−2.798	−2.566	−2.317	−1.953	−1.641	−1.282	1.349	1.752	2.112	2.543	2.846	3.134
.40	−2.796	−2.568	−2.322	−1.962	−1.651	−1.293	1.384	1.802	2.177	2.628	2.947	3.251
.50	−2.807	−2.581	−2.337	−1.979	−1.669	−1.309	1.426	1.861	2.253	2.727	3.063	3.385
.60	−2.829	−2.605	−2.362	−2.003	−1.692	−1.330	1.475	1.929	2.339	2.839	3.195	3.536
.70	−2.862	−2.638	−2.395	−2.035	−1.722	−1.355	1.530	2.005	2.436	2.963	3.340	3.703
.80	−2.904	−2.680	−2.436	−2.073	−1.757	−1.385	1.590	2.088	2.542	3.099	3.498	3.884
.90	−2.955	−2.729	−2.484	−2.117	−1.796	−1.418	1.656	2.176	2.656	3.244	3.667	4.078
1.00	−3.013	−2.786	−2.538	−2.167	−1.840	−1.454	1.727	2.275	2.777	3.399	3.848	4.284
1.25	−3.188	−2.954	−2.697	−2.310	−1.967	−1.559	1.921	2.538	3.109	3.820	4.336	4.840
1.50	−3.396	−3.152	−2.883	−2.476	−2.113	−1.678	2.135	2.628	3.472	4.278	4.866	5.442
1.75	−3.630	−3.374	−3.091	−2.660	−2.274	−1.809	2.364	3.136	3.857	4.763	5.426	6.077
2.00	−3.885	−3.615	−3.316	−2.859	−2.447	−1.949	2.603	3.456	4.259	5.268	6.008	6.737
2.50	−4.440	−4.139	−3.804	−3.288	−2.820	−2.251	3.104	4.131	5.097	6.318	7.217	8.104
3.00	−5.037	−4.702	−4.326	−3.747	−3.218	−2.571	3.624	4.828	5.963	7.402	8.463	9.512
3.50	−5.663	−5.291	−4.873	−4.225	−3.633	−2.905	4.156	5.540	6.847	8.507	9.733	10.95
4.00	−6.308	−5.898	−5.435	−4.717	−4.059	−3.247	4.696	6.262	7.743	9.626	11.02	12.40
4.50	−6.968	−6.518	−6.010	−5.219	−4.493	−3.596	5.241	6.991	8.647	10.75	12.31	13.86
5.00	−7.638	−7.147	−6.593	−5.728	−4.933	−3.950	5.789	7.725	9.557	11.89	13.62	15.32
6.00	−9.001	−8.426	−7.777	−6.762	−5.826	−4.667	6.894	9.202	11.39	14.17	16.24	18.28
7.00	−10.38	−9.724	−8.977	−7.809	−6.730	−5.393	8.006	10.69	13.23	16.47	18.87	21.25
8.00	−11.78	−11.03	−10.19	−8.864	−7.642	−6.125	9.122	12.18	15.08	18.77	21.51	24.22
9.00	−13.18	−12.35	−11.41	−9.926	−8.558	−6.860	10.24	13.67	16.93	21.08	24.16	27.21
10.00	−14.59	−13.67	−12.63	−10.99	−9.478	−7.598	11.36	15.17	18.78	23.39	26.81	30.19

70 Degrees of freedom

					Confidence					
S(y)	.0025	.005	.01	.025	.05	.10	.90	.95	.975	.99	.995	.9975
.10	−2.836	−2.595	−2.337	−1.962	−1.643	−1.278	1.298	1.676	2.010	2.405	2.679	2.937
.20	−2.810	−2.574	−2.322	−1.954	−1.639	−1.278	1.318	1.706	2.050	2.458	2.743	3.013
.30	−2.798	−2.567	−2.318	−1.955	−1.643	−1.284	1.346	1.745	2.101	2.526	2.823	3.106
.40	−2.800	−2.572	−2.326	−1.965	−1.655	−1.295	1.380	1.794	2.163	2.607	2.919	3.216
.50	−2.814	−2.588	−2.344	−1.984	−1.673	−1.312	1.421	1.851	2.236	2.701	3.030	3.343
.60	−2.840	−2.614	−2.371	−2.011	−1.698	−1.334	1.468	1.916	2.320	2.808	3.155	3.486
.70	−2.876	−2.650	−2.406	−2.044	−1.729	−1.360	1.521	1.989	2.413	2.927	3.293	3.644
.80	−2.921	−2.695	−2.449	−2.084	−1.765	−1.390	1.580	2.070	2.515	3.057	3.444	3.816
.90	−2.975	−2.747	−2.499	−2.130	−1.806	−1.425	1.644	2.157	2.625	3.197	3.606	4.001
1.00	−3.036	−2.807	−2.556	−2.181	−1.852	−1.462	1.713	2.251	2.742	3.345	3.778	4.197
1.25	−3.219	−2.981	−2.720	−2.328	−1.981	−1.569	1.903	2.507	3.063	3.750	4.246	4.727
1.50	−3.435	−3.187	−2.913	−2.499	−2.131	−1.690	2.112	2.789	3.414	4.191	4.755	5.304
1.75	−3.677	−3.416	−3.127	−2.688	−2.296	−1.824	2.335	3.089	3.788	4.660	5.293	5.913
2.00	−3.940	−3.664	−3.358	−2.892	−2.472	−1.967	2.569	3.403	4.178	5.147	5.854	6.545
2.50	−4.511	−4.202	−3.858	−3.330	−2.853	−2.273	3.060	4.059	4.992	6.163	7.019	7.859
3.00	−5.125	−4.779	−4.393	−3.798	−3.258	−2.599	3.570	4.740	5.835	7.213	8.223	9.215
3.50	−5.767	−5.382	−4.952	−4.286	−3.680	−2.938	4.091	5.435	6.696	8.285	9.450	10.60
4.00	−6.428	−6.004	−5.526	−4.788	−4.113	−3.285	4.621	6.141	7.569	9.370	10.69	11.99
4.50	−7.104	−6.638	−6.113	−5.299	−4.554	−3.639	5.155	6.854	8.450	10.47	11.95	13.40
5.00	−7.790	−7.281	−6.708	−5.818	−5.002	−3.998	5.694	7.572	9.337	11.57	13.21	14.82
6.00	−9.185	−8.589	−7.916	−6.870	−5.909	−4.726	6.779	9.017	11.12	13.78	15.74	17.67
7.00	−10.60	−9.914	−9.141	−7.935	−6.828	−5.462	7.871	10.47	12.92	16.01	18.29	20.53
8.00	−12.02	−11.25	−10.38	−9.010	−7.754	−6.203	8.967	11.93	14.72	18.25	20.85	23.41
9.00	−13.46	−12.60	−11.62	−10.09	−8.685	−6.949	10.07	13.39	16.53	20.49	23.41	26.29
10.00	−14.90	−13.95	−12.86	−11.17	−9.620	−7.697	11.17	14.86	18.34	22.74	25.98	29.17

80 Degrees of freedom

						Confidence						
$S(y)$	.0025	.005	.01	.025	.05	.10	.90	.95	.975	.99	.995	.9975
.10	−2.831	−2.591	−2.334	−1.961	−1.643	−1.278	1.297	1.674	2.005	2.397	2.669	2.924
.20	−2.807	−2.573	−2.321	−1.954	−1.640	−1.279	1.316	1.703	2.044	2.449	2.730	2.995
.30	−2.799	−2.568	−2.320	−1.957	−1.645	−1.285	1.343	1.740	2.093	2.513	2.806	3.084
.40	−2.803	−2.575	−2.330	−1.969	−1.658	−1.297	1.376	1.787	2.153	2.591	2.898	3.189
.50	−2.820	−2.594	−2.349	−1.989	−1.677	−1.315	1.416	1.842	2.224	2.682	3.004	3.310
.60	−2.848	−2.623	−2.378	−2.017	−1.703	−1.337	1.462	1.906	2.304	2.785	3.124	3.447
.70	−2.887	−2.661	−2.415	−2.052	−1.735	−1.364	1.514	1.977	2.394	2.900	3.257	3.599
.80	−2.935	−2.708	−2.460	−2.093	−1.772	−1.395	1.572	2.056	2.493	3.025	3.402	3.764
.90	−2.991	−2.762	−2.512	−2.140	−1.814	−1.430	1.635	2.141	2.600	3.160	3.559	3.942
1.00	−3.056	−2.824	−2.571	−2.193	−1.860	−1.468	1.702	2.232	2.715	3.304	3.725	4.130
1.25	−3.245	−3.004	−2.740	−2.344	−1.993	−1.577	1.888	2.483	3.027	3.696	4.177	4.642
1.50	−3.467	−3.215	−2.938	−2.518	−2.145	−1.700	2.093	2.758	3.369	4.125	4.669	5.198
1.75	−3.716	−3.451	−3.157	−2.711	−2.313	−1.836	2.313	3.052	3.734	4.580	5.192	5.787
2.00	−3.986	−3.705	−3.393	−2.918	−2.493	−1.981	2.543	3.359	4.115	5.054	5.735	6.399
2.50	−4.571	−4.255	−3.903	−3.365	−2.880	−2.292	3.025	4.003	4.911	6.045	6.868	7.672
3.00	−5.198	−4.844	−4.448	−3.841	−3.291	−2.622	3.527	4.671	5.736	7.068	8.038	8.987
3.50	−5.853	−5.459	−5.017	−4.337	−3.719	−2.965	4.040	5.354	6.579	8.113	9.233	10.33
4.00	−6.528	−6.092	−5.602	−4.846	−4.158	−3.317	4.562	6.047	7.434	9.173	10.44	11.69
4.50	−7.218	−6.738	−6.199	−5.365	−4.605	−3.675	5.089	6.747	8.297	10.24	11.66	13.05
5.00	−7.918	−7.393	−6.804	−5.892	−5.059	−4.038	5.619	7.453	9.166	11.32	12.89	14.43
6.00	−9.339	−8.724	−8.032	−6.959	−5.978	−4.774	6.689	8.873	10.92	13.48	15.36	17.20
7.00	−10.78	−10.07	−9.276	−8.040	−6.909	−5.518	7.765	10.30	12.68	15.66	17.85	19.99
8.00	−12.23	−11.43	−10.53	−9.130	−7.846	−6.268	8.845	11.74	14.45	17.85	20.34	22.78
9.00	−13.69	−12.80	−11.79	−10.23	−8.789	−7.022	9.929	13.18	16.22	20.04	22.84	25.58
10.00	−15.16	−14.17	−13.06	−11.33	−9.735	−7.779	11.01	14.62	17.99	22.24	25.34	28.39

100 Degrees of freedom

					Confidence					
S(y)	.0025	.005	.01	.025	.05	.10	.90	.95	.975	.99	.995	.9975
.10	−2.823	−2.586	−2.331	−1.960	−1.642	−1.279	1.295	1.670	1.999	2.387	2.655	2.906
.20	−2.805	−2.572	−2.321	−1.955	−1.641	−1.280	1.314	1.697	2.035	2.434	2.711	2.971
.30	−2.800	−2.571	−2.323	−1.960	−1.648	−1.287	1.339	1.733	2.081	2.494	2.782	3.053
.40	−2.809	−2.582	−2.336	−1.974	−1.662	−1.301	1.371	1.777	2.138	2.567	2.867	3.150
.50	−2.831	−2.604	−2.358	−1.996	−1.683	−1.319	1.409	1.830	2.205	2.653	2.967	3.264
.60	−2.863	−2.636	−2.390	−2.026	−1.711	−1.342	1.454	1.891	2.282	2.750	3.079	3.392
.70	−2.906	−2.678	−2.430	−2.063	−1.744	−1.370	1.504	1.960	2.368	2.859	3.205	3.534
.80	−2.958	−2.728	−2.478	−2.107	−1.783	−1.403	1.560	2.035	2.462	2.978	3.342	3.690
.90	−3.018	−2.786	−2.533	−2.156	−1.826	−1.438	1.621	2.117	2.565	3.107	3.490	3.857
1.00	−3.087	−2.851	−2.595	−2.211	−1.874	−1.478	1.686	2.205	2.674	3.244	3.648	4.035
1.25	−3.286	−3.041	−2.772	−2.368	−2.012	−1.589	1.866	2.447	2.974	3.618	4.077	4.518
1.50	−3.519	−3.261	−2.977	−2.548	−2.169	−1.716	2.066	2.713	3.303	4.027	4.546	5.045
1.75	−3.779	−3.506	−3.204	−2.747	−2.341	−1.855	2.279	2.997	3.655	4.463	5.044	5.605
2.00	−4.059	−3.769	−3.448	−2.961	−2.526	−2.003	2.503	3.295	4.022	4.919	5.564	6.188
2.50	−4.665	−4.338	−3.974	−3.420	−2.921	−2.321	2.974	3.920	4.791	5.870	6.648	7.402
3.00	−5.314	−4.946	−4.536	−3.908	−3.342	−2.657	3.463	4.569	5.590	6.856	7.770	8.659
3.50	−5.991	−5.580	−5.121	−4.416	−3.780	−3.007	3.965	5.233	6.406	7.863	8.917	9.941
4.00	−6.688	−6.232	−5.722	−4.938	−4.228	−3.366	4.474	5.908	7.235	8.885	10.08	11.24
4.50	−7.399	−6.896	−6.334	−5.470	−4.685	−3.731	4.989	6.590	8.072	9.916	11.25	12.55
5.00	−8.120	−7.570	−6.956	−6.008	−5.148	−4.100	5.508	7.277	8.915	10.95	12.43	13.87
6.00	−9.583	−8.938	−8.216	−7.100	−6.086	−4.849	6.555	8.661	10.61	13.05	14.81	16.53
7.00	−11.07	−10.32	−9.492	−8.206	−7.036	−5.607	7.607	10.05	12.32	15.15	17.20	19.20
8.00	−12.56	−11.72	−10.78	−9.320	−7.992	−6.370	8.665	11.45	14.04	17.26	19.60	21.88
9.00	−14.07	−13.13	−12.07	−10.44	−8.953	−7.136	9.725	12.85	15.76	19.38	22.01	24.56
10.00	−15.58	−14.54	−13.37	−11.56	−9.918	−7.906	10.79	14.26	17.48	21.50	24.42	27.26

120 Degrees of freedom

					Confidence					
$S(y)$	.0025	.005	.01	.025	.05	.10	.90	.95	.975	.99	.995	.9975
.10	−2.819	−2.582	−2.329	−1.959	−1.643	−1.279	1.294	1.668	1.995	2.380	2.645	2.893
.20	−2.804	−2.571	−2.322	−1.956	−1.642	−1.281	1.312	1.693	2.028	2.424	2.698	2.954
.30	−2.803	−2.573	−2.326	−1.963	−1.650	−1.289	1.336	1.727	2.073	2.481	2.765	3.031
.40	−2.815	−2.587	−2.341	−1.978	−1.666	−1.303	1.367	1.770	2.127	2.551	2.846	3.123
.50	−2.839	−2.612	−2.365	−2.002	−1.688	−1.322	1.404	1.822	2.192	2.633	2.940	3.231
.60	−2.875	−2.647	−2.399	−2.034	−1.716	−1.346	1.448	1.881	2.266	2.726	3.048	3.353
.70	−2.920	−2.691	−2.442	−2.072	−1.751	−1.375	1.497	1.947	2.349	2.831	3.169	3.489
.80	−2.976	−2.744	−2.492	−2.118	−1.791	−1.408	1.552	2.020	2.441	2.945	3.300	3.638
.90	−3.039	−2.805	−2.549	−2.169	−1.836	−1.445	1.611	2.100	2.540	3.069	3.442	3.798
1.00	−3.111	−2.873	−2.613	−2.225	−1.885	−1.485	1.675	2.186	2.646	3.201	3.594	3.968
1.25	−3.318	−3.069	−2.796	−2.386	−2.026	−1.599	1.851	2.421	2.937	3.563	4.007	4.432
1.50	−3.559	−3.296	−3.007	−2.571	−2.186	−1.728	2.046	2.681	3.257	3.959	4.459	4.940
1.75	−3.827	−3.548	−3.240	−2.775	−2.362	−1.869	2.255	2.958	3.599	4.382	4.941	5.479
2.00	−4.116	−3.819	−3.490	−2.993	−2.550	−2.020	2.475	3.250	3.957	4.824	5.444	6.041
2.50	−4.738	−4.402	−4.029	−3.461	−2.953	−2.343	2.937	3.861	4.707	5.748	6.494	7.214
3.00	−5.404	−5.024	−4.603	−3.959	−3.381	−2.684	3.418	4.496	5.487	6.706	7.583	8.430
3.50	−6.097	−5.673	−5.200	−4.477	−3.826	−3.039	3.911	5.148	6.284	7.687	8.696	9.671
4.00	−6.811	−6.339	−5.813	−5.008	−4.282	−3.403	4.412	5.809	7.095	8.681	9.824	10.93
4.50	−7.538	−7.018	−6.438	−5.549	−4.746	−3.773	4.918	6.478	7.913	9.687	10.96	12.20
5.00	−8.275	−7.707	−7.072	−6.097	−5.216	−4.147	5.429	7.152	8.738	10.70	12.11	13.48
6.00	−9.771	−9.103	−8.356	−7.208	−6.169	−4.906	6.458	8.509	10.40	12.74	14.42	16.06
7.00	−11.29	−10.52	−9.656	−8.332	−7.132	−5.673	7.495	9.876	12.07	14.79	16.75	18.65
8.00	−12.82	−11.94	−10.97	−9.464	−8.103	−6.446	8.536	11.25	13.75	16.85	19.09	21.25
9.00	−14.35	−13.37	−12.28	−10.60	−9.078	−7.223	9.579	12.63	15.44	18.92	21.42	23.85
10.00	−15.89	−14.81	−13.61	−11.74	−10.06	−8.002	10.63	14.01	17.12	20.98	23.77	26.47

160 Degrees of freedom

						Confidence						
$S(y)$	.0025	.005	.01	.025	.05	.10	.90	.95	.975	.99	.995	.9975
.10	−2.813	−2.579	−2.327	−1.958	−1.643	−1.280	1.293	1.664	1.989	2.371	2.633	2.877
.20	−2.803	−2.572	−2.323	−1.958	−1.644	−1.283	1.309	1.688	2.020	2.411	2.680	2.932
.30	−2.807	−2.578	−2.330	−1.967	−1.654	−1.292	1.332	1.720	2.061	2.464	2.742	3.003
.40	−2.824	−2.595	−2.348	−1.985	−1.671	−1.307	1.362	1.761	2.113	2.529	2.817	3.088
.50	−2.852	−2.624	−2.376	−2.011	−1.694	−1.327	1.398	1.810	2.174	2.606	2.906	3.188
.60	−2.892	−2.663	−2.413	−2.045	−1.724	−1.352	1.440	1.866	2.244	2.694	3.007	3.303
.70	−2.943	−2.711	−2.459	−2.086	−1.761	−1.382	1.487	1.930	2.324	2.793	3.121	3.430
.80	−3.002	−2.768	−2.512	−2.133	−1.803	−1.416	1.540	2.001	2.411	2.902	3.245	3.569
.90	−3.070	−2.832	−2.573	−2.187	−1.849	−1.454	1.597	2.077	2.506	3.019	3.379	3.720
1.00	−3.146	−2.904	−2.640	−2.246	−1.901	−1.495	1.659	2.160	2.608	3.145	3.522	3.881
1.25	−3.365	−3.110	−2.831	−2.413	−2.046	−1.612	1.830	2.387	2.887	3.490	3.914	4.319
1.50	−3.618	−3.348	−3.051	−2.605	−2.211	−1.745	2.019	2.637	3.194	3.869	4.345	4.800
1.75	−3.898	−3.610	−3.293	−2.815	−2.392	−1.890	2.223	2.906	3.524	4.273	4.804	5.312
2.00	−4.199	−3.891	−3.552	−3.040	−2.586	−2.044	2.437	3.188	3.869	4.697	5.284	5.847
2.50	−4.846	−4.496	−4.109	−3.522	−2.999	−2.374	2.887	3.781	4.594	5.585	6.290	6.966
3.00	−5.535	−5.139	−4.701	−4.034	−3.438	−2.723	3.356	4.399	5.348	6.507	7.334	8.127
3.50	−6.253	−5.809	−5.316	−4.565	−3.893	−3.085	3.837	5.032	6.121	7.452	8.402	9.315
4.00	−6.991	−6.496	−5.947	−5.110	−4.359	−3.456	4.327	5.675	6.906	8.411	9.486	10.52
4.50	−7.742	−7.196	−6.590	−5.665	−4.834	−3.833	4.822	6.326	7.699	9.381	10.58	11.74
5.00	−8.503	−7.906	−7.241	−6.226	−5.314	−4.215	5.321	6.982	8.499	10.36	11.69	12.96
6.00	−10.05	−9.344	−8.561	−7.364	−6.287	−4.988	6.328	8.305	10.11	12.33	13.91	15.43
7.00	−11.61	−10.80	−9.896	−8.515	−7.271	−5.770	7.341	9.637	11.74	14.31	16.15	17.92
8.00	−13.18	−12.26	−11.24	−9.674	−8.262	−6.557	8.360	10.97	13.37	16.30	18.40	20.42
9.00	−14.77	−13.74	−12.59	−10.84	−9.258	−7.348	9.381	12.32	15.00	18.30	20.65	22.92
10.00	−16.36	−15.22	−13.95	−12.01	−10.26	−8.141	10.41	13.66	16.64	20.29	22.90	25.42

200 Degrees of freedom

					Confidence					
$S(y)$	.0025	.005	.01	.025	.05	.10	.90	.95	.975	.99	.995	.9975
.10	−2.811	−2.577	−2.326	−1.958	−1.643	−1.280	1.292	1.662	1.986	2.365	2.625	2.868
.20	−2.804	−2.573	−2.324	−1.959	−1.646	−1.284	1.307	1.685	2.015	2.403	2.669	2.919
.30	−2.811	−2.581	−2.334	−1.970	−1.656	−1.294	1.330	1.716	2.054	2.453	2.728	2.985
.40	−2.830	−2.602	−2.354	−1.989	−1.674	−1.309	1.358	1.755	2.103	2.515	2.799	3.066
.50	−2.862	−2.633	−2.384	−2.017	−1.699	−1.330	1.393	1.802	2.162	2.589	2.884	3.161
.60	−2.905	−2.674	−2.423	−2.052	−1.731	−1.356	1.434	1.857	2.230	2.674	2.981	3.270
.70	−2.959	−2.725	−2.471	−2.095	−1.768	−1.387	1.481	1.919	2.307	2.769	3.090	3.392
.80	−3.022	−2.785	−2.527	−2.144	−1.811	−1.422	1.532	1.988	2.392	2.873	3.209	3.526
.90	−3.093	−2.852	−2.590	−2.199	−1.859	−1.460	1.588	2.062	2.484	2.987	3.338	3.670
1.00	−3.172	−2.926	−2.659	−2.260	−1.912	−1.503	1.649	2.143	2.583	3.109	3.474	3.824
1.25	−3.399	−3.140	−2.856	−2.432	−2.060	−1.622	1.816	2.364	2.854	3.442	3.855	4.246
1.50	−3.661	−3.384	−3.082	−2.628	−2.229	−1.757	2.002	2.609	3.154	3.810	4.271	4.710
1.75	−3.949	−3.654	−3.330	−2.844	−2.414	−1.904	2.201	2.872	3.475	4.203	4.715	5.204
2.00	−4.258	−3.943	−3.596	−3.073	−2.611	−2.061	2.411	3.148	3.812	4.615	5.181	5.722
2.50	−4.922	−4.562	−4.165	−3.565	−3.032	−2.396	2.853	3.729	4.520	5.479	6.157	6.806
3.00	−5.629	−5.221	−4.770	−4.086	−3.478	−2.750	3.315	4.334	5.257	6.378	7.172	7.932
3.50	−6.364	−5.905	−5.398	−4.627	−3.940	−3.118	3.789	4.956	6.013	7.299	8.212	9.085
4.00	−7.119	−6.608	−6.042	−5.182	−4.414	−3.494	4.270	5.588	6.782	8.235	9.267	10.26
4.50	−7.887	−7.323	−6.697	−5.746	−4.896	−3.876	4.758	6.227	7.559	9.182	10.33	11.44
5.00	−8.666	−8.047	−7.361	−6.317	−5.383	−4.263	5.250	6.871	8.343	10.14	11.41	12.63
6.00	−10.24	−9.515	−8.706	−7.474	−6.371	−5.045	6.241	8.170	9.922	12.06	13.58	15.03
7.00	−11.84	−11.00	−10.06	−8.644	−7.369	−5.837	7.240	9.479	11.51	13.99	15.76	17.45
8.00	−13.45	−12.50	−11.44	−9.822	−8.375	−6.634	8.243	10.79	13.11	15.94	17.95	19.88
9.00	−15.07	−14.00	−12.81	−11.01	−9.385	−7.435	9.250	12.11	14.71	17.89	20.15	22.31
10.00	−16.69	−15.51	−14.19	−12.19	−10.39	−8.238	10.26	13.43	16.32	19.84	22.35	24.74

250 Degrees of freedom

						Confidence					
$S(y)$	.0025	.005	.01	.025	.05	.10	.90	.95	.975	.99	.995	.9975
.10	−2.809	−2.576	−2.325	−1.958	−1.643	−1.280	1.291	1.661	1.983	2.360	2.619	2.860
.20	−2.805	−2.574	−2.326	−1.961	−1.647	−1.285	1.306	1.682	2.010	2.396	2.660	2.907
.30	−2.815	−2.585	−2.337	−1.973	−1.658	−1.295	1.327	1.712	2.048	2.443	2.715	2.970
.40	−2.837	−2.608	−2.359	−1.993	−1.677	−1.311	1.355	1.749	2.095	2.503	2.784	3.047
.50	−2.872	−2.641	−2.391	−2.023	−1.703	−1.333	1.389	1.795	2.152	2.574	2.865	3.138
.60	−2.918	−2.685	−2.433	−2.060	−1.736	−1.360	1.429	1.849	2.218	2.656	2.958	3.242
.70	−2.974	−2.738	−2.483	−2.104	−1.775	−1.391	1.475	1.909	2.293	2.748	3.063	3.359
.80	−3.040	−2.800	−2.540	−2.154	−1.819	−1.427	1.525	1.976	2.375	2.849	3.178	3.488
.90	−3.114	−2.870	−2.605	−2.211	−1.868	−1.466	1.580	2.049	2.465	2.959	3.303	3.627
1.00	−3.196	−2.947	−2.676	−2.273	−1.921	−1.509	1.640	2.128	2.561	3.077	3.437	3.776
1.25	−3.430	−3.167	−2.879	−2.450	−2.073	−1.631	1.804	2.344	2.826	3.401	3.803	4.184
1.50	−3.700	−3.418	−3.111	−2.650	−2.245	−1.768	1.986	2.584	3.118	3.759	4.208	4.633
1.75	−3.996	−3.695	−3.365	−2.870	−2.433	−1.917	2.183	2.842	3.432	4.142	4.640	5.113
2.00	−4.313	−3.990	−3.636	−3.104	−2.634	−2.077	2.389	3.113	3.762	4.544	5.093	5.615
2.50	−4.993	−4.624	−4.217	−3.604	−3.061	−2.416	2.825	3.683	4.455	5.387	6.044	6.669
3.00	−5.716	−5.296	−4.834	−4.135	−3.514	−2.775	3.279	4.278	5.179	6.266	7.034	7.765
3.50	−6.467	−5.995	−5.473	−4.685	−3.983	−3.147	3.746	4.889	5.920	7.168	8.048	8.888
4.00	−7.238	−6.711	−6.129	−5.248	−4.464	−3.528	4.221	5.512	6.675	8.084	9.079	10.03
4.50	−8.022	−7.440	−6.796	−5.821	−4.952	−3.915	4.702	6.140	7.438	9.010	10.12	11.18
5.00	−8.816	−8.178	−7.472	−6.401	−5.447	−4.306	5.187	6.774	8.207	9.944	11.17	12.34
6.00	−10.43	−9.673	−8.839	−7.576	−6.448	−5.098	6.166	8.053	9.759	11.83	13.29	14.69
7.00	−12.05	−11.19	−10.22	−8.763	−7.459	−5.899	7.151	9.342	11.32	13.72	15.42	17.04
8.00	−13.69	−12.71	−11.62	−9.958	−8.478	−6.705	8.142	10.64	12.89	15.63	17.57	19.41
9.00	−15.34	−14.24	−13.02	−11.16	−9.501	−7.515	9.135	11.94	14.47	17.54	19.71	21.79
10.00	−17.00	−15.78	−14.42	−12.37	−10.53	−8.327	10.13	13.24	16.05	19.45	21.86	24.16

300 Degrees of freedom

					Confidence						
S(y)	.0025	.005	.01	.025	.05	.10	.90	.95	.975	.99	.995	.9975
.10	−2.808	−2.575	−2.325	−1.958	−1.644	−1.281	1.290	1.659	1.981	2.357	2.614	2.854
.20	−2.806	−2.575	−2.327	−1.962	−1.648	−1.285	1.305	1.680	2.007	2.391	2.654	2.899
.30	−2.818	−2.588	−2.340	−1.975	−1.660	−1.296	1.326	1.709	2.043	2.437	2.707	2.959
.40	−2.843	−2.612	−2.363	−1.997	−1.680	−1.313	1.353	1.746	2.090	2.494	2.773	3.033
.50	−2.879	−2.648	−2.397	−2.027	−1.707	−1.335	1.387	1.790	2.145	2.563	2.851	3.121
.60	−2.927	−2.693	−2.440	−2.065	−1.740	−1.362	1.426	1.843	2.210	2.643	2.942	3.222
.70	−2.986	−2.748	−2.491	−2.110	−1.779	−1.394	1.471	1.902	2.283	2.733	3.044	3.336
.80	−3.053	−2.812	−2.550	−2.162	−1.824	−1.430	1.520	1.968	2.363	2.832	3.156	3.461
.90	−3.130	−2.884	−2.617	−2.220	−1.874	−1.470	1.575	2.040	2.451	2.939	3.278	3.597
1.00	−3.214	−2.963	−2.690	−2.283	−1.929	−1.514	1.633	2.117	2.545	3.054	3.408	3.742
1.25	−3.454	−3.188	−2.897	−2.463	−2.083	−1.637	1.795	2.330	2.805	3.372	3.767	4.140
1.50	−3.729	−3.444	−3.132	−2.666	−2.257	−1.776	1.976	2.566	3.093	3.722	4.162	4.578
1.75	−4.031	−3.726	−3.391	−2.889	−2.448	−1.927	2.169	2.820	3.402	4.098	4.586	5.047
2.00	−4.354	−4.026	−3.667	−3.127	−2.651	−2.088	2.373	3.088	3.726	4.493	5.030	5.539
2.50	−5.047	−4.670	−4.257	−3.634	−3.084	−2.431	2.804	3.650	4.409	5.322	5.965	6.573
3.00	−5.782	−5.354	−4.882	−4.171	−3.541	−2.794	3.253	4.238	5.122	6.186	6.935	7.646
3.50	−6.545	−6.062	−5.531	−4.728	−4.016	−3.169	3.715	4.842	5.853	7.073	7.931	8.747
4.00	−7.328	−6.789	−6.195	−5.298	−4.501	−3.553	4.186	5.456	6.598	7.975	8.944	9.867
4.50	−8.124	−7.528	−6.871	−5.877	−4.995	−3.944	4.662	6.077	7.350	8.887	9.969	11.00
5.00	−8.930	−8.277	−7.555	−6.464	−5.494	−4.339	5.142	6.704	8.110	9.806	11.00	12.14
6.00	−10.56	−9.793	−8.940	−7.652	−6.505	−5.137	6.111	7.968	9.641	11.66	13.09	14.44
7.00	−12.22	−11.33	−10.34	−8.852	−7.527	−5.945	7.087	9.242	11.18	13.53	15.18	16.76
8.00	−13.88	−12.87	−11.75	−10.06	−8.555	−6.758	8.068	10.52	12.74	15.40	17.29	19.08
9.00	−15.55	−14.42	−13.17	−11.28	−9.588	−7.574	9.052	11.81	14.29	17.29	19.40	21.42
10.00	−17.23	−15.97	−14.59	−12.49	−10.62	−8.394	10.04	13.10	15.85	19.17	21.52	23.75

400 Degrees of freedom

						Confidence					
S(y)	.0025	.005	.01	.025	.05	.10	.90	.95	.975	.99	.995	.9975
.10	−2.807	−2.575	−2.325	−1.959	−1.644	−1.281	1.289	1.658	1.978	2.352	2.608	2.847
.20	−2.808	−2.578	−2.329	−1.964	−1.649	−1.287	1.303	1.677	2.003	2.384	2.645	2.888
.30	−2.823	−2.593	−2.344	−1.978	−1.663	−1.298	1.324	1.705	2.037	2.427	2.695	2.945
.40	−2.850	−2.619	−2.370	−2.001	−1.684	−1.315	1.350	1.740	2.082	2.482	2.758	3.015
.50	−2.890	−2.657	−2.405	−2.033	−1.711	−1.338	1.383	1.784	2.135	2.549	2.833	3.099
.60	−2.941	−2.706	−2.450	−2.073	−1.746	−1.366	1.421	1.835	2.198	2.625	2.920	3.196
.70	−3.002	−2.763	−2.504	−2.120	−1.786	−1.399	1.465	1.892	2.268	2.712	3.018	3.305
.80	−3.073	−2.830	−2.565	−2.173	−1.832	−1.436	1.514	1.957	2.347	2.808	3.126	3.425
.90	−3.153	−2.904	−2.634	−2.233	−1.884	−1.477	1.567	2.027	2.432	2.912	3.244	3.555
1.00	−3.240	−2.986	−2.709	−2.298	−1.940	−1.521	1.624	2.102	2.524	3.024	3.370	3.695
1.25	−3.489	−3.218	−2.922	−2.481	−2.097	−1.646	1.783	2.310	2.777	3.332	3.717	4.080
1.50	−3.772	−3.481	−3.164	−2.690	−2.275	−1.788	1.960	2.542	3.058	3.673	4.101	4.504
1.75	−4.083	−3.770	−3.428	−2.917	−2.469	−1.941	2.150	2.791	3.360	4.039	4.512	4.959
2.00	−4.415	−4.079	−3.710	−3.159	−2.675	−2.105	2.351	3.053	3.678	4.424	4.944	5.436
2.50	−5.125	−4.738	−4.313	−3.676	−3.115	−2.453	2.775	3.605	4.346	5.233	5.853	6.439
3.00	−5.877	−5.436	−4.951	−4.223	−3.581	−2.820	3.218	4.183	5.045	6.077	6.800	7.484
3.50	−6.658	−6.160	−5.612	−4.790	−4.062	−3.200	3.673	4.776	5.762	6.944	7.772	8.557
4.00	−7.458	−6.902	−6.290	−5.369	−4.555	−3.590	4.137	5.380	6.492	7.826	8.761	9.647
4.50	−8.272	−7.656	−6.989	−5.958	−5.056	−3.985	4.606	5.991	7.231	8.719	9.762	10.75
5.00	−9.095	−8.420	−7.676	−6.555	−5.562	−4.385	5.060	6.608	7.976	9.619	10.77	11.86
6.00	−10.76	−9.965	−9.086	−7.761	−6.588	−5.194	6.035	7.852	9.480	11.44	12.81	14.11
7.00	−12.45	−11.53	−10.51	−8.981	−7.623	−6.011	6.999	9.106	10.99	13.26	14.86	16.37
8.00	−14.15	−13.10	−11.95	−10.21	−8.666	−6.834	7.967	10.37	12.52	15.10	16.91	18.63
9.00	−15.85	−14.68	−13.39	−11.44	−9.713	−7.660	8.938	11.63	14.05	16.94	18.98	20.91
10.00	−17.57	−16.27	−14.84	−12.68	−10.76	−8.489	9.911	12.90	15.57	18.79	21.05	23.19

600 Degrees of freedom

							Confidence					
S(y)	.0025	.005	.01	.025	.05	.10	.90	.95	.975	.99	.995	.9975
.10	−2.806	−2.575	−2.325	−1.959	−1.645	−1.282	1.288	1.656	1.975	2.348	2.602	2.839
.20	−2.811	−2.581	−2.332	−1.966	−1.651	−1.288	1.302	1.674	1.998	2.377	2.636	2.876
.30	−2.829	−2.599	−2.349	−1.982	−1.666	−1.300	1.321	1.700	2.031	2.417	2.682	2.928
.40	−2.861	−2.628	−2.377	−2.007	−1.688	−1.318	1.347	1.734	2.072	2.469	2.741	2.994
.50	−2.904	−2.769	−2.415	−2.041	−1.717	−1.342	1.378	1.776	2.124	2.532	2.812	3.074
.60	−2.959	−2.721	−2.463	−2.082	−1.753	−1.371	1.416	1.825	2.184	2.606	2.895	3.166
.70	−3.024	−2.781	−2.519	−2.131	−1.795	−1.404	1.458	1.881	2.252	2.689	2.989	3.269
.80	−3.098	−2.851	−2.583	−2.186	−1.842	−1.442	1.506	1.944	2.328	2.781	3.092	3.384
.90	−3.182	−2.929	−2.654	−2.248	−1.895	−1.484	1.558	2.012	2.411	2.881	3.205	3.509
1.00	−3.273	−3.014	−2.732	−2.315	−1.953	−1.530	1.614	2.085	2.500	2.989	3.326	3.643
1.25	−3.512	−3.254	−2.953	−2.504	−2.114	−1.658	1.769	2.288	2.745	3.286	3.661	4.012
1.50	−3.825	−3.527	−3.202	−2.718	−2.296	−1.802	1.942	2.514	3.018	3.616	4.031	4.420
1.75	−4.147	−3.825	−3.474	−2.952	−2.494	−1.958	2.129	2.757	3.312	3.971	4.428	4.858
2.00	−4.489	−4.143	−3.764	−3.200	−2.705	−2.125	2.326	3.013	3.622	4.345	4.847	5.319
2.50	−5.221	−4.821	−4.383	−3.728	−3.154	−2.479	2.741	3.553	4.274	5.131	5.727	6.289
3.00	−5.995	−5.537	−5.037	−4.287	−3.628	−2.853	3.176	4.119	4.956	5.953	6.647	7.301
3.50	−6.797	−6.280	−5.714	−4.865	−4.119	−3.239	3.624	4.700	5.657	6.797	7.592	8.341
4.00	−7.619	−7.040	−6.406	−5.456	−4.620	−3.634	4.080	5.293	6.371	7.657	8.553	9.399
4.50	−8.454	−7.813	−7.110	−6.057	−5.130	−4.035	4.542	5.892	7.094	8.527	9.526	10.47
5.00	−9.299	−8.595	−7.823	−6.665	−5.645	−4.441	5.008	6.497	7.823	9.405	10.51	11.55
6.00	−11.01	−10.18	−9.264	−7.895	−6.688	−5.262	5.948	7.718	9.294	11.18	12.49	13.73
7.00	−12.74	−11.78	−10.72	−9.137	−7.741	−6.091	6.896	8.949	10.78	12.96	14.48	15.92
8.00	−14.48	−13.39	−12.19	−10.39	−8.801	−6.925	7.849	10.19	12.27	14.75	16.49	18.13
9.00	−16.23	−15.00	−13.66	−11.64	−8.965	−7.763	8.806	11.43	13.77	16.56	18.50	20.34
10.00	−17.98	−16.63	−15.14	−12.90	−10.93	−8.604	9.764	12.67	15.26	18.36	20.51	22.55

1000 Degrees of freedom

					Confidence					
$S(y)$	.0025	.005	.01	.025	.05	.10	.90	.95	.975	.99	.995	.9975
.10	−2.807	−2.576	−2.326	−1.960	−1.645	−1.282	1.288	1.654	1.972	2.343	2.597	2.832
.20	−2.815	−2.584	−2.335	−1.969	−1.653	−1.289	1.300	1.671	1.993	2.370	2.627	2.866
.30	−2.837	−2.605	−2.355	−1.987	−1.669	−1.302	1.318	1.696	2.024	2.407	2.670	2.914
.40	−2.872	−2.638	−2.385	−2.014	−1.693	−1.321	1.343	1.728	2.064	2.457	2.725	2.975
.50	−2.919	−2.682	−2.426	−2.049	−1.723	−1.346	1.374	1.769	2.113	2.516	2.793	3.050
.60	−2.977	−2.737	−2.476	−2.092	−1.760	−1.375	1.410	1.816	2.171	2.586	2.871	3.137
.70	−3.046	−2.801	−2.535	−2.143	−1.804	−1.410	1.452	1.870	2.237	2.666	2.961	3.236
.80	−3.124	−2.874	−2.602	−2.200	−1.853	−1.449	1.498	1.931	2.310	2.755	3.060	3.345
.90	−3.212	−2.955	−2.676	−2.264	−1.907	−1.492	1.549	1.997	2.390	2.851	3.168	3.464
1.00	−3.307	−3.043	−2.757	−2.333	−1.966	−1.538	1.603	2.068	2.476	2.955	3.285	3.593
1.25	−3.576	−3.293	−2.985	−2.528	−2.131	−1.669	1.755	2.266	2.714	3.243	3.606	3.947
1.50	−3.881	−3.575	−3.242	−2.748	−2.318	−1.816	1.925	2.486	2.980	3.562	3.964	4.340
1.75	−4.213	−3.882	−3.522	−2.987	−2.514	−1.976	2.108	2.723	3.266	3.906	4.348	4.763
2.00	−4.567	−4.210	−3.820	−3.241	−2.736	−2.145	2.301	2.974	3.567	4.269	4.753	5.208
2.50	−5.321	−4.907	−4.455	−3.782	−3.194	−2.505	2.709	3.503	4.204	5.033	5.607	6.146
3.00	−6.118	−5.643	−5.125	−4.353	−3.677	−2.885	3.136	4.057	4.870	5.833	6.500	7.126
3.50	−6.943	−6.405	−5.818	−4.943	−4.177	−3.278	3.576	4.627	5.555	6.656	7.418	8.134
4.00	−7.787	−7.185	−6.527	−5.547	−4.688	−3.680	4.025	5.208	6.254	7.494	8.353	9.160
4.50	−8.644	−7.977	−7.247	−6.160	−5.206	−4.087	4.478	5.796	6.961	8.342	9.300	10.20
5.00	−9.511	−8.778	−7.976	−6.779	−5.731	−4.499	4.937	6.390	7.674	9.198	10.26	11.25
6.00	−11.27	−10.40	−9.449	−8.033	−6.791	−5.333	5.863	7.588	9.115	10.93	12.19	13.37
7.00	−13.04	−12.03	−10.94	−9.299	−7.863	−6.173	6.796	8.797	10.57	12.67	14.13	15.50
8.00	−14.82	−13.68	−12.43	−10.57	−8.940	−7.020	7.735	10.01	12.03	14.42	16.08	17.64
9.00	−16.62	−15.34	−13.94	−11.85	−10.02	−7.870	8.676	11.23	13.49	16.18	18.04	19.80
10.00	−18.41	−17.00	−15.45	−13.14	−11.11	−8.723	9.620	12.45	14.96	17.94	20.00	21.94

GLOSSARY OF MEASUREMENT TERMINOLOGY*

GENERAL TERMS

Accuracy [EPA, rev.]   The degree of agreement between an observed value and an accepted reference value. Accuracy includes a combination of two types of components: random error (precision) and systematic error (bias), both types can be affected by any or all of the parts of the measurement process. An alternative definition of accuracy is "absence of bias." EPA recommends that *accuracy* not be used and that *precision* and *bias* be used to convey the information usually associated with accuracy. *See also* Precision *and* Bias.

Assessment monitoring   Investigative monitoring that is initiated after the presence of a contaminant has been detected in groundwater above a relevant criterion at one or more locations. The objective of the monitoring program is to determine if there is a statistically significant exceedance of a standard or criterion at a potential area of concern (PAOC) or at the groundwater venting to surface-water interface, and/or to quantify the rate and extent of migration of constituents detected in groundwater above residential criteria.

---

*Abbreviations:
AG-MF—Agilent Technologies/Measurement Forum (formerly part of Hewlett-Packard), see http://metrologyforum.tm.agilent.com/terminology.shtml; EPA—United States Environmental Protection Agency; see http://www.epa.gov; ICH—International Conference on Harmonisation, see http://www.ifpma.org; IEEE—Institute of Electrical and Electronics Engineers; see http://www.ieee.org; M—Mensor, see http://www. mensor.com; MG—Measurements Group, see http://www.measurementsgroup.com/mg.htm.

324 GLOSSARY OF MEASUREMENT TERMINOLOGY

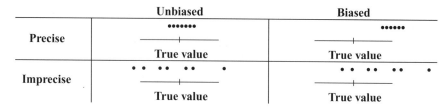

Figure G.1 Bias and precision.

Bias [EPA, rev.] Systematic or persistent error due to distortion of a measurement process, which deprives the result of representativeness (i.e., the expected sample measurement is different than the sample's true value; see Figures G.1 and G.2).

Calibrate To determine, by systematic experimentation, the transformation (mathematical or algorithmic) that relates the response of a measurement system to a true value over a range of true values (see Figure G.3).

Capable/capability In a measurement system, meeting the needs of the measurement customer. Capability is demonstrated by evaluating the system and providing objective evidence that the requirements for a specific intended use are satisfied.

Compliance monitoring Instituted when hazardous constituents have been detected above a relevant criterion at the compliance point during detection monitoring. Groundwater samples are collected at the compliance point, property-boundary of the facility and at upgradient monitoring wells for analysis of hazardous constituents to determine if they are leaving the regulated unit at concentrations that are statistically significant.

Corrective action monitoring Instituted when hazardous constituents from a regulated unit have been detected at statistically significant concentrations between the compliance point and the downgradient property-boundary of the facility. Corrective action monitoring is conducted throughout the corrective action program that is implemented to address groundwater contamination.

Data quality objectives (DQOs) [EPA, rev.] Qualitative and quantitative statements of the overall level of uncertainty that a decision maker is willing to accept

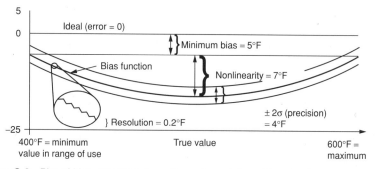

Figure G.2 Plot of bias, precision, linearity, and resolution of temperature measurement.

GENERAL TERMS 325

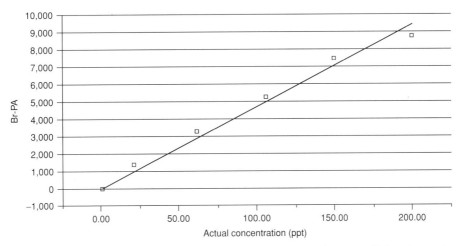

Figure G.3 Straight-line calibration [Br-PA = −48.454 + 46.414 x (ppt actual)]: bromine peak areas from ion chromatography versus actual concentration.

in results or decisions derived from measurements. DQOs provide the statistical framework for planning and managing measurement plans consistent with the data user's needs.

Detection/detection limit The lowest level for which there is simultaneous high confidence that (a) detection will occur if the true value is at the detection limit; (b) there will *not* be detection if the true value is zero (or any specified "null" value; see Figure G.4).

Detection monitoring A program of monitoring for the express purpose of determining whether or not there has been a release of a contaminant to groundwater.

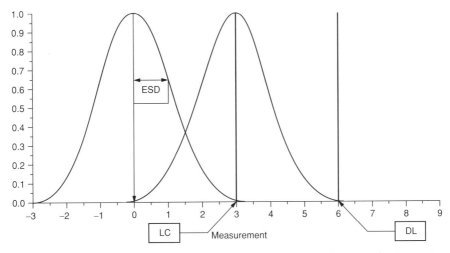

Figure G.4 Distributions of unbiased measurements at true = 0 and true = LC = 3, showing measurement errors and detection limit = 6.

Detection monitoring involves collection of groundwater samples from compliance point and upgradient monitoring wells on a semiannual basis for analysis of hazardous constituents of concern. Results are evaluated to determine if there is a statistically significant exceedance of background.

Error (measurement) [EPA, rev.] The difference between an observed measurement value and a specified, supposedly correct value (i.e., the true value or accepted value).

Error standard deviation (ESD) The standard deviation of measurement error, commonly used to quantify precision. ESD is estimated by making repeat measurements, or measurements of replicate samples. Note that ESD can be defined for different contexts, such as within-laboratory, or between laboratory, measurement. ESD may depend on the true value. *See also* Precision function.

Gauge repeatability and reproducibility study (gauge R&R study) A basic experimental plan to estimate repeatability and some version of reproducibility (such as between-analyst, or between-instrument, or both). A simple gauge R&R study might be used to demonstrate the need for a more extensive and expensive, customized measurement capability study. There are several versions of gauge R&R studies: a gauge potential study might involve two people measuring each of 10 parts/samples twice; a short-term study might involve at least one person making five measurements of each of 25 parts/samples; a long-term study might involve at least one person making 25 measurements of each of 8 parts/samples (see Figures G.5 to G.7).

In control [EPA] A condition indicating that performance of the quality control system is within the specified control limits (i.e., that a stable system of chance is operating and resulting in statistical control). Also: a condition where time-series plots of key variables show random scatter within constant limits; there are no trends, outliers, shifts, cycles, or changes in variability.

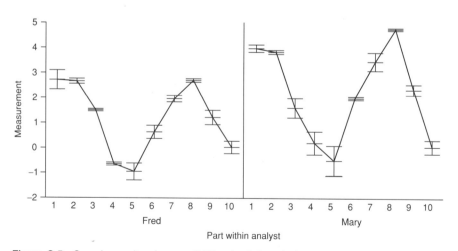

Figure G.5 Sample results of gauge R&R potential study (two people × two measurements × 10 parts.

Gage R&R		
		%Tolerance
Measurement Unit Analysis	1.9912890	497.82
Repeatability	3.5250412	881.26
Operator *Part Variation (IV)	1.8419473	460.49
Gage R&R (RR)	4.4479116	1111.98
Part Variation (PV)	7.9666755	1991.67
Total Variation (TV)	9.1242444	
Sigma Multiple	5.1500000	
Tolerance	0.4000000	

Gage R&R Report assumes that column 'Analyst' represents Operator, and column 'part' represents Part.

Figure G.6 Summary statistics of gauge R&R potential study.

Measurement capability study (MCS) An experiment designed to provide estimates of measurement performance such as precision, bias, response time, or robustness to extraneous factors. The gauge R&R study is a special case of a MCS.

Measurement uncertainty [AS/MF] A figure of merit associated with the actual measured value; the boundary limits within which the "true" value lies. Contributors to this "potential for inaccuracy" include the performance of the equipment used to make the measurement, the test process or technique itself, and environmental effects. Additional imprecision may result from behavior of the measurand or part/sample.

Method/procedure/standard operating procedure (SOP) The formal step-by-step procedure to be followed to make a measurement. This may include, but is not limited to, the sample, the reference standard and the reagents preparations, use of the apparatus, generation of the calibration curve, and use of the formulas for the calculation. May be standardized by organizations and agencies such as ASTM, ISO, MIL-Standards, NIST, or EPA, or by individual corporations or laboratories.

Method detection limit (MDL) A critical level defined by EPA in 40 Code of Federal Regulations (CFR), Part 136, Appendix B (in its basic version) as $3.14s'$, where s' is the sample standard deviation of measurement of seven trace-level subsamples. 40 CFR136 B also describes variants of this definition, involving more samples.

Variance Component Estimates					
Component	Var. Comp. Est.	% of Total	Cum. Total	Cum. %	Sqrt(VC)
Analyst	0.4685047	14.93	0.469	14.9	0.684
Part	2.3929840	76.24	2.861	91.2	1.547
Analyst*part	0.1279204	4.08	2.989	95.2	0.358
Error	0.1495045	4.76	3.139	100.0	0.387

Figure G.7 Variance components analysis of gauge R&R potential study.

EPA's claim is that the MDL is the minimum concentration of an analyte which in a given matrix and with a specific method has a 99% probability of being identified, qualitatively or quantitatively measured, and reported to be greater than zero. Technically, the MDL is a critical level, *not* a detection limit. *See also* Detection limit.

Noise [EPA]   The sum of random errors (i.e., time-varying errors) in the response of a measuring instrument.

Precision [EPA]   The degree to which a set of observations or measurements of the same property, usually obtained under similar conditions, conform to themselves; a data quality indicator. Precision is usually expressed as standard deviation, variance, or range, in either absolute or relative terms. *See also* Accuracy and Bias (in this section); Standard deviation (in the section "Technical Terms").

Precision/tolerance ratio (*P/T* ratio)   A simple figure of merit for measurement systems, indicating what proportion of the tolerance (i.e., a required specification range) is "eaten up" by the imprecision (representing uncertainty). Bias is assumed to be zero. $P/T = 2(3s)/(USL - LSL)$, where s is the standard deviation of measurement error, USL is the upper specification limit (for the product or process), and LSL is the lower specification limit. Generally, it is recommended to ensure $P/T < 0.25$, or $P/T < 0.1$.

QA/QC   Shorthand for quality assurance and quality control systems or activities, combined.

Quality assurance (QA) [EPA]   An integrated system of activities involving planning, quality control, quality assessment, reporting, and quality improvement to ensure that a product or service meets defined standards of quality with a stated level of confidence.

Quality control (QC) [EPA]   The overall system of technical activities whose purpose is to measure and control the quality of a product or service so that it meets the needs of users. The aim is to provide quality that is satisfactory, adequate, dependable, and economical.

Quantitation/quantitation limit (QL)   The lowest level at which a specified (estimated) relative standard deviation is achieved, typically 10%, 20%, or 30%. (*Note*: EPA uses an ambiguous, almost-circular definition of QL: the maximum or minimum levels or quantities of a target variable that can be quantified with the certainty required by the data user.)

Random [EPA]   Lacking a definite plan, purpose, or pattern; due to chance.

Random error [EPA, rev.]   The deviation of an observed value from a true value, which behaves like a variable in that any particular value occurs as though chosen at random from a probability distribution of such errors. The distribution of random error is often assumed to be normal or lognormal.

Repeatability [EPA]   The degree of agreement between mutually independent test results produced by the same analyst using the same test method and equipment on random aliquots of the same sample within a short period of time. Repeatability is usually quantified by the repeatability of standard deviation.

Reproducibility [EPA]   The extent to which a method, test, or experiment yields the same or similar results when performed on subsamples of the same sample by different analysts or laboratories. [*Note*: The term *reproducibility* should be avoided, because it has been used in such a wide variety of ways. Instead, it is better simply to state what factors were allowed (or forced) to vary and what factors were held constant. Alternatively, one can deliberately design a MCS to provide data for a variance components analysis, which will partition the observed variation into bins associated with different factors. Note that by one definition, reproducibility includes repeatability variation; by another, it does not.]

Sample [EPA] A part of a larger whole or a single item of a group; a finite part or subset of a statistical population. A sample serves to provide data or information concerning the properties of the entire group or population.

Significant digit/significant figure [EPA] Any of the digits 0 through 9, excepting leading zeros and some trailing zeros, which is used with its place value to denote a numerical quantity to a desired rounded number. Also: a digit that has some information content (after rounding).

Standard [EPA] A substance or material with a property quantified with sufficient accuracy to permit its use to evaluate the same property in a similar substance or material. Standards are generally prepared by placing a reference material in a matrix.

Standard reference material (SRM) [EPA] A certified reference material produced by the U.S. National Institute of Standards and Technology and characterized for absolute content, independent of analytical method.

Traceability [AG/MF, rev.] The ability to link measuring equipment to a standard of higher accuracy. For any parameter/range we should be able to illustrate this type of hierarchical relationship:

National standard	Accurate to 0.002%
Calibration laboratory	0.01%
Company "master" item	0.07%
Company production equipment	1.0%
Produced product	10.0%

True value The theoretically correct value, never knowable, but for which a measurement is an estimate. Also known as the *actual value*. *See also* Accepted value (in the section "Technical Terms").

TECHNICAL TERMS

Accepted value [M] An assigned value accepted as valid by the parties affected; a historically determined value of a process parameter. *See also* True value (in the section "General Terms").

Blank (sample) [EPA, rev.] A clean sample or a sample of matrix processed so as to measure artifacts in the measurement (sampling and analysis) process. There are many kinds of blanks, including field blanks, trip blanks, and reagent blanks.

Blind sample [EPA]   A subsample submitted for analysis with a composition and identity known to the submitter but unknown to the analyst and used to test the analyst's or laboratory's proficiency in the execution of the measurement process.

Certified reference material (CRM) [EPA]   A reference material that has one or more of its property values established by a technically valid procedure and is accompanied by or traceable to a certificate or other documentation issued by a certifying body.

Check sample/check standard [EPA]   An uncontaminated sample matrix spiked with known amounts of analytes usually from the same source as the calibration standards. It is generally used to establish the stability of the analytical system but may also be used to assess the performance of all or a portion of the measurement system.

[M]   A stable, well-characterized in-house standard that is requalified or remeasured at periodic intervals to determine that the device is within acceptable limits.

Coefficient of variation (COV, CV) [EPA]   A measure of relative dispersion (precision), based on a set of measurements made at one concentration. It is equal to the ratio of the standard deviation of the measurements divided by the arithmetic mean for the known concentration. *See also* Relative standard deviation.

Critical level (L_c or LC)   A threshold used to determine whether or not detection has occurred. The threshold can be applied either to raw responses or reported measurement values. Not to be confused with a detection limit (as is commonly the case with the EPA's MDL).

[Digital] resolution [AG/MF]   The discrimination that the instrument can show (not necessarily equal to the uncertainty); the smallest discernible difference between two measurements.

[M]   The limits between which the input to an instrument can vary without causing a change in the instrument output.

Drift   Instability in a measurement process, typically apparent with the passage of time of 1 second or longer (if shorter, typically such instability is called and treated as "noise"). In general, drift may be due to a factor within any component of the measurement system: "man" (i.e., operator or analyst), machine (i.e., instrumentation), material (i.e., sample being measured, or calibration sample), or method (i.e., procedure), although typically drift is blamed on the machine.

Duplicate samples [EPA]   Two samples taken from and representative of the same population and carried through all steps of the sampling and analytical procedures in an identical manner. Duplicate samples are used to assess the variance of the total method, including sampling and analysis.

Dynamic range [IEEE, rev.]   The ratio, usually expressed in decibels, of the maximum to the minimum true measurand values over which a measurement system can operate within some specified range of performance.

Error, random   *See* Random error (in the section "General Terms").

Error, systematic [EPA] A consistent or predictable deviation in the results of sampling and/or analytical processes from the expected or known value. Such error is caused by human and methodological bias. May be a function of the true value. Also called "bias."

False hit/false positive [EPA] Determining (incorrectly) that an analyte is present when it is actually not present. Also called a *Type I error* or α *error*.

False negative (result) [EPA] Determining (incorrectly) that an analyte is not present when it actually is present (at some potentially detectable level). Also called a *Type II error* or β *error*. The risk of a false negative is a function of the true value.

Filtering A process applied to a signal to reduce temporal or spatial noise or to extract the essential part of the signal (such as a particular band of wavelengths). For example, simple filters include successive average or median, running average (i.e., moving average), exponentially weighted moving average. More complex filters include low-bandpass filters and multidimensional filters.

Gain As determined through calibration; the conversion rate between response units and measurement units (e.g., displacement distance in millimeters per kilogram). Synonym for the slope of the calibration function. (*Note*: Not presented here is the unrelated definition relating gain to signal amplification.) In Figure G.3 the gain is 46.414 peak area counts per ppt bromine.

Guardband [AG/MF] Particularly employed when Test Accuracy Ratios (TARs) are low, a guardband is a safety margin having the purpose of tightening an acceptance (pass) limit when testing a specified product. The guardband limit might, simplistically, be set at a point equal to the specification minus the uncertainty but is often "tuned" to recreate the same confidence that would result from using a 4:1 TAR with the acceptance point set at the specification limits. Guardbands are also employed in manufacturing where routine testing may only be a subset of the product's full (customer) specifications in extent or environment, yet assurance of compliant shipments is desired.

Hysteresis [MG, rev.] The dependence of measurement not only on the instantaneous true value of the measurand, but also on the previous history of the measurand.

Intercept A synonym for *offset*, as determined through calibration; the response predicted when the true value is zero (i.e., the intersection of the calibration line or curve with the vertical axis).

Interference [EPA] A positively or negatively valued effect on a measurement caused by a variable other than the one being investigated.

Interlaboratory method validation study (IMVS) [EPA] The formal study of a sampling and/or analytical method, conducted with replicate, representative matrix samples, following a specific study protocol and utilizing a specific written method, by a minimum of seven laboratories, for the purpose of estimating interlaboratory precision, bias, and analytical interferences. Also called a between-laboratory method validation study.

Interlaboratory round robin (IRR) A study in which non-consumable and non-degrading standards are sent to one qualified laboratory (in a group) after another, and each laboratory makes and reports at least one measurement of each standard, using a prescribed method. Alternatively, large "master" samples may be split and simultaneously distributed to all participating laboratories. The objectives of an IRR generally include the following:

1. To identify individual laboratories with performance problems, such as high bias or variability.
2. To identify types of samples or analyte concentration ranges (or levels) for which the method is especially accurate or inaccurate.
3. Quantify bias and precision for individual laboratories or for aggregated laboratories' measurements.
4. Identify characteristics of the method that need to be improved, perhaps "robustified."

One approach to an IRR is described by ASTM's D2777—a Standard Guide to determining bias and precision. ASTM's Standard Practices, D6091 (IDE) and D6512 (IQE), provide other alternative designs (both Youden and non-Youden) for an IRR.

International System of Units (SI) [M] Seven basic units of measurement recommended by the Système International d'Unités that have been internationally recognized and agreed upon. The seven basic units are the meter (m), kilogram (kg), second (s), kelvin (K), ampere (a), candela (cd), and mole (mol).

Inverse prediction Mathematical inversion of the results of a calibration fit of a (dependent variable) response on a (independent variable) true value, so that a new response value is transformed to a measurement (i.e., an estimate of the true value).

Limit of detection (LOD) *See* Detection limit (in the section "General Terms").

Limit of quantitation (LOQ) *See* Quantitation limit (in the section "General Terms"). (*Note*: EPA uses a definition that is different than that for QL, and similar to the accepted UIPAC/ISO definition of detection limit: the concentration of analyte in a specific matrix for which the probability of producing analytical values above the method detection limit is 99%).

Linearity [EPA] The degree of agreement between the calibration curve of a method and a straight-line assumption can be expressed as a percentage of full-scale deviation.

Lognormal distribution A frequency distribution for a variable whose logarithm follows a normal distribution.

Lower confidence limit (LCL) A statistical estimate of the lower bound for the true mean concentration (or a percentile of the concentration distribution) with specified level of confidence (e.g., 95%) based on m samples.

Lower prediction limit (LPL) A statistical estimate of the minimum concentration that will provide a lower bound for the next series of k measurements from that distribution, or the mean of m new measurements for each of k sets of samples, with specified level of confidence (e.g., 95%) based on a sample of n background measurements.

Mean The Arithmetical average: the sum of a set of values divided by the number of values. *See also* Median.

Measurand [MG] A physical parameter being quantified by measurement (e.g., force, yield strength, percent of copper by weight).

Measurement assurance [M] The ability to quantify the total uncertainty of the measurements (both random error and systematic components of error) with respect to national or other designated standards, and to demonstrate that the total uncertainty is sufficiently small to meet the users' requirements.

Measurement assurance program (MAP) [M] A quality assurance program for a measurement process that quantifies the total uncertainty of the measurements with respect to a national or other designated standard.

Median The value for which one-half (50%) of the observations (when ranked) will lie above that value and one-half will lie below that value. When the number of values in the sample is even, the median is computed as the average of the two middle values.

Metrology The study of measurement systems.

Nondestructive testing (NDT)/nondestructive evaluation (NDE) Any one of a noninvasive measurement technique, such as ultrasonics, radiography, penetrants, magnetics, and eddy currents, used to determine the integrity of a structure, component, or material without destroying the usefulness of the part.

Nonparametric Description of a statistical technique in which the distribution of the constituent in the population is unknown and is not restricted to be of a specified form.

Normal distribution [EPA, rev.] An idealized probability density function that approximates the distribution of many random variables associated with measurements of natural phenomena and averages of measurements. The density takes the form of a symmetric "bell-shaped curve". Mathematically expressed as $f(z) = (2\pi)^{1/2} \exp\{-Z^{1/2}\}$ (see Figure G.8). A characteristic property of the normal

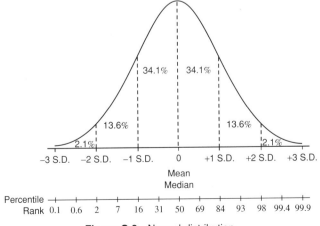

Figure G.8 Normal distribution.

distribution is that 68% of all of its observations fall within a range of ±1 standard deviation from the mean, and a range of ±2 standard deviations includes 95% of the scores.

Offset *See* Intercept.

Outlier [EPA] An observed value that appears to be discordant from the other observations in a sample. The declaration of an outlier is dependent on the significance level of the identification test applied.

Phase I environmental site assessment (ESA) A nonintrusive investigation that identifies potential areas of concern (PAOCs) which may require further investigation in subsequent phases of work.

Phase II environmental site assessment (ESI) An intrusive survey to confirm or deny existence of a release into the environment at a PAOC at levels which may adversely affect public health or the environment.

Potential area of concern (PAOC) An area with a documented release or likely presence of a hazardous substance that could pose an unacceptable risk to human health or the environment.

Precision function A mathematical function relating the estimated precision of a measurement system (e.g., the predicted measurement standard deviation) to the true value or mean measured value. For example, a precision function might be $s = g + hT$, or $s = \sqrt{g^2 + (hT)^2}$, where s is the predicted standard deviation, T is the true value, and g and h are either theoretical or estimated constants.

Prediction interval An interval that will, with a specified degree of confidence, contain one or more future randomly selected measurements.

Proficiency testing [EPA] A systematic program in which one or more standardized samples is analyzed by one or more laboratories to determine the capability of each participant. *See also* Interlaboratory round robin.

Random (sub)sample [EPA] A subset of a population or a subset of a sample, selected according to the laws of chance with a randomization procedure.

Range The interval between the upper and lower levels for which it has been demonstrated that the measurement procedure has a suitable level of precision, accuracy, and linearity.

Reference method [EPA] A sampling and/or measurement method which has been officially specified by an organization as meeting its data quality requirements.

Reference standard/reference material [EPA] A material or substance, one or more properties of which are sufficiently well established to be used for the calibration of an apparatus, for assessment of a measurement method, or for assigning values to materials.

Relative standard deviation (RSD) [EPA] The standard deviation of a group of measurements of a single concentration expressed as a percentage of the mean recovery (or the true concentration) (i.e., RSD is the coefficient of variation multiplied by 100).

Replicate samples [EPA] Two or more samples representing the same population characteristic, time, and place, which are carried independently through all steps of the sampling and measurement process in an identical manner. Replicate samples are used to assess total (sampling and analysis) method variance. Often incorrectly used in place of the term *replicate analysis*.

Representative sample [EPA] A sample taken so as to reflect the variable(s) of interest in the population as accurately and precisely as specified. To ensure representativeness, the sampling procedure may be either completely random or stratified, depending on the conceptualized population and the sampling objective (i.e., on the decision to be made).

Response time (time constant) The time required for a measurement system (or sensor) to register (i.e., change by) a specified percentage (typically 63.2%) of the difference between an old level and a new level after application of a step change or disturbance, under a specified set of conditions. Three time constants are required for the sensor to reach 95% of the step change value. For example, it may take a thermometer 2 seconds for its reading to travel 63% of the way from room temperature (25°C) to 200°C when plunged in boiling water [i.e., after 2 seconds, it will read $135° = 25° + 0.63(200 - 25)°C$]. Note that response time and precision can often be traded off through signal averaging (e.g., moving average, or exponentially weighted moving average).

Robustness A measure of the capacity of a measurement system to remain unaffected by small (possibly deliberate) variations in method or environmental parameters; provides an indication of its reliability during normal usage.

Sample preparation The act of producing a manageable sample, suitable for measurement while ensuring that the prepared sample is homogeneous and fully representative of the original material. Typically, sample preparation is required in environmental measurement.

Sample standard deviation [EPA] A measure of the dispersion of a set of values. The square root of the sum of the squares of the difference between the individual values of a set and the arithmetic mean of the set, divided by the square root of 1 less than the number of values in the set. The sample standard deviation is the square root of the sample variance, $S = [\Sigma(x - \bar{x})^2/(n - 1)]^{1/2}$

Sample variance [EPA] A measure of the dispersion of a set of values. The sum of the squares of the difference between the individual values of a set and the arithmetic mean of the set, divided by 1 less than the number of values in the set. The sample variance is the square of the sample standard deviation.

Sampling error [EPA] The difference between an estimate of a population value and its true value. Sampling error is due to observing only a limited number of the total possible values and is distinguished from errors due to imperfect selection, bias in response, errors of observation, measurement or recording, and so on.

Selectivity [EPA] In analytical chemistry, the capability of a method or instrument to respond to a target substance or constituent in the presence of nontarget substances.

Sensitivity [AG/MF] The smallest change in the input (stimulus) that causes a discernible change in the output.

Signal processing The processing (such as detection, averaging, smoothing, transforming, and filtering) of signals, which results in their transformation into other forms.

Signal-to-noise ratio (SNR) [multiple possible definitions]   The ratio between the amplitude of a desired signal and the amplitude of noise signals, often expressed in decibels. Alternatively, the reciprocal of the coefficient of variation.

Significance level [EPA]   The magnitude of the acceptable probability of rejecting a true null hypothesis or of accepting a false null hypothesis.

Single-operator precision [EPA, rev.]   The degree of variation among the individual measurements of a series of measurements by the same analyst or operator, all other conditions being equal (equivalent to repeatability).

Slope   Synonym for the gain as determined through calibration; the conversion rate between response units and measurement units (e.g., displacement distance in millimeters per kilogram).

Specificity [ICH, for analytical measurements]   The ability to assess unequivocally the analyte in the presence of components that may be expected to be present. Typically, these might include impurities, degradants, matrix, and so on.

Standard deviation (σ)   A parameter of a distribution or population defining the degree of variability; estimated by the sample standard deviation. The standard deviation is the square root of the variance.

Subsample [EPA]   A representative portion of a sample. A subsample may be taken from any laboratory or field sample.

Test   A measurement procedure that produces a quantitative or qualitative result. Alternatively, a procedure used to produce an indication of fitness or suitability, possibly with a result in arbitrary units, or possibly a binary result, such as pass/fail.

Test accuracy ratio (TAR) [AG/MF]   A test specification divided by uncertainty. Higher is better.

Upper confidence limit (UCL)   A statistical estimate of the upper bound for the true mean concentration (or a percentile of the concentration distribution or some other parameter) with specified level of confidence (e.g., 95%) based on m samples.

Upper prediction limit (UPL)   A statistical estimate of the maximum concentration that will not be exceeded by the next series of k measurements from that distribution, or the mean of m new measurements for each of k sets of samples, with specified level of confidence (e.g., 95%) based on a sample of n background measurements.

Validation/verification sample   A sample/part taken and measured or tested to confirm or refute an unexpected measurement result.

Variance   A parameter of a distribution or population defining the degree of variability; estimated by the sample variance. The variance is the square of the standard deviation.

Variance component [S]   Used in the context of experimental designs with random effects to denote the estimate of the (amount of) variance that can be attributed to those effects.

Weighted least squares (WLS) An estimation procedure for linear regression models that differentially weights observations based on the magnitude of their variance.

Youden-pair design/Youden study An interlaboratory round-robin study where most or all samples sent to any particular laboratory are *not* replicated but are designed as pairs, where for each pair, the values are within 20% of the mean.

Youden plot A plot of the results of a Youden pair study: first sample versus second sample, where the plotting symbol is different for each laboratory. The benefit is that biased or noisy laboratories stand out.

MATHEMATICAL SYMBOLS

b_{0w} WLS estimate of the intercept of the recovery curve

b_{1w} WLS estimate of the slope of the recovery curve

H_L factor developed by Land (1971) to obtain the lower $100(\alpha)\%$ confidence limit for the mean of a lognormal distribution

H_U factor developed by Land (1971) to obtain the upper $100(\alpha)\%$ confidence limit for the mean of a lognormal distribution

k number of future comparisons for a single monitoring event (e.g., the number of downgradient monitoring wells multiplied by the number of constituents to be monitored) for which statistics are to be computed

k_i variance of sample i at true concentration x (i.e., $s_{x_i}^2$) used as weight in the WLS regression

$K_{0.95,\,0.99}$ one-sided normal tolerance limit factor for 99% coverage and 95% confidence based on n measurements

m number of on-site or downgradient measurements used in computing the on-site mean concentration

n number of background (off-site or upgradient) measurements

p number of unknown parameters in the standard deviation versus concentration model

s sample-based standard deviation of a constituent computed from n background measurements

s^2 sample-based variance of a constituent computed from n background measurements

MATHEMATICAL SYMBOLS

s_x	standard deviation of instrument response or measured concentration at true concentration x
s_w^2	weighted residual variance
t	$100(1 - \alpha)$ percentage point of Student's t-distribution on $n - 1$ degrees of freedom
$\text{Var}(\hat{y})$	variance of a predicted measured concentration or instrument response
$\bar{x}$	sample-based mean or average concentration of a constituent computed from n background measurements
y_C	measured concentration or instrument response at the critical level
y_D	measured concentration or instrument response at the detection limit
y_Q	measured concentration or instrument response at the quantification limit
α	false positive rate for an individual comparison (i.e., one sampling location and constituent)
α^*	site-wide false positive rate covering all sampling locations and constituents
μ	true population mean of a constituent
σ^2	true population variance of a constituent

WEB REFERENCES

American Society for Testing and Materials (ASTM)
http://www.astm.org

Organized in 1898, ASTM is one of the largest voluntary standards development organizations in the world. ASTM is a not-for-profit organization that provides a forum for the development and publication of voluntary consensus standards for materials, products, systems, and services. More than 32,000 members representing producers, users, ultimate consumers, and representatives of government and academia from over 100 countries develop documents that serve as a basis for manufacturing, procurement, and regulatory activities. ASTM develops standard test methods, specifications, practices, guides, classifications, and terminology in 130 areas, covering subjects such as metals, paints, plastics, textiles, petroleum, construction, energy, the environment, consumer products, medical services and devices, computerized systems, electronics, and others. About ASTM 10000 standards are (re-)published each year.

British Standards Institute (BSI)
http://www.bsi-global.com/index.html

The BSI group: (1) facilitates writing the standards that industry and business use to increase efficiency and safety and to trade internationally, (2) oversees the implementation of management systems, (3) inspects commodities and tests products to ensure that they are what they claim to be and do what they claim to do efficiently and safely, and (4) provides related support services and learning opportunities.

Cal Lab—The International Journal of Metrology
http://www.callabmag.com

This useful publication is dedicated to metrology and calibration. Articles are written by practitioners in the field from national laboratories, research facilities, and leading companies, and cover all aspects of calibration theory and practice, traceability, maintenance of standards, measurement methods, error quantification, metrology education, interlaboratory cooperation, and legal, regulatory, and business aspects of metrology. There are numerous links to standards and metrological organizations, laboratories and universities.

Eurachem/Cooperation on International Traceability in Analytical Chemistry
http://www.measurementuncertainty.org

This Web site has an excellent glossary and "Eurachem/Citac Guide: Quantifying Uncertainty in Analytical Measurement," 2nd ed. The focus is on characterizing measurement uncertainty in analytical measurement.

National Institute of Standards and Technology (NIST)
http://www.nist.gov

NIST is an agency of the U.S. Department of Commerce Technology Administration. Established in 1901, NIST works with industry to develop and apply technology, measurements, and standards through four major programs: Measurements and Standards Laboratories (which provides traceability and first-principles-based methods), Advanced Technology Program, Manufacturing Extension Partnership, and Baldrige National Quality Award. NIST has an annual budget of about $720 million, employs about 3200 people, and operates primarily in two locations: Gaithersburg, Maryland, and Boulder, Colorado.

United States Environmental Protection Agency (USEPA)
http://www.epa.gov

This extensive site contains news items, laws and regulations, programs, data, and documents on every imaginable environmental topic.

ANNOTATED BIBLIOGRAPHY

Abramowitz, M.; Stegun, I. A., Eds. (1964). *Handbook of Mathematical Functions.* Applied Mathematics Series 55. National Bureau of Standards, U.S. Government Printing Office, Washington DC.
Excellent source for a variety of tabled values of mathematical functions.

ACS (1985). *Trace Residue Analysis: Chemometric Estimations of Sampling, Amount, and Error.* ACS Symposium Series 284. Kurtz, D. A., Ed. American Chemical Society, Washington, DC.

ACS (1988a). *Detection in Analytical Chemistry: Importance, Theory, and Practice.* ACS Symposium Series 361. Currie, L. A., Ed. American Chemical Society, Washington, DC.
Referenced in Currie (1995).

ACS (1988b). *Principles of Environmental Sampling.* American Chemical Society, Washington, DC.
Lists components of measurement error.

Adams, P. B.; Passmore, W. O.; Campbell, D. E. (1966). *Symposium on Trace Characterization: Chemical and Physical.* Paper 14. National Bureau of Standards, Washington, DC.
Referenced in Currie (1968) and Gibbons (1994). Defines a *minimum working concentration* as that for which the relative standard deviation was 10%.

Agterdenbos, J. (1957a). Uber die genaue Berücksichtigung des Blindwertes bei colorimetrischen Bestimmungen. *Fresenius Z. Anal. Chem.*, **157**, 161–165.
Not in English. Referenced in Agterdenbos (1979). Suggests that the total standard deviation may sometimes be found by (quadratic) addition of an SD and an RSD component.

Agterdenbos, J. (1957b). Veränderung des optimalen Extinktionsgebietes photoelektrischer colorimetrischer Bestimmungen durch nichtphotometrische Einflusse. *Fresenius Z. Anal. Chem.*, **154**, 401–406.

Not in English. Referenced in Agterdenbos (1979). Suggests that the total standard deviation may sometimes be found by (quadratic) addition of an SD and an RSD component.

Agterdenbos, J. (1958). Indirekte photometrische Mikrobestimmung von Barium nach Fällung als Chromat, mit Fehlerdiskussion. *Fresenius Z. Anal. Chem.*, **159**, 202–205.

Not in English. Referenced in Agterdenbos (1979). Suggests that the total standard deviation may sometimes be found by (quadratic) addition of an SD and an RSD component.

Agterdenbos, J. (1979). Calibration in quantitative analysis: 1. General considerations. *Anal. Chim. Acta*, **108**, 315–323.

Discussion of general calibration issues. Claims that the standard deviation is independent of x if $x < 22p$, where p is the limit of detection, and the relative standard deviation is independent of x if $x > 50p$. The total standard deviation may sometimes be found by (quadratic) addition of an SD and an RSD component (similar to Rocke and Lorenzato). Mentions that with weighting there is no good procedure for calculation of a confidence region for the analyte content in the unknown sample.

Agterdenbos, J.; Maessen, F. J. M. J.; Balke, J. (1981). Calibration in quantitative analysis: 2. Confidence regions for the sample content in the case of linear calibration relations. *Anal. Chim. Acta.*, **132**, 127–137.

Detection limits are calculated based on confidence (our prediction) limits around the calibration curve.

Aitchison, J. (1955). On the distribution of a positive random variable having a discrete probability mass at the origin. *J. Am. Stat. Assoc.*, **50**, 901–908.

Useful method for the analysis of censored data where nondetects are treated as zero.

Aitchison, J.; Brown, J. A. C. (1957). *The Lognormal Distribution*. Cambridge: Cambridge University Press.

Excellent source on details of the lognormal distribution and corresponding applications, although not in environmental problems.

Ajhar, R. M.; Dalager, P. D.; Davison, A. L. (1976). Multielement analysis with an inductively coupled plasma/optical emission system. *Am. Lab.*, Mar., pp. 71–78.

The coefficient of variation decreases as concentration increases for ICP data.

Alkemade, C. Th. J.; Snelleman, W.; Boutilier, G. D.; Pollard, B. D.; Winefordner, J. D.; Chester, T. L.; Omenetto, N. (1978). A review and tutorial discussion of noise and signal-to-noise ratios in analytical spectrometry: I. *Spectrochim. Acta B*, **33B**, 383–399.

Provides signal-to-noise-based expressions for calculating detection limits.

Altshuler, B.; Pasternack, B. (1963). Statistical measures of the lower limit of detection of a radioactivity counter. *Health Phys.*, **9**, 293–298.

Before Currie (1968). Detection limits connected with errors of the first (Type I, false positive) and second (Type II, false negative) kind. Points out the inadequacy of methods that do not account for false negative errors. Terminology: minimum significant measured activity (Lc) and minimum detectable true activity (Ld).

Analytical Methods Committee, Royal Society of Chemistry (1987). Recommendations for the definition, estimation, and use of the detection limit. *Analyst*, **112**, 199–204.

Clarification of the IUPAC definition: $X_L = X_B + Ks_B$, where X_L is the detection limit, X_B is the mean blank value, K is recommended to be set at 3, and s_B is the standard deviation of a blank (suggests a minimum of 10 replicates). Suggests that the limit of quantification is not necessary. Discusses instrument detection limits and lists reasons that they are unrealistically low. Notes that precision decreases as concentration increases (nonconstant variance). Discusses the presence of bias in the analytical method. Notes the variety of definitions for detection limits in the literature. Suggests that there is little point in determining detection limits to great precision since the results cannot be interpreted in terms of confidence intervals.

ASTM (1964). *Methods for Emission Spectrochemical Analysis*, 4th ed. American Society for Testing and Materials, Philadelphia.
    Referenced in Gabriels (1970). Detection limit calculated as a function of the standard deviation on the blank value.

ASTM (1974). *Testing Thermal Conductivity Detectors Used in Gas Chromatography.* E 516-74. American Society for Testing and Materials, Philadelphia.
    Referenced in Foley and Dorsey (1984). Discusses the concept of minimum detectability.

ASTM (1977). *Testing Flame Ionization Detectors Used in Gas Chromatography.* E 594-77. American Society for Testing and Materials, Philadelphia.
    Referenced in Foley and Dorsey (1984). Discusses the concept of minimum detectability.

ASTM (1979a). *Testing Fixed Wavelength Photometric Detectors Used in Liquid Chromatography.* E 685-79. American Society for Testing and Materials, Philadelphia.
    Referenced in Foley and Dorsey (1984). Discusses the concept of minimum detectability.

ASTM (1979b). *Use of Electron Capture Detectors Used in Gas Chromatography.* E 697-79. American Society for Testing and Materials, Philadelphia.
    Referenced in Foley and Dorsey (1984). Discusses the concept of minimum detectability.

ASTM (1985). *Standards on Precision and Bias for Various Applications*, 2nd ed. American Society for Testing and Materials, Philadelphia. Referenced in Clayton et al. (1987).
    Points out the inadequacy of methods that do not account for Type II (false negative) errors. Quote: "[If the] true concentration is equal to the criterion of detection and if all analytical results below the criterion of detection were reported as such, then the probability of discerning the substance would be 0.5 or 50%."

ASTM (1987). *American Society for Testing and Materials 1987 Book of Standards.* American Society for Testing and Materials, Philadelphia.
    Points out the incorrectness of the PQL.

ASTM (1989). *Standard Practice for Interlaboratory Quality Control Procedures and a Discussion on Reporting Low-Level Data.* D 4210-89. American Society for Testing and Materials, Philadelphia.
    Referenced in Currie (1995). Provides support for identifying the "detection limit" with L_d instead of L_c.

ASTM (1997). *Standard Practice for 99%/95% Interlaboratory Detection Estimate (IDE) for Analytical Methods with Negligible Calibration Error.* D 6091-97. American Society for Testing and Materials, Philadelphia.
    Uses a 99% confidence, 90% coverage tolerance interval (K) to calculate L_c and a 95% confidence, 90% coverage tolerance interval to calculate L_d. Models the standard deviation as a function of concentration using either a constant model, a straight-line model, or an exponential model. Provides guidance for the appropriate use and application.

ASTM (1998). *Standard Practice for Determination of Precision and Bias of Applicable Test Methods of Committee D-19 on Water.* D 2777-98. American Society for Testing and Materials, Philadelphia.

ASTM (2000). *Standard Practice for Interlaboratory Quantitation Estimate.* D 6512-00. American Society for Testing and Materials, Philadelphia.

Bailey, C. J.; Cox, E. A.; Springer, J. A. (1978). High pressure liquid chromatographic determination of the intermediates/side reaction products in FD&C Red No. 2 and FD&C Yellow No. 5: statistical analysis of instrument response. *J. Assoc. Off. Anal. Chem.*, **61**, 1404–1414.

Calculates limits of detection using the Hubaux and Vos (1970) method (unweighted prediction intervals). Confuses Y_{Ld} and Y_{Lq}. Reports recoveries up to 97% for spiked concentrations of dye that were below the detection limit (as referenced in Owens 1988).

Barnett, V.; Lewis, T. (1984). *Outliers in Statistical Data*, 2nd ed. Wiley, New York.
Comprehensive text on detecting outliers in data.

Beckman, R. J.; Cook, R. D. (1983). Outlier. . . . . s. *Technometrics*, **25**, 119–152.
Review article on statistical methods for detecting outliers.

Berthouex, P. M. (1993). A study of the precision of lead measurements at concentrations near the method limit of detection. *Water Environ. Res.*, **65**, 620–629.

Presents results of an interlaboratory study of lead measurement. Results show that a wide range of calculated MDLs are possible both between laboratories and within a single laboratory, depending on the data used. Comment that the accuracy of these low-level measurements may be affected more by bias than by poor precision (note that EPA's MDL reflects only precision, not bias). Stresses that the detection limit is a statistical, not a chemical concept.

Berthouex, P. M. (1994). *Statistics for Environmental Engineers*. Lewis Publishers, Boca Raton., FL.

Chapter 9 is a discussion of the limit of detection. Chapter 10 is a discussion of "simple methods for analyzing data that are below the limit of detection."

Berthouex, P. M.; Gan, D. R. (1993). A model of measurement precision at low concentrations. *Water Environ. Res.*, **65**, 759–763.

Describes a new approach to understanding the precision of measurements at low concentrations; originally proposed by Pallesen (1985). Authors like the Rocke and Lorenzato (1995) two-component model. Error consists of background noise (b_i), which is constant, and analytical error (a_i), which is proportional to concentration. Defines the characteristic limit as the concentration where the variances of background noise and analytical error are equal, and the limit of guaranteed purity, which defines how large the true concentration might be in light of a given measured value.

Bocek, P.; Novak, J. (1970). Statistical processing of calibration data in quantitative analysis by gas chromatography. *J. Chromatogr.*, **51**, 375–383.

Reports that gas chromatographic analysis may involve nonuniform variance.

Bock, R. D. (1975). *Multivariate Statistical Methods in Behavioral Research*. McGraw-Hill, New York.

Excellent source on multivariate statistical methods.

Boque, R.; Rius, F. X. (1996). Multivariate detection limits estimators. *Chemometr. Intell. Lab. Syst.*, **32**, 11–23.

From the abstract: "This paper critically reviews the theoretical approaches advanced up to the present, describes the hypotheses and theoretical backgrounds on which they are based, and discusses the advantages and limitations of the different techniques and derived estimators. The connection with the derivation of the detection limits based on the variance associated to the predicted response of an individual observation in regression theory is introduced and finally, some suggestions as to potential areas of interest for future development are envisaged."

Boque, R.; Rius, F. X. (1997a). Computing detection limits in multicomponent spectroscopic analysis. *TRAC—Trends Anal. Chem.*, **16**, 432–436.

Boque, R.; Rius, F. X. (1997b). Detection limits in GC-MS multivariate analysis. *Quim. Anal.*, **16**, 81–86.

From the abstract: "A new technique to calculate multivariate detection limits is applied to CC-MS analysis. The approach, based on a principal component analysis followed by the application of the Clayton's method, takes advantage of the high correlation among the instrumental responses recorded. In this way, all information given by the technique is employed in addition to the minimisation of the instrumental noise. The described procedure has been used to calculate the MDL of several commercial pesticides analysed using a GCMS system with electronic impact ionisation. The calculated detection limits values are in the range 0.3–1.2 ppm. Characteristic curves of detection are also provided."

Boque, R.; Larrechi, M. S.; Rius, F. X. (1999). Multivariate detection limits with fixed probabilities of error. *Chemometr. Intell. Lab. Syst.*, **45**, 397–408.

From the abstract: "In this paper, a new approach to calculate multivariate detection limits (MDL) for the commonly used inverse calibration model is discussed. The derived estimator follows the latest recommendations of the International Union of Pure and Applied Chemistry (IUPAC) concerning the detection capabilities of analytical methods. Consequently, the new approach: (a) is based on the theory of hypothesis testing and takes into account the probabilities of false positive and false negative decisions, and (b) takes into account all the different sources of error, both in calibration and prediction steps, which affect the final result. The MDL is affected by the presence of other analytes in the sample to be analyzed; therefore, it has a different value for each sample to be tested and so the proposed approach attempts to find whether the concentration derived from a given response can be detected or not at the fixed probabilities of error. The estimator has been validated with and applied to real samples analysed by NIR spectroscopy."

Boumans, P. W. J. M. (1978). A tutorial review of some elementary concepts in the statistical evaluation of trace element measurements. *Spectrochim. Acta B*, **33B**, 625–634.

Fairly complete review of basic detection limit calculation procedures.

Boumans, P. W. J. M. (1980). *Line Coincidence Tables for Inductively Coupled Plasma Atomic Emission Spectrometry*. Pergamon Press, Oxford.

Review of detection limit calculations.

Boumans, P. W. J. M.; deBoer, F. J. (1972). Studies of flame and plasma torch emission for simultaneous multi-element analysis: I. *Spectrochim. Acta B*, **27B**, 391–414.

The appendix provides a discussion of detection limit calculations. Equation: detection limit $= z s_{rel}/b_1$, where s_{rel} is the relative standard deviation of the background (in response units), b_1 the slope, and z is recommended to be set at 2. The reason that the relative standard deviation is used is unclear. Assumes constant standard deviation.

Boumans, P. W. J. M.; Maessen, F. J. M. J. (1966). Evaluation of detection limits in emission spectroscopy. *Z. Anal. Chem.*, **220**, 241–260.

Detection limits for photographic emission spectroscopy are calculated, as well as both "practical" detection limits attained with a given background structure and "ideal" detection limits attained with a smooth background structure. Provides a brief review of Kaiser's theory.

Boutilier, G. D.; Pollard, B. D.; Winefordner, J. D.; Chester, T. L.; Omenetto, N. (1978). A review and tutorial discussion of noise and signal-to-noise ratios in analytical spectrometry: II. *Spectrochim. Acta B*, **33B**, 401–415.

Signal-to-noise-based expressions for calculating detection limits.

Box, G. E. P.; Jenkins, G. M. (1976). *Time Series Analysis: Forecasting and Control*, 2nd ed. Holden-Day, San Francisco.

One of the leading books on statistical time series.

Brodsky, A. (1986). *Accuracy and Detection Limits for Bioassay Measurements in Radiation Protection*. NUREG-1156. Nuclear Regulatory Commission, Washington, DC.

Referenced in Chambless et al. (1992). Recommends that the MDC should be defined as the 95% upper confidence limit of the distribution of M, where M is obtained when converting L_d into activity or concentration units.

Burke, J. A. (1976). In *Pesticide Residues: A Contribution to Their Interpretation, Relevance, and Legislation*. Frehse, H.; Geissbuhler, H., Eds. Pergamon Press, Oxford.

Referenced in Horwitz (1980). Provides data (pesticide residues) supporting the hypothesis that the coefficient of variation is related to concentration.

Carlson, R. F., MacCormick, A. J. A.; Watts, D. G. (1970). Applications of linear random models to four annual streamflow series. *Water Resour. Res.*, **6**, 1070–1078.

Carroll, R. J.; Ruppert, D. (1988). *Transformation and Weighting in Regression*. London: Chapman & Hall.

Excellent advanced-level source on weighted least squares regression models.

Caulcutt, R.; Boddy, R. (1983). *Statistics for Analytical Chemists*. Chapman & Hall, London.

Cetorelli, J. J.; Winefordner, J. D. (1967). Estimation of the minimum detectable concentration and optimum instrumental conditions in precision absorption spectrophotometry of substances in the condensed phase. *Talanta*, **14**, 705–713.

Defines theoretical minimum detectable concentration. C_{mt} and practical minimum detectable concentration. C_{mp}. Equation for molecular absorption spectroscopy: $C_{mp} = 0.002/(ab)$, where a is the molar absorptivity and b is the thickness of the absorbing layer. Shows that the signal-to-noise (RMS) ratio is a linear function of the analyte concentration near the detection limit.

Cetorelli, J. J.; McCarthy, W. J.; Winefordner, J. D. (1968). The selection of optimum conditions for spectrochemical methods. *J. Chem. Educ.*, **45**, 98–102.

Uses previously developed methods for calculating the limiting detectable sample concentration to evaluate molecular absorption spectrometry, molecular fluorescence spectrometry, and phosphorescence spectrometry.

Chambless, D. A.; Dubose, S. S.; Sensintaffar, E. L. (1992). Detection limit concepts: foundations, myths, and utilization. *Health Phys.*, **63**, 338–340.

General discussion of detection limits. Terminology: minimum detectable concentration (MDC). MDC is "the result of expressing L_d in activity concentration units." Discusses several myths concerning detection limits. Stresses that the detection limit concept was developed for the specific purpose of permitting scientists to compare analytical instruments and methodologies to assist in developing more powerful techniques. Also provides some discussion of censoring issues.

Chatfield, C. (1984). *The Analysis of Time Series: An Introduction*, 3rd ed. Chapman & Hall, London.

Introductory text on time series analysis.

Cheeseman, R. V.; Wilson, A. L. (1978). *Manual on Analytical Quality Control for the Water Industry*. TR66. Water Resource Center, Berkeley, CA.

Referenced in Kirchmer (1983). Suggests reporting the actual analytical results together with their 95% confidence limits.

Chew, V. (1968). Simultaneous prediction intervals. *Technometrics*, **10**, 323–331.

Chou, Y. M.; Owen, D. B. (1986). One-sided distribution-free simultaneous prediction limits for p future samples. *J. Qual. Technol.*, **18**, 96–98.

Clark, M. J. R.; Whitfield, P. H. (1994). Conflicting perspectives about detection limits and about the censoring of environmental data. *Water Resour. Bull.*, **30**, 1063–1079.

   The criterion of detection (COD, equal to L_c), limit of detection (LOD), and limit of quantitation (LOQ) are all concentration ranges, not sharp boundaries. Stevenson and Winefordner (1991) accept LODs as being equivalent if within a factor of 2 or 3. Recommends that experimental concentrations be 20 times the LOD. Data censoring issues also discussed.

Clayton, C. A.; Hines, J. W.; Elkins, P. D. (1987). Detection limits with specified assurance probabilities. *Anal. Chem.*, **59**, 2506–2514.

   Prediction interval-based critical levels (L_c) and detection limits (L_d) for various analytes recovered from spiked sediment are calculated. Uses central t (the usual Student's t) to calculate L_c, but must use noncentral t to calculate L_d. Provides a table of delta values (includes central and noncentral t). Replaces $(L_d - \overline{X})^2$ with $(\overline{X})^2$ to simplify the equations. Uses internal standard corrected data and a square-root transformation to correct for nonconstant variance [$y = (Y)^{0.5}$, $x = (X + 0.1)^{0.5} - (0.1)^{0.5}$]. Evaluates calculated detection limits by analyzing spiked sediment samples. *Note:* Can use a smaller concentration range if the detection limit calculation is the sole purpose, but if analyte concentration estimation is also needed, a larger concentration range may be required.

Cohen, A. C. (1959). Simplified estimators for the normal distribution when samples are singly censored or truncated. *Technometrics*, **1**, 217–237.

   First of two seminal papers on maximum likelihood estimation of the mean and variance of a censored normal distribution.

Cohen, A. C. (1961). Tables for maximum likelihood estimates: singly truncated and singly censored samples. *Technometrics*, **3**, 535–541.

   Second of two seminal papers on maximum likelihood estimation of the mean and variance of a censored normal distribution.

Cohen, A. C. (1991). *Truncated and Censored Samples: Theory and Applications.* Marcel Dekker, New York.

   Text devoted to the analysis of censored data.

Coleman, D. (1993a). Comparison of different types of method detection limits (MDLs). Alcoa memorandum. Mar. 26.

   Referenced in Gibbons (1994). Comments on Clayton method. $\overline{X}$ replaces $(X - \overline{X})^2$. As n increases, the difference between t and the noncentral t becomes negligible.

Coleman, D. (1993b). Some simulation results of different types of method detection limits (MDLs). Alcoa Technical Memorandum, May 20.

   Referenced in Coleman (1995). "The MDL procedure . . . nominally provides a FP rate of 1%, but since it does not account for calibration error, the actual FP rate is commonly 2% or higher."

Coleman, D. (1993c). Technical links between calibration, relative prediction error (RPE), detection limits, signal-to-noise, and ROCs. Alcoa Technical Memorandum. Apr. 7.

   Referenced in Coleman (1995). "For quantitation, the Hubaux–Vos (1970) approach has been extended to compute a PQL defined as the lowest concentration at which measurement with one significant digit can be made . . . with high confidence."

Coleman, D. (1995a). Comment on "Method detection limits in solid waste analysis." *Environ. Sci. Technol.*, **29**, 279–280.

   Correspondence commenting on Kimbrough and Wakakuwa (1993). Takes issue with the authors on three points: (1) detection definitions, (2) accuracy, and (3) quantitation.

Coleman, D. (1995b). *Proceedings of the International Conference on Environmetrics and Chemometrics*, Las Vegas, NV.

Referenced in Gibbons et al. (1997a)." . . . a 10% or even 20% RSD may not always be achievable for a given constituent and analytical method."

Coleman, D.; Auses, J.; Grams, N. (1997). Regulation: from an industry perspective or relationships between detection limits, quantitation limits, and significant digits. *Chemometr. Intell. Lab. Syst.*, **37**, 71–80.

Recommends a data reporting format that includes the standard deviation and the number of degrees of freedom. Terminology: relative measurement error (RME). RME = (estimated measurement uncertainty)/(estimated level). Links RME to the number of significant digits (w). $\frac{1}{2} \times 10^{-w} \leq \text{RME} < \frac{1}{2} \times 10^{-w+1}$, or $w \geq \text{SNR} > w - 1$, where SNR = log(signal/noise) = log(measurement/total error). Using unweighted prediction intervals: at the detection limit SNR = 0, RME = 50%, and $w = 0$; at the quantitation limit, SNR $\geq$ 1, RME $\leq$ 5%, $w = 1$.

Commission on Spectrochemical and Other Optical Procedures for Analysis (Analytical Chemistry Division) (1976a). Nomenclature, symbols, units, and their usage in spectrochemical analysis: II. *Pure Appl. Chem.*, **45**, 99–103.

Model adopted by the IUPAC in 1975 for limit of detection calculations. Quote: "The limit of detection, expressed as a concentration, cL, or the quantity, qL, is derived from the smallest measure, X_L, that can be detected with reasonable certainty for a given analytical procedure." Equation: $X_L = X_{b1} + ks_{b1}$, where X_{bl} is the mean of the blank measures, s_{b1} is the standard deviation of the blank measures, and k is a numerical factor chosen according to the confidence level desired. Strongly recommend $k = 3$.

Commission on Spectrochemical and Other Optical Procedures for Analysis (Analytical Chemistry Division) (1976b). Nomenclature, symbols, units and their usage in spectrochemical analysis: III. *Pure Appl. Chem.*, **45**, 105–123.

Same detection limit discussion as in Part II.

Commission on Spectrochemical and Other Optical Procedures for Analysis (Analytical Chemistry Division) (1978a). Nomenclature, symbols, units, and their usage in spectrochemical analysis: II. *Spectrochim. Acta B*, **33B**, 242–245.

Model adopted by the IUPAC in 1975 for limit of detection calculations. Quote: "The limit of detection, expressed as a concentration, c_L, or the quantity, q_L, is derived from the smallest measure, X_L, that can be detected with reasonable certainty for a given analytical procedure." Equation: $X_L = X_{b1} + ks_{b1}$, where X_{b1} is the mean of the blank measures, s_{b1} is the standard deviation of the blank measures, and k is a numerical factor chosen according to the confidence level desired. Strongly recommend $k = 3$.

Commission on Spectrochemical and Other Optical Procedures for Analysis (Analytical Chemistry Division) (1978b). Nomenclature, symbols, units, and their usage in spectrochemical analysis: III. *Spectrochim. Acta B*, **33B**, 248–269.

Same detection limit discussion as in Part II.

Conover, W. J. (1980) *Practical Nonparametric Statistics*, 2nd ed. Wiley, New York.

General text on nonparametric statistical methods.

Currie, L. A. (1968). Limits for qualitative detection and quantitative determination. *Anal. Chem.*, **40**, 586–593.

Propose and define *Lc*, *Ld*, *Lq*, and errors of the first (Type I) and second (Type II) kind. The critical level (*Lc*): "a 'decision limit' at which one may decide whether or not the result of an analysis indicates detection." Concerned with the signal or measured concentration. The detection limit (Ld): "a 'detection limit' at which a given analytical procedure may be relied upon to lead to detection"-concerned with the true concentration. The determination limit (Lq): "a 'determination limit' at which a given procedure will be sufficiently precise to yield a satisfactory quantitative estimate." Includes discussions of both the general analytical case and radioactivity data (Poisson distribution). Nonconstant variance is discussed for the radioactivity data.

Currie, L. A. (1972). The limit of precision in nuclear and analytical chemistry. *Nucl. Instrum. Methods*, **100**, 387–395.

Referenced in Long and Winefordner (1983). Review of detection limit calculations.

Currie, L. A. (1982). Quality of analytical results, with special reference to trace analysis and social-chemical problems. *Pure Appl. Chem.*, **54**, 715–754.

Discusses detection theory and hypothesis testing (false positive/false negative issues).

Currie, L. A. (1984a). *Lower Limit of Detection: Definition and Elaboration of a Proposed Position for Radiological Effluent and Environmental Measurements.* NUREG/CR-4007. Nuclear Regulatory Commission, Washington, DC.

Referenced in Chambless et al. (1992). Recommends that the MDC be defined as the 95% upper confidence limit of the distribution of M, where M is obtained when converting Ld into activity or concentration units.

Currie, L. A. (1984b). In *Chemometrics, Mathematics and Statistics in Chemistry.* Kowalski, B. R., Ed. D. Reidel, Dordrecht, The Netherlands.

Broad discussion of detection limits.

Currie, L. A. (1985). In *Trace Residue Analysis, Chemometric Estimations of Sampling, Amount, and Error.* ACS Symposium Series 284. Kurtz, D. A., Ed. American Chemical Society, Washington, DC.

Calculates a weighted prediction interval-based critical level and a z-based detection limit (see the appendix) for fenvalerate data. Models the weights as a function of concentration using a linear model $s_x = a_0 + a_1 X$. No data at very low concentrations (the lowest standard is greater than 40 times the standard deviation of the blank, so it is much larger than the detection limit). The lowest concentration exceeded the calculated detection limit by more than an order of magnitude. Discusses the instrumental "threshold" problem. Discusses detection in higher dimensions. *Note:* "It can be shown that with an inadequate design the detection limit may not even exist!"

Currie, L. A. (1995). Nomenclature in evaluation of analytical methods including detection and quantification capabilities. *Pure Appl. Chem.*, **67**, 1699–1723.

IUPAC recommendations for nomenclature in evaluation of analytical methods. See Section 3.7. Discusses and defines Lc, Ld, and Lq. Nomenclature: L = the generic symbol, S = the signal, and X = the concentration. Discusses the effect of an instrumental threshold to discriminate against small signal—this is essentially a de facto Lc. See Section 3.7.8. Must account for nonconstant variance.

Dagnall, R. M.; Taylor, M. R. G.; West, T. S. (1968). Some detection limits in atomic fluorescence spectroscopy with microwave excited electrodeless discharge tubes as excitation sources. *Spectrosc. Lett.*, **1**, 397–405.

Detection limits compared for a number of metals in atomic fluorescence and absorption spectroscopy.

D'Agostino, R. B. (1971). Linear estimation of the Weibull parameters. *Technometrics*, **13**, 171–182.

David, D. J. (1960). The application of atomic absorption to chemical analysis. *Analyst*, **85**, 779–791.

Discusses the limit of detection for atomic absorption analyses. Uses the terms *detection limit* and *sensitivity* interchangeably.

Davis, C. B. (1994). In *Handbook of Statistics. Environmental Statistics*, Vol. 12 Patil, G. P.; Rao, C. R., Eds. Elsevier, Amsterdam.

Referenced in Gibbons (1995). Detection limit review.

Davis, C. B.; McNichols, R. J. (1987). One-sided intervals for at least p of m observations from a normal population on each of r future observations. *Technometrics*, **29**, 359–370.

One-sided prediction intervals for at least p of m observations on each of r future occasions are developed and applied to groundwater quality monitoring. Selected tables are presented.

Davis, C. B.; McNichols, R. J. (1994). Ground-water monitoring statistics update: I. Progress since 1988. *Ground Water Monitor. Remediation*, **14**, 148–158.

Review of statistical methods for ground water monitoring.

Davis, C.; McNichols, R. (1999). Simultaneous nonparametric prediction limits. *Technometrics*, **41**, 89.

Excellent statistical review with several commentaries on the derivation of nonparametric prediction limits and relevant environmental applications.

DeGalan, L. (1970). On sensitivity and detection limit with reference to flame spectrometry. *Spectrosc. Lett.*, **3**, 123–127.

Attempts to resolve the confusion between the terms *detection limit* and *sensitivity*. Defines sensitivity as the slope of the calibration curve; the detection limit is based on the standard deviation of the blank. The equation for calculating the detection limit requires the sensitivity to convert between the analytical signal and the concentration. Since the two values are inversely proportional, the sensitivity increases as the detection limit decreases. Also point out that the two values have different units, so it is meaningless to compare them.

Dixon, W. J. (1953). Processing data for outliers. *Biometrics*, **9**, 74–89.

Seminal paper in the area of statistical outlier detection.

Dixon, W. J.; Massey, F. J., Jr. (1969). *Introduction to Statistical Analysis*. McGraw-Hill, New York.

Referenced in Grant et al. (1991). Mentions that the size of a and b risks can be varied to fit the problem requirements.

Doerffel, K. (1966). *Reinststoffprobleme: II. Reinststoffanalytik*. Akademie-Verlag, Berlin.

Referenced in Gabriels (1970). Calculates the detection limit as a function of the standard deviation on the blank value.

Draper, N. R.; Smith, H. (1981). *Applied Regression Analysis*, 2nd ed. Wiley, New York.

Excellent reference for regression theory. Includes unweighted confidence and prediction intervals only.

Draper, W. M.; Dhoot, J. S.; Dhaliwal, J. S.; Remoy, J. W.; Perera, S. K.; Baumann, F. J. (1998). Detection limits of organic contaminants in drinking water. *J. Am. Water Works Assoc.*, **90**, 82–90.

Discusses the reasons that MDLs vary: technique, instruments, contamination, method, spike level, bias, gross error, and random error. Shows specific examples of varying MDLs (e. g., MDLs reported for benzene analyzed by EPA method 624 varied over a 10,000-fold range). ". . . MDLs would be more uniform among laboratories if (1) uniform spike concentrations were used in MDL determination; (2) analytical methods were more uniform as to procedures; and (3) tighter guidelines were established for conducting MDL experiments and handling MDL data."

Dunnett, C. W. (1955). A multiple comparisons procedure for comparing several treatments with a control. *J. Am. Stat. Assoc.*, **50**, 1096–1121.

Describes Dunnett's test, a general method for comparing the means of several experimental groups to a common control.

Dunnett, C. W.; Sobel, M. (1955). Approximations to the probability integral and certain percentage points of a multivariate analogue of Student's *t*-distribution. *Biometrika*, **42**, 258–260.

Basic statistical foundation for Dunnett's test.

Eaton, A. (1993). Estimation of interlaboratory MDLs and RQLs. *Environ. Lab.*, Apr.–May. Referenced in Sanders et al. (1996).

Erhlich, G. (1968). Zur objektiven Bewertung des Nachweisvermögens in der Emissionsspektrographie. *Z. Anal. Chem.*, **232**, 1–17.

Not in English. Referenced in Liteanu and Rica (1975). Study of the relationship between the background noise and the analytical signal in the frame of the problem of the detection limit.

Ehrlich, G.; Mai, H. (1966). Zur objektiven Bewertung des Nachweisvermögens in der HF-Funken-Massenspektrographie. *Z. Anal. Chem.*, **218**, 1–17.

Not in English. Referenced in Liteanu and Rica (1975). Study of the relationship between the background noise and the analytical signal in the frame of the problem of the detection limit.

El-Shaarawi, A. H. (1989). Inferences about the mean from censored water quality data. *Water Resour. Res.*, **25**, 685–690.

El-Shaarawi, A. H.; Niculescu, S. P. (1992). On Kendall's tau as a test of trend in time. *Environmentrics*, **3**, 385–412.

Describes the application of trend tests to environmental data.

Falk, H. (1980). Analytical capabilities of atomic spectrometric methods using tunable lasers: a theoretical approach. *Prog. Anal. At. Spectrosc.*, **3**, 181–208.

Extensive calculations of detection limits for atomic spectrometric methods.

Feigl, F. (1949). *Chemistry of Specific, Selective and Sensitive Reactions.* Academic Press, New York.

The limit of detection is not a well-defined concentration or amount but instead, represents a relatively wide region of uncertainty. Existence of a broad "region of uncertain reaction" near the detection limit.

Filliben, J. J. (1975). The probability plot correlation coefficient test for normality. *Technometrics*, **17**, 111–117.

Describes a test of normality that is suitable for censored data.

Finney, D. J. (1941). On the distribution of a variate whose logarithm is normally distributed. *J. R. Stat. Soc. Ser. B*, **7**, 155–161.

Seminal paper on the lognormal distribution and estimation of its parameters.

Finney, D. J. (1976). Radioligand assay. *Biometrics*, **32**, 721–740.

Uses weighted least-squares regression for radioligand assay calibration. Mentions variance modeling. Also covers transformations briefly.

Finney, D. J.; Phillips, P. (1977). The form and estimation of a variance function, with particular reference to radioimmunoassay. *Appl. Stat.*, **26**, 312–320.

Variance is not constant with concentration for radioimmunoassay data. Concerned with maximum likelihood estimation of the variance function, using either a quadratic or an exponential model.

Firestone, D.; Horwitz, W. (1979). Oils and fats. IUPAC gas chromatographic method for determination of fatty acid composition: collaborative study. *J. Assoc. Off. Anal. Chem.*, **62**, 709–721.

Provides data (methyl esters of fatty acids) supporting the hypothesis that the coefficient of variation is related to concentration.

Fisher, R. A. (1929). Moments and product-moments of sampling distributions. *Proc. London Math. Soc.*, **30**(2), 199.

Fisher, R. A. (1930). The moments of the distribution for normal samples of measures of departure from normality. *Proc. R. Soc. Ser. A*, **130**, 16.

Seminal paper on testing for normality by the founder of modern statistics.

Fligner, M. A.; Wolfe, D. A. (1979). Nonparametric prediction limits for a future sample median. *J. Am. Stat. Assoc.*, **30**, 78–85.

Florian, K. (1990). Qualitative or quantitative characterization of spectrographic methods? The detection and determination limits. *Chem. Anal. (Warszawa)*, **35**, 129–139.

Referenced in Clark and Whitfield (1994). "Many protocols define Limits of Quantitation as simple multiples of the Limits of Detection in spite of some evidence to the contrary."

Foley, J. P.; Dorsey, J. G. (1984). Clarification of the limit of detection in chromatography. *Chromatographia*, **18**, 503–511.

Points out current problems with limit-of-detection concepts in chromatography. Proposes two possible models for calculating the LOD. The first model is based on the standard deviation of the blank: $LOD = 3s_B/S$, where S is the slope of the calibration curve. The second model is based on error propagation: $LOD = 3[s_B^2 + s_i^2 + (i/S)^2 s_s^2]^{1/2}/S$, where s_i is the standard deviation of the intercept, i is the intercept, and s_s is the standard deviation of the slope.

Franke, J. P.; de Zeeuw, R. A.; Hakkert, R. (1978). Evaluation and optimization of the standard addition method for absorption spectrometry and anodic stripping voltammetry. *Anal. Chem.*, **50**, 1374–1380.

Variance is not constant with concentration. Applies weighted least squares regression and a transformation to the standard addition method commonly used in atomic absorption spectrometry and anodic stripping voltammetry.

Fuller, F. C.; Tsokos, C. P. (1971). Time series analysis of water pollution data. *Biometrics*, **27**, 1017–1034.

Gabriels, R. (1970). A general method for calculating the detection limit in chemical analysis. *Anal. Chem.*, **42**, 1439–1440.

Assumes constant variance. Propose calculating the detection limit (D, in units of instrument response) as $D = b_0 + s(2/n)^{0.5}(t_1 + t_2)$, where s is the standard deviation of a blank (assume equal to the standard deviation at D), t_1 corresponds to a significance level of alpha, t_2 corresponds to a significance level of $2(1 - P)$; test statistics are two-sided. The value D must be converted to a concentration.

Garden, J. S.; Mitchell, D. G.; Mills, W. N. (1980). Nonconstant variance regression techniques for calibration-curve-based analysis. *Anal. Chem.*, **52**, 2310–2315.

Calibration modeling using weighted least-squares correction for nonconstant variance. Incorrect use of an unweighted least-squares analysis could cause gross errors in the estimation of trace concentrations. Discusses sources of noise for various instruments.

Garfield, F. M. (1984). *Quality Assurance Principles for Analytical Laboratories.* Association of Official Analytical Chemists, Arlington, VA.
   Lists AOAC quality assurance procedures.

Garton, F. W. J.; Ramsden, W.; Taylor, R.; Webb, R. J. (1956). The assesment of performance of spectrographic methods. *Spectrochim. Acta,* **8**, 94–101.
   Suggests multiplying the standard deviation by either 2 or 3 to determine the sensitivity of detection (or limit of detection).

Geary, R. C. (1935). The ratio of the mean deviation to the standard deviation as a test of normality. *Biometrika,* **27**, 310–332.

Geary, R. C. (1936). Moments of the ratio of the mean deviation to the standard deviation for normal samples. *Biometrika,* **28**, 295–307.

Gibbons, R. D. (1987a). Statistical prediction intervals for the evaluation of ground-water quality. *Ground Water,* **25**, 455–465.
   Early paper on the use of prediction limits for analysis of water quality data.

Gibbons, R. D. (1987b). Statistical models for the analysis of volatile organic compounds in waste disposal sites. *Ground Water,* **25**, 572–580.
   Early paper on analysis of constituents with low detection frequency. Introduces Poisson tolerance and prediction limits.

Gibbons, R. D. (1989). A comment on "Statistical methods for evaluating ground-water monitoring from hazardous waste facilities," final rule. *Ground Water,* **27**, 252–254.

Gibbons, R. D. (1990). A general statistical procedure for ground-water detection monitoring at waste disposal facilities. *Ground Water,* **28**, 235–243.
   First paper introducing use of nonparametric prediction limits for environmental applications.

Gibbons, R. D. (1991a). Some additional nonparametric prediction limits for ground-water detection monitoring at waste disposal facilities. *Ground Water,* **29**, 729–736.
   Introduces exact results for the earlier paper and generalization to alternative resampling plans.

Gibbons, R. D. (1991b). Statistical tolerance limits for ground-water monitoring. *Ground Water,* **29**, 563–570.

Gibbons, R. D. (1991c). Statistical tolerance limits for ground-water monitoring. *Water Resour. J.,* **171**, 21–28. Reprinted by the United Nations Economic and Social Commission for Asia and the Pacific.

Gibbons, R. D. (1994). *Statistical Methods for Groundwater Monitoring.* Wiley, New York.
   Comprehensive environmental statistics book with chapters on detection and quantification limit estimators.

Gibbons, R. D. (1995). Some statistical and conceptual issues in the detection of low-level environmental pollutants. *Environ. Ecol. Stat.,* **2**, 125–167.
   Review article for environmental statisticians.

Gibbons, R. D. (1996). Some conceptual and statistical issues in analysis of groundwater monitoring data. *Environmetrics,* **7**, 185–199.
   Overview of statistical methods used in analysis of groundwater monitoring data.

Gibbons, R. D. (1998a). Some conceptual and statistical issues in analysis of ground-water monitoring data. In *Encyclopedia of Environmental Analysis and Remediation.* Wiley, New York.

Gibbons, R. D. (1998b). Some conceptual and statistical issues in detection and quantification of environmental pollutants. In *Encyclopedia of Environmental Analysis and Remediation*. Wiley, New York.

Gibbons, R. D. (1998c). False positives in groundwater statistics. *Waste Age*, **29**, 32–37.

Gibbons, R. D. (1999a). Use of combined Shewhart–CUSUM control charts for groundwater monitoring applications. *Ground Water*, **37**, 682–691.

Explores use of resampling and alternative multipliers for combined Shewhart– CUSUM control charts.

Gibbons, R. D. (1999b). Discussion of simultaneous nonparametric prediction limits. *Technometrics*, **41**, 104–105.

Gibbons, R. D. (2000). Detection and quantification of environmental pollutants. In *Encyclopedia of Analytical Chemistry*. Wiley, New York.

Recent overview of the literature on detection and quantification.

Gibbons, R. D.; Baker, J. (1991). The properties of various statistical prediction limits. *J. Environ. Sci. Health*, **A26-4**, 535–553.

Gibbons, R. D.; Bhaumik, D. (2001). Weighted random-effects regression models with application to inter-laboratory calibration. *Technometrics*, in press.

Statistical paper on the estimating interlaboratory calibration curves and uncertainty intervals.

Gibbons, R. D.; Coleman, D. E. (1996). QL assumptions questioned. *J. Am. Water Works Assoc.*, **88**, 4–8.

Gibbons, R. D.; Jarke, F. H.; Stoub, K. P. (1988). Method detection limits. *Proceedings of the 5th Annual USEPA Waste Testing and Quality Assurance Symposium*, Vol. 2, pp. 292–319.

Review of literature on detection limits from a statistical perspective.

Gibbons, R. D.; Jarke, F. H.; Stoub, K. P. (1991). In *Waste Testing and Quality Assurance*. ASTM STP 1075. Friedman, D., Ed. American Society for Testing and Materials, Philadelphia.

Weighted least-squares approach to computing tolerance limit–based detection limits.

Gibbons, R. D.; Grams, N. E.; Jarke, F. H.; Stoub, K. P. (1992). Practical quantitation limits. *Chemometr. Intell. Lab. Syst.*, **12**, 225–235.

Early qantification limit estimator using transformation to bring about homoscedasticity.

Gibbons, R. D.; Coleman, D. E.; Maddalone, R. F. (1997a). An alternative minimum level definition for analytical quantification. *Environ. Sci. Technol.*, **31**, 2071–2077.

Introduces a new quantification limit estimator based on WLS.

Gibbons, R. D.; Coleman, D. E.; Maddalone, R. F. (1997b). Response to comment on "An alternative minimum level definition for analytical quantification." *Environ. Sci. Technol.*, **31**, 3729–3731.

Gibbons, R. D.; Coleman, D. E.; Maddalone, R. F. (1998). Response to comment on "An alternative minimum level definition for analytical quantification." *Environ. Sci. Technol.*, **32**, 2349–2353.

Gibbons, R. D.; Coleman, D. E.; Maddalone, R. F. (1999a). Response to comment on "An alternative minimum level definition for analytical quantification." *Environ. Sci. Technol.*, **33**, 1313–1314.

Gibbons, R. D.; Dolan D. G.; May H.; O'Leary, K.; O'Hara, R. (1999b). Statistical comparison of leachate from hazardous, co-disposal, and municipal solid waste landfills. *Ground Water Monitor. Remediation*, **19**, 57–72.

Paper describing what types of compounds are detected in which types of facilities. A model is developed for the classification of leachate data as hazardous, co-disposal, or municipal.

Gibson, K. S.; Balcom, M. M. (1947). *J. Res. Nat. Bur. Stand.*, **38**, 601–601.

Referenced in Cetorelli and Winefordner (1967). The minimum detectable sample concentration in molecular absorption spectrophotometry is generally defined as that sample concentration resulting in an absorbance of 0.002.

Gilbert, R. O. (1987). *Statistical Methods for Environmental Pollution Monitoring.* Van Nostrand Reinhold, New York.

Excellent book on environmental statistics that is particularly strong on experimental sampling and design issues.

Gilliom, R. J.; Helsel, D. R. (1986). Estimation of distributional parameters for censored trace level water quality data: 1. Estimation techniques. *Water Resourc. Res.*, **22**, 135–146.

Early paper on analysis of censored samples in environmental data.

Glaser, J. A.; Foerst, D. L.; McKee, G. D.; Quave, S. A.; Budde, W. L. (1981). Trace analyses for wastewaters. *Environ. Sci. Technol.*, **15**, 1426–1435.

Basis for the EPA's method detection limit (MDL). Calculates MDLs for organic priority pollutants by various methods. Assumes that the standard deviation is a function of concentration, then that the standard deviation is constant with concentration. Calculates a confidence interval on the MDL and proposes a more detailed iterative approach. Quotes: "The error variance of the analytical method changes as the analyte concentration increases above the MDL." "To ensure that the estimate of the MDL is a good estimate, the analyst must determine that a lower concentration of analyte will not result in a significantly lower calculated MDL." "When economically feasible, the full regression treatment should be used." "Experience has shown that when the relative standard deviation is at or near 10%, the calculated MDL values can be below instrumental detection limits." Does this make sense? A 10% RSD should be near the limit of quantification.

Gleit, A. (1985). Estimation for small normal data sets with detection limits. *Environ. Sci. Technol.*, **19**, 1201–1206.

Early paper on analysis of censored samples in environmental data.

Gottschalk, G. (1962). *Statistik in der quantitativ chemischen Analyse Ferd.* Enke, Stuttgart, Germany.

Referenced in St. John et al. (1967). Reference to the limiting detectable sample concentration, C_m, in spectrochemical methods.

Gottschalk, G.; Dehmel, P. (1958). Statistik in der chemischen Analyse: II. *Z. Anal. Chem.*, **160**, 161–169.

Not in English. Referenced in Liteanu and Rica (1975b). Study of the relationship between the background noise and the analytical signal in the frame of the problem of the detection limit.

Grant, C. L.; Hewitt, A. D.; Jenkins, T. F. (1991). Experimental comparison of EPA and USATHAMA detection and quantitation capability estimators. *Am. Lab.*, Feb., pp. 15–33.

Compares the certified reporting limit (CRL) specified by the U.S. Army Toxic and Hazardous Materials Agency (USATHAMA) and the MDL. ". . . It has been suggested that thase two procedures yield widely divergent estimates for the same analyses." The CRL uses confidence bands as described by Hubaux and Vos—slight variation. Suggests sequential truncation of highest concentration values to remove nonlinearity. Good discussion of nonconstant variance (p. 18).

Grubbs, F. E. (1969). Procedures for detecting outlying observations in samples. *Technometrics*, **11**, 1–21.
Introduces a widely used method for testing for outliers.

Grubbs, F. E.; Beck, G. (1972). Extension of sample sizes and percentage points for significance tests of outlying observations. *Technometrics*, **14**, 847–854.

Guenther, W. C. (1975). The inverse hypergeometric: a useful model. *Stat. Neerlandica*, **29**, 129–144.
Statistical foundational paper useful in deriving nonparametric prediction intervals.

Guilbaud, O. (1983). Nonparametric prediction intervals for sample medians in the general case. *J. Am. Stat. Assoc.*, **78**, 937–941.

Gupta, A. K. (1952). Estimation of the mean and standard deviation of a normal population from a censored sample. *Biometrika*, **39**, 260–273.

Gupta, S. S. (1963). Probability integrals of multivariate normal multivariate *t*. *Ann. Math. Stat.*, **34**, 792–828.

Gupta, S. S.; Panchapakesan, S. (1979). *Multiple Decision Procedures*. Wiley, New York.

Guttman, I. (1970). *Statistical Tolerance Regions: Classical and Bayesian*. Hafner, Darien, CT.
Develops and provides tables of tolerance interval test statistics (K); see Table 4.6 for values of one-sided K's. The original reference for this table is Owen (1963).

Haas, C. N.; Scheff, P. A. (1990). Estimation of averages in truncated samples. *Environ. Sci. Technol.*, **24**, 912–919.

Hahn, G. J. (1970). Additional factors for calculating prediction intervals for samples from a normal distribution. *J. Am. Stat. Assoc.*, **65**, 1668–1676.
Seminal paper on statistical prediction.

Hahn, G. J.; Meeker, W. Q. (1991). *Statistical Intervals: A Guide for Practitioners*. Wiley, New York.
Excellent text on the application of statistical prediction, tolerance, and confidence intervals.

Haisch, U. (1970). Uber die Bestimmung kleinster Konzentrationen chemischer Elemente in der Emissionsspektralanalyse mit photoelektrischer Strahlungsmessung: I. *Spectrochim. Acta*, **25B**, 597–612.
Not in English. Referenced in Boumans and deBoer (1972). An earlier emission spectrography method for calculating the limit of detection.

Haisch, U.; Laqua, K.; Hagenah, W. D. (1971). Uber die Bestimmung kleinster Konzentrationen chemischer Elemente in der Emissionsspektralanalyse mit photoelektrischer Strahlungsmessung: II. *Spectrochim. Acta*, **26B**, 651–667.
Not in English. Referenced in Boumans and deBoer (1972). An earlier emission spectrography method for calculating the limit of detection.

Hall, P.; Selinger, B. (1989). A statistical justification relating interlaboratory coefficients of variation with concentration levels. *Anal. Chem.*, **61**, 1465–1466.
Uses the term *Horwitz trumpet* to describe the behavior of the coefficient of variation with concentration. Equation: $CV(p) = p^{-0.15}/50$.

Hall, I. J., Prarie, R. R.; Motlagh, C. K. (1975). Non-parametric prediction intervals. *J. Qual. Technol.*, **7**, 109–114.
Early work on nonparametric prediction limits.

Hashimoto, L. K.; Trussell, R. R. (1983). Evaluating water quality data near the detection limit. *Proceedings of the American Water Works Association Advanced Technology Conference*, Las Vegas, NV, June 5–9.

Helsel, D. R. (1990). Less than obvious: statistical treatment of data below the detection limit. *Environ. Sci. Technol.*, **24**, 1766–1774.
  Review of methods for dealing with censored environmental data.

Helsel, D. R.; Cohn, T. A. (1988). Estimation of descriptive statistics for multiply censored water quality data. *Water Resour. Res.*, **24**, 1997–2004.

Herrmann, R.; Alkemade, C. T. J. (1963). *Chemical Analysis by Flame Photometry.* Interscience, New York.
  Adoptes Kaiser's definition of the detection limit for flame photometry.

Hertz, C. D.; Brodovsky, J.; Marrollo, L.; Harper, R. E. (1992). In *Proceedings of the Water Quality and Technology Conference*, Toronto.
  Referenced in Kimbrough (1997).

Hewett, P.; Ganser, G. H. (1997). Simple procedures for calculating confidence intervals around the sample mean and exceedance fraction derived from lognormally distributed data. *Appl. Occupat. Environ. Hygiene*, **12**, 132–142.
  Method for approximating Land's coefficients.

Hobbs, D. J.; Iny, A. (1970). The "limit of detection" in spectrographic analysis: further thoughts and experiments. *Appl. Spectrosc.*, **24**, 522–526.
  Defines the region of uncertain detectability (RUD) that relates concentration to the number of not-detected results.

Hobbs, D. J.; Smith, D. M. (1965). *Proceedings of the International Colloquium on Spectroscopy.* Hilger & Watts, London.
  Referenced in Boumans and Maessen (1966). Discussion of detection limits.

Hochberg, Y.; Tamhane, A. C. (1987). *Multiple Comparison Procedures.* Wiley, New York.
  Devoted to problems in multiple comparisons, such as the comparison of multiple sampling locations to a common background.

Hocking, R. R.; Bremer, R. H.; Green, J. W. (1989). Variance component estimation with model-based diagnostics. *Technometrics*, **31**, 227–239.

Holden, A. V. (1975). In *Environmental Quality and Safety.* Coulston, F.; Korte, F., Eds. George Thieme, Stuttgart.
  Provides data (pesticide residues) supporting the hypothesis that the coefficient of variation is related to concentration.

Holland, D. M.; McElroy, F. F. (1986). Analytical method comparison by estimates of precision and lower detection limit. *Environ. Sci. Technol.*, **20**, 1157–1161.
  Calculates detection limits for nitrogen dioxide analyzed by chemiluminescence.

Horwitz, W. (1977). The variability of AOAC methods of analysis as used in analytical pharmaceutical chemistry. *J. Assoc. Off. Anal. Chem.*, **60**, 1355–1363.
  Provides data (pharmaceutical preparations) supporting the hypothesis that the coefficient of variation is related to concentration.

Horwitz, W. (1982). Evaluation of analytical methods used for regulation of foods and drugs. *Anal. Chem.*, **54**, 67A–76A.
  Discussion concerns interlaboratory data. Describes changes in precision from one method intended for a given range of conentrations to another method intended for a different range. Terminology: "Limit of Reliable Measurement... the smallest amount (or concentration) of a material that can be measured with a stated degree of confidence."

Figure showing percent coefficient of variation versus concentration for interlaboratory data. Equation: % CV $= 2^{1 - 0.5 \log C}$.

Horwitz, W.; Kamps, L. R.; Boyer, K. W. (1980). Quality assurance in the analysis of foods for trace constituents. *J. Assoc. Off. Anal. Chem.*, **63**, 1344–1354.

Discussion concerns interlaboratory data. Describes changes in precision from one method intended for a given range of concentrations to another method intended for a different range. Figure showing percent coefficient of variation versus concentration for interlaboratory data. Within-laboratory error is approximately one-half to two-thirds that of the total (between-laboratory) variability. Analytes: aflatoxins, drugs in feeds, pesticides, trace elements, and trace organics.

Hoyle, M. H. (1968). The estimation of variances after using a Gaussianating transformation. *Ann. Math. Stat.*, **39**, 1125–1143.

Early paper on estimating the variance of the lognormal distribution.

Hsu, D. A.; Hunter, J. S. (1976). Time series analysis and forecasting for air pollution concentrations with seasonal variations. *Proceedings of the EPA Conference on Environmental Modeling and Simulation*, PB-257 142. National Technical Information Service, Springfield, VA, pp. 673–677.

Hubaux, A.; Vos, G. (1970). Decision and detection limits for linear calibration curves. *Anal. Chem.*, **42**, 849–855.

Unweighted least-squares prediction intervals (Lc and Ld). Ways to improve decision and detection limits: improve precision, increase n, sufficiently high range ($R = 20$), modes of repartition of standards, and replication of the unknown.

Hunt, D. T. E.; Wilson, A. L. (1986). *The Chemical Analysis of Water: General Principles and Techniques.* Alden Press, Oxford.

Random error reported as coefficient of variation ranges from about 25% to 100% at the limit of detection.

Hunt, W. F., Jr.; Akland, G.; Cox, W.; Curran, T.; Frank, N.; Goranson, S.; Ross, P.; Sauls, H.; Suggs, J. (1981). *U.S. Environmental Protection Agency Intra-agency Task Force Report on Air Quality Indicators.* EPA-450/4-81-015. U.S. Environmental Protection Agency, National Technical Information Service, Springfield, VA.

Ingle, J. D., Jr. (1974). Sensitivity and limit of detection in quantitative spectrometric methods. *J Chem. Educ.*, **51**, 100–105.

Definitions of the limit of detection using z or t test statistics—apply to emission, luminescence, and absorbance measurements. Assume constant variance—claim that at concentrations near the limit of detection, the standard deviation is independent of concentration. Concerned only with false positives. Currie (1984) mentions that this paper implies that detection limit definitions concerned with errors of the second type are too complex for common understanding and use. The coefficient of variation ranges from about 25% to 100% at the limit of detection. Suggests that quantitative analysis be performed at 10 times the limit of detection.

Ingle, J. D., Jr.; Wilson, R. L. (1976). Difficulties with determining the detection limit with nonlinear calibration curves in spectrometry. *Anal. Chem.*, **48**, 1641–1642.

Equation: $S_t - S_b = z s_b$, where S_t is the total signal at the detection limit, S_b is the blank signal, and s_b is the standard deviation for a blank measurement. For cases of nonlinear regression (where the calibration curve levels off at low concentration) the intercept and the mean blank response can sometimes be very different. The paper suggests using the extrapolated blank signal (intercept of the nonlinear curve) rather than the measured blank signal for S_b in the equation.

IUPAC (1972). *Appendices on Tentative Nomenclature, Symbols, Units and Standards: II. Terms and Symbols Related to Analytical Functions and Their Figures of Merit.* Number 28. International Union of Pure and Applied Chemistry, Oxford.
   Referenced in MacDougall and Crummett (1980). Provide a definition of the limit of detection.

Jacquez, J. A.; Norusis, M. (1973). Sampling experiments on the estimation of parameters in heteroscedastic linear regression. *Biometrics*, **29**, 771–780.
   Referenced in Rodbard et al. (1976). Points out that use of empirically determined weights may result in a decreased performance of a linear curve-fitting procedure.

Jindal, R. P. (1988). Attaining the ideal detection limit using random multiplication. *J. Appl. Phys.*, **63**, 2824–2827.
   Evaluates detection abilities of different detectors; however, there is no discussion of calculating the detection limit. Some discussion of signal-to-noise determination.

Johnson, G. L. (1971). In *Developments in Applied Spectroscopy*, Vol. 9. Grove, E. L.; Perkins, A. J., Eds. Plenum Press, New York.
   Provides a discussion of detection limits.

Johnson, N. L.; Kotz, S. (1969). *Discrete Distributions*. Wiley, New York.
   Excellent reference text. Part of a four-volume set that includes tests on continuous and multivariate distributions.

Jolodovsky, P. D. (1970). Some comments on Kaiser's views on the problem of limits of detection. *Spectrosc. Lett.*, **3**, 311–315.
   Points out that some authors fail to distinguish between one- and two-tailed confidence levels.

Kahn, H. D.; Telliard, W. A.; White, C. E. (1998). Comment on "An alternative minimum level definition for analytical quantification." *Environ. Sci. Technol.*, **32**, 2346–2348.

Kaiser, H. (1947). Die Berechnung der Nachweisempfindlichkeit. *Spectrochim. Acta*, **3**, 40–67.
   Not in English. Referenced in St. John et al. (1967). Reference to the limiting detectable sample concentration, C_m, in spectrochemical methods.

Kaiser, H. (1964). *Optik*, **21**, 309–309.
   Referenced in Parsons (1969). Development of the detection limit definition.

Kaiser, H. (1965). Zum Problem der Nachweisgrenze. *Fresenius Z. Anal. Chem.*, **209**, 1–18.
   Not in English.

Kaiser, H. (1966). Zur Definition der Nachweisgrenze, der Garantiegrenze und der dabei benutzen Begriffe: Fragen und Ergebnisse der Diskussion. *Fresenius Z. Anal. Chem.*, **216**, 80–94.
   Not in English.

Kaiser, H. (1969). *Two Papers on the Limit of Detection of a Complete Analytical Procedure.* Hafner, New York.
   Referenced in Currie (1985). Defines Ld in accordance with the statistical theory of hypothesis testing.

Kaiser, H. (1970a). Quantitation in elemental analysis. *Anal. Chem.*, **42**, 24A–41A.
   Nothing relating directly to detection limits in Part I; see Part II (Kaiser, 1970b).

Kaiser, H. (1970b). Quantitation in elemental analysis: II. *Anal. Chem.*, **42**, 26A–59A.
   Terminology: limit of detection (corresponds to Lc) and limit of guarantee of purity (corresponds to Ld). Equations: $x = x_b + ks_b$, where x is the limit of detection (in units of instrument response), x_b is the mean blank response, k is recommended to be set at 3,

and s_b is the standard deviation of a blank; $x_G = x_b + ks_b$, where x_G is the limit of guarantee of purity (in units of instrument response), and k is recommended to be set at 6. No mention of nonconstant variance.

Kaiser, H. (1973). Guiding concepts relating to trace analysis. *Pure Appl. Chem.*, **34**, 35–61.
Brief discussion of previously described detection limit concepts.

Kaiser, H. (1978). Foundations for the critical discussion of analytical methods. *Spectrochim. Acta.*, **33B**, 551–576.
Terminology: limit of detection (corresponds to Lc). Equation: $x = x_b + ks_b$, where x is the limit of detection (in units of instrument response), x_b is the mean blank response, k is recommended to be set at 3, and s_b is the standard deviation of a blank. Provides examples of chance fluctuations of the blank (instrument noise): impurities in reagents, losses through adsorption on the walls of the vessels, errors of weighing or titration, secondary reactions, temperature fluctuations of the light source in spectrochemical analysis, etc.

Kaiser, H.; Menzies, A. C. (1968). *The Limit of Detection of a Complete Analytical Procedure.* Adam Hilger, London.
Referenced in Glaser et al. (1981).

Kaiser, H.; Specker, H. (1956). Bewertung und Vergleich von Analysenverfahren. *Fresenius Z. Anal. Chem.*, **149**, 46–66.
Not in English.

Karger, B. L.; Martin, M.; Guiochon, G. (1974). Role of column parameters and injection volume on detection limits in liquid chromatography. *Anal. Chem.*, **46**, 1640–1647.
Defines the minimum detectable concentration as five times the signal-to-noise ratio. Examines the influence of various chromatographic parameters on the detection limit.

Keith, L., Ed. *Principles of Environmental Sampling*, American Chemical Society, Washington, DC.

Keith, L. H. (1991). *Environmental Sampling and Analysis: A Practical Guide.* Lewis Publishers, Chelsea, MI.
Referenced in Sanders et al. (1996). Comments that the MDL varies from laboratory to laboratory and provides protection only against false positive results.

Keith, L. H. (1992a). Revising definitions: low-level analyses. *Environ. Lab.*, June–July, pp. 58–61.
Discuss an American Chemical Society Committee on Environmental Improvement syposium entitled "Regulatory Problems and Solutions with Method Detection Limits." Suggests setting the reliable detection level (RDL) at two times MDL and the reliable quantitation level at two times RDL.

Keith, L. H. (1992b). ACS, EPA, etc. to redefine MDL. *EnvirofACS*, **40**, 1–5.
Referenced in Sanders et al. (1996).

Keith, L. H. (1992c). Update on efforts to redefine EPA's MDL. *EnvirofACS*, **40**, 4–4.
Referenced in Sanders et al. (1996).

Keith, L. H. (1993a). EPA's MDL issue: where is it going? *EnvirofACS*, **41**, 3–3.
Referenced in Sanders et al. (1996).

Keith, L. H. (1993b). Chaos revisited: EPA's MDL. *EnvirofACS*, **41**, 9–9.
Referenced in Sanders et al. (1996).

Keith, L. H. (1994a). EPA's Office of Water surges toward one MDL solution: the ML. *EnvirofACS*, **42**, 8–8.
Referenced in Sanders et al. (1996).

Keith, L. H. (1994b). Throwaway data. *Environ. Sci. Technol.*, **28**, 389A–390A.
   Discusses the issue of data censorship. Concludes that the analytical laboratory should ask data users whether or not they want the data censored.

Keith, L. H.; Lewis, D. L. (1992). *Revised Concepts for Reporting Data near Method Detection Levels: Proceedings of the 203rd Meeting of the American Chemical Society, Committee on Environmental Improvement.* American Chemical Society, San Francisco.
   Referenced in Kimbrough (1997). Reference for the reliable detection limit (RDL).

Keith, L. H.; Crummett, W.; Deegan, J., Jr.; Libby, R. A.; Taylor, J. K.; Wentler, G. (1983). Principles of environmental analysis. *Anal. Chem.*, **55**, 2210–2218.
   American Chemical Society (ACS) Committee on Environmental Improvement (CEI) guidelines or principles for analytical measurements of environmental samples. Defines the limit of detection (LOD) as LOD = $S_b + 3s$, where S_b is the blank signal and s is the standard deviation. "LOD is numerically equivalent to the MDL as S_b approaches zero." Define the limit of quantitation (LOQ) as LOQ = $10s$. Quote: ". . . quantitative interpretation, decision-making, and regulatory actions should be limited to data at or above the limit of quantitation." Revision of MacDougall and Crummett (1980).

Kendall, M. G. (1975) *Rank Correlation Methods*, 4th ed. Charles Griffin, London.
   Useful source on nonparametric trend analysis.

Kennedy, E. E. (1963). The impact on the analytical chemist of government regulations pertaining to tissue residues. *J. Agric. Food Chem.*, **11**, 393–395.
   Proposes a comparison of control samples with fortified samples to arrive at a sensitivity range.

Kimbrough, D. E. (1997). Comment on "An alternative minimum level definition for analytical quantification." *Environ. Sci. Technol.*, **31**, 3727–3728.
   Note that data quality objectives (DQOs) of the data user are also important. Gibbons et al. (1997) example results in a bias of 96% and a relative standard deviation of 286%. This may or may not be acceptable, depending on the DQOs of the user. Incorrectly states that the Gibbons approach is the basis for the MDL. Refers to the EPA's Office of Drinking Water (ODR) approach, the minimum reporting level (MRL), the lowest concentration that meets the data quality objectives of the information collection rule. An analyst must spike at the MRL and recover it within a fixed percentage.

Kimbrough, D. E.; Wakakuwa, J. R. (1991). In *Proceedings of the 7th Annual U.S. EPA Symposium on Solid Waste Testing and Quality Assurance*. U.S. Government Printing Office, Washington, DC.
   Referenced in Kimbrough and Wakakuwa (1994). Questions the usefulness of the MDL procedure on both theoretical and practical grounds.

Kimbrough, D. E.; Wakakuwa, J. (1993). Method detection limits in solid waste analysis. *Environ. Sci. Technol.*, **27**, 2692–2699.
   Calculates MDLs for five analytes in soil: arsenic, cadmium, molybdenum, selenium, and thallium and compares the calculated MDLs with the actual method performance on real spiked soils with these analytes at concentrations above and below the calculated MDL. Found that a wide range of MDLs are obtained, depending on which estimation procedure is used. Also, the calculated MDL can predict concentrations that produce no measurable result or are strongly biased.

Kimbrough, D. E.; Wakakuwa, J. R. (1994). Quality control level: an alternative to detection levels. *Environ. Sci. Technol.*, **28**, 338–345.

Defines the quality control level (QCL) as the lowest concentration that meets the data quality objectives of the data user. The QCL is defined in terms of the minimum acceptable precision and accuracy. Found that at high concentrations, the accuracy is high and consistent; however, below a certain concentration, the bias increased dramatically (i.e., the accuracy decreased). Also found that the percent relative standard deviation is low at high concentrations and increases at lower concentration.

Kimbrough, D. E.; Suffet, I. H.; Hertz, C. D. (1996). In *Proceedings of the Water Quality and Technology Conference*, Vol. 2., Boston.

Referenced in Kimbrough (1997). Statistically based reporting methods do not incorporate the need for a measure of accuracy in the reporting limit.

Kirchmer, C. J. (1983). Quality control in water analyses. *Environ. Sci. Technol.*, **17**, 174A–181A.

Review of Currie's method for calculating detection and quantification limits.

Koch, R. C. (1960). *Activation Analysis Handbook*. Academic Press, New York.

Uses the term *sensitivity* for the detection limit.

Koch, O. G.; Koch-Dedic, G. A. (1964). *Handbuch der Spurenanalyse*. Springer-Verlag, Berlin.

Calculates the detection limit as a function of the standard deviation on the blank value.

Koehn, J. W.; Zimmermann, A. G. (1990a). Method detection limits: how low can you really go? *Environ. Lab.*, Feb.–Mar.

Provides background information on the classical approach and calibration curve regression theory for calculating detection limits.

Koehn, J. W.; Zimmermann, A. G. (1990b). Method detection limits: how low can you really go? II. *Environ. Lab.*, June–July, pp. 35–60.

Application of a modified version of Hubaux and Vos (1970) provided by the U.S. Army Toxic and Hazardous Materials Agency (USATHMA); this may be the same method as that referred to by Coleman et al. (1997). Step 1: calibrate the instrument. Step 2: method certification—spike samples and process through the entire analytical method over several days. Use step 1 to convert instrument responses in step 2 to measured concentrations.

Koorse, S. J. (1989). False positives, detection limits, and other laboratory imperfections: the regulatory implications. *Environ. Law Rep.*, **19**, 10211–10222.

Review of the legal implications of EPA's operational definitions.

Kravchenko, M. S.; Fumarova, M. S.; Bugaevski, A. A. (1988). Estimation of detection limits of test methods for water analysis. *Int. J. Environ. Anal. Chem.*, **33**, 257–267.

Referenced in Clark and Whitfield (1994). Some of the basic detection limit assumptions are dubious.

Krull, I.; Swartz, M. (1998). Determining limits of detection and quantitation. *Liq. Chromatogr. Gas Chromatogr.*, **16**, 922–924.

LOD and LOQ are commonly determined from signal-to-noise ratios. For example, LOD has an S/N of 2:1 or 3:1, and LOQ has an S/N of 10:1. Note that it is very difficult to measure the noise in liquid and gas chromatography. Also discusses a "new convention" that is calibration based: LOD = 3.3 (s/S) and LOQ = 10 (s/S), where S is the slope and s is the response standard deviation of the blank, the residual standard deviation of the regression line, or the standard deviations of the y intercept. Suggests analyzing two or more replicates of five standard concentrations from the reporting level to 120% of the control limit.

Kurtz, D. A. (1983). The use of regression and statistical methods to establish calibration graphs in chromatography. *Anal. Chim. Acta.*, **150**, 105–114.

Calibration modeling using log-log correction for nonconstant variance. Very similar to subsequent Kurtz et al. (1985) paper.

Kurtz, D. A.; Rosenberger, J. L.; Tamayo, G. J. (1985). In *Trace Residue Analysis*: *Chemometric Estimations of Sampling, Amount, and Error.* ACS Symposium Series 284. Kurtz, D. A., Ed. American Chemical Society, Washington, DC.

Calculates confidence intervals from regression on transformed data; does not calculate detection limits. Analyzed for various pesticides using gas chromatography. Discusses the issue of nonconstant variance: ". . . we found that the calibration graph response data obtained from gas chromatography seldom have constant variance along the length of the graph." Many references for performing transformations and subsequent calculations.

Kurtz, D. A.; Taylor, J. K.; Sturdivan, L.; Crummett, W. B.; Midkiff, C. R., Jr.; Watters, R. L., Jr.; Wood, L. J.; Hanneman, W. W.; Horwitz, W. (1988). In *Detection in Analytical Chemistry: Importance, Theory, and Practice.* ACS Symposium Series 361. Currie, L. A., Ed. American Chemical Society, Washington, DC.

Discusses several real-world detection limit problems including basic data quality, calibration errors, public trust, forensic analyses, blanks, and regulatory limits.

Laird, N. M.; Ware, J. H. (1982). Random-effects models for longitudinal data. *Biometrics*, **38**, 963–974.

Fundamental paper on mixed-effects regression models for analysis of longitudinal data.

Lambert, D.; Peterson, B.; Terpenning, I. (1991). Nondetects, detection limits, and the probability of detection. *J. Am. Stat. Assoc.*, **86**, 266–277.

Terminology: *Probability of acceptance curve*, $p(m)$: relates measured concentration (m) to the probability of detection and "shows how likely a measurement of m is to be accepted by chemists and reported as a detect." *Probability of detection curve*, $pi(c)$: "describes how likely a field concentration of c is to give a measurement reported as detect." *Censoring limit* = 90th percentile of $p(m)$. *Minimum reliably detected concentration* = 90th percentile of $pi(c)$. Discusses infinite detection limits: happen "whenever the regression line is so flat relative to the scatter around the line that the lower prediction limit never reaches" the intercept. This method is also discussed in Gibbons (1995).

Land, C. E. (1971). Confidence intervals for linear functions of the normal mean and variance. *Ann. Math. Stat.*, **42**, 1187–1205.

Original paper on exact confidence limits for a lognormal mean.

Land, C. E. (1972). An evaluation of approximate confidence interval estimation methods for lognormal means. *Technometrics*, **14**, 145–158.

Land, C. E. (1975). Tables of confidence limits for linear functions of the normal mean and variance. In *Selected Tables in Mathematical Statistics*, Vol. III. American Mathematical Society, Providence RI, pp. 385–419.

Original source of Land's coefficients for lognormal confidence limits.

Laqua, K. (1966). Zum heutigen Stand der spektrochemischen Spurenanalyse. *Z. Anal. Chem.*, **221**, 44–68.

Not in English. Referenced in Boumans and deBoer (1972). An earlier emission spectrography method for calculating the limit of detection.

Laqua, K.; Hagenah, W. D.; Waechter, H. (1967). Spektrochemische Spurenanalyse mit spektralaufgelösten Analysenlinien. *Z. Anal. Chem.*, **225**, 142–174.

Not in English. Referenced in Boumans and deBoer (1972). An earlier emission spectrography method for calculating the limit of detection.

Lieberman, G. J.; Miller, R. G. (1963). Simultaneous tolerance intervals in regression. *Biometrika*, **50**, 155–168.

Proposes four different simultaneous tolerance interval calculation techniques. The fourth is the simplest but works as well as the others.

Lilliefors, H. W. (1967). On the Kolmogorov–Smirnov test for normality with mean and variance unknown. *J. Am. Stat. Assoc.*, **62**, 399–402.

Lilliefors, H. W. (1969). Correction to the paper "On the Kolmogorov–Smirnov test for normality with mean and variance unknown. " *J. Am. Stat. Assoc.*, **64**, 1702.

Lindstedt, J. (1993). LOD, LOQ, MDL, and PQL: the driving force of the next generation of regulations. *Met. Finish.*, **91**, 64–70.

Link, C. L. (1986). *An Equation for One-Sided Tolerance Limits for Normal Distributions*. FPL 458 1-4. U.S. Department of Agriculture, Forest Products Laboratory.

Provides an equation for calculating a one-sided tolerance limit for any sample size n that does not require the use of tables.

Linning, F. J.; Mandel, J. (1964). Which measure of precision? The evaluation of the precision of analytical methods involving linear calibration curves. *Anal. Chem.*, **36**, 25A–32A.

Not related directly to detection limits. Reference for unweighted prediction intervals.

Liteanu, C.; Florea, I. (1966). Sur une nouvelle méthode statistique pour la détermination de la sensibilité des réactions d'identification. *Mikrochim. Acta*, **1966**, 983–999.

Not in English. Referenced in Liteanu and Rica (1975a). Study of the relationship between the background noise and the analytical signal in the frame of the problem of the detection limit.

Liteanu, C.; Rica, I. (1973). Uber die Definition der Nachweisgrenze. *Mikrochim. Acta*, **1973**, 745–757.

Not in English. Referenced in Liteanu and Rica (1975). Study of the relationship between the background noise and the analytical signal in the frame of the problem of the detection limit.

Liteanu, C.; Rica, I. (1975a). On the detection limit. *Pure Appl. Chem.*, **44**, 535–553.

Equations: $Y_k = Y_0 + 3s_0$, $Y_D = Y_k + 3s_0$, $L_D = (Y_D - a)/b$. Also calculate a confidence interval around L_D. Use a parametrical test (Student), a nonparametrical test (Wilcoxon), and a sequential test (Wald) to determine detection. Assume constant variance.

Liteanu, C.; Rica, I. (1975b). On the frequentometric estimation of the detection limit. *Mikrochim. Acta*, **1975**, 311–323.

Use a two-step model to estimate the detection limit. First, the discrimination of the analyte signal is made relative to the background signal, and second, the discrimination is made relative to the detection level.

Liteanu, C.; Rica, I. (1980). *Statistical Theory and Methodology of Trace Analysis*. Wiley, New York.

Referenced in Currie (1995). Provides support for identifying the detection limit with Ld instead of Lc. Type II error $f(x)$. Referenced in Gibbons (1994). Review of the principles of experimental design of method detection limit studies.

Long, G. L.; Winefordner, J. D. (1983). Limit of detection: a closer look at the IUPAC definition. *Anal. Chem.*, **55**, 712A–724A.

Simple and general discussion of methods for calculating limits of detection. Evaluates IUPAC method ($Ld = 3s$), a graphical approach, and a propagation of errors method. Recommend IUPAC or propagation of errors approach.

Lucas, J. M. (1982). Combined Shewhart–CUSUM quality control schemes. *J. Qual. Technol.*, **14**, 51–59.

   Original source for combined Shewhart–CUSUM control charts, which are widely used for intrawell groundwater detection monitoring.

MacDougall, D.; Crummett, W. B. (1980). Guidelines for data acquisition and data quality evaluation in environmental chemistry. *Anal. Chem.*, **52**, 2242–2249.

   American Chemical Society (ACS) Committee on Environmental Improvement (CEI) guidelines or principles for analytical measurements of environmental samples. Defines the limit of detection (LOD) as LOD = S_b + $3s$, where S_b is the blank signal and s is the standard deviation. Defines the limit of quantitation (LOQ) as LOQ = $10s$. These guidelines are revised in Keith et al. (1983).

Maddalone, R. F.; Scott, J. W.; Frank, J. (1988). *Round-Robin Study of Methods for Trace Metal Analysis.* Vols. 1–3 EPRI CS-5910. Electric Power Research Institute, Palo Alto, CA.

   Referenced in Gibbons et al. (1997a). ". . . a 10% or even 20% RSD may not always be achievable for a given constituent and analytical method."

Maindonald, J. H. (1984). *Statistical Computation.* Wiley, New York.

   Good text for a variety of computational statistical problems, including approximating a variety of distributions.

Mandel, J. (1967). *The Statistical Analysis of Experimental Data.* Wiley, New York.

   Referenced in Boumans (1978). Provides more elaborate statistical calculations to take into account the uncertainty of the calibration function.

Mandel, J.; Stiehler, R. D. (1964). *J. Res. Nat. Bur. Stand.*, **A53**, 155.

   Referenced in Skogerboe and Grant (1970). Proposes that the sensitivity be defined as the ratio of the curve slope to the standard deviation of measurement.

Mandelstam, S. L.; Nedler, V. V. (1961). On the sensitivity of emission spectrochemical analysis. *Spectrochim. Acta*, **17**, 885–894.

   Refers to the minimum detectable sample concentration, $T_{L(min)}$, in spectrochemical methods. Equation: $T_{L(min)} \sim 10 DT_P$, where DT_P is the variation in the blank signal calculated using a complex analysis-specific expression.

Mann, H. B. (1945). Nonparametric tests against trend. *Econometrica*, **13**, 245–259.

   Source of the Mann–Kendall trend test.

Marsden, P.; Tsang, S. F. (1996). Achieving lower detection limits. *Environ. Test. Anal.*, Mar.–Apr., pp. 17–18.

   Ways to decrease detection limits: (1) increase the sample size, (2) decrease the background signal, (3) decrease the final extract volume, (4) derivatize specific analytes to increase signal-to-noise ratio, (5) increase the instrument sensitivity, (6) decrease laboratory contamination, (7) demonstrate (laboratory or analyst) proficiency, and (8) establish specific QA/QC procedures.

Maskarinec, M. P.; Holladay, S. K. (1987). *Quality Assurance/Quality Control in Waste Site Characterization and Remedial Action.* ORNL/TM-10600. Oak Ridge National Laboratory, Oak Ridge, TN.

   Referenced in Grant et al. (1991). Suggests that the MDL and certified reporting limit (CRL) yield widely divergent estimates for the same analyses.

Massart, D. L.; Vandingste, B. G. M.; Deming, S. N.; Michotte, Y.; Daufman, L. (1988). *Chemometrics: A Textbook.* Elsevier, Amsterdam.

   Referenced in Rocke and Lorenzato (1995) (states that values below the PQL do not yield useful quantitative information about the concentration of the analyte). Referenced in Currie (1995). Provides support for identifying the detection limit with L_d instead of L_c.

McCleary, R.; Hay, R. A. (1980). *Applied Time Series Analysis for the Social Sciences*. Sage Publications, Beverly Hills, CA.

McCollister, G. M.; Wilson, K. R. (1975). Linear stochastic models for forecasting daily maxima and hourly concentrations of air pollutants. *Atmos. Environ.*, **9**, 417–423.

McNichols, R. J.; Davis, C. B. (1988). Statistical issues and problems in ground water detection monitoring at hazardous waste facilities. *Ground Water Monitor. Rev.*, **8**, 135–150.

Early overview of methods for the analysis of groundwater monitoring data.

Meijers, C. A. M.; Hulsman, J. A. R. J.; Huber, J. F. K. (1972). Bestimmung von Cortisol in Serum mit Hilfe der Saulen-Flussigkeits-Chromatographie. *Z. Anal. Chem.*, **261**, 347–353.

Referenced in Karger et al. (1974). Discusses problems of achieving low detection limits in liquid chromatography through the optimization of chromatographic operation.

Midgley, A. R.; Niswender, G. D.; Rebar, R. W. (1969). Principles for the assessment of the reliability of radioimmunoassay methods (precision, accuracy, sensitivity, specificity). *Acta Endocrinol. (Suppl.)*, **63**, 163–184.

Defines sensitivity as "the smallest amount of unlabeled hormone which can be distinguished from no hormone." Notes that radioimmunoassays show nonuniformity of variance. Suggests using weighted least squares in situations of nonconstant variance.

Miller, R. G. (1966). *Simultaneous Statistical Inference*. McGraw-Hill, New York.

Discusses simultaneous intervals. Page 1: Explains simultaneous—"simultaneously bracket the expected values . . . for all values of x." Page 111: Conversion to nonsimultaneous intervals—t replaces $(2F)^{0.5}$.

Milliken, G. A.; Johnson, D. E. (1984). *Analysis of Messy Data*, Vol. 1, *Designed Experiments*. Van Nostrand Reinhold, New York.

Mitchell, D. G.; Garden, J. S. (1982). Measuring and maximizing precision in analyses based on use of calibration graphs. *Talanta*, **29**, 921–929.

Advocates estimating the minimum reportable concentration (MRC) on the basis of the calibration equation. Defines the MRC as the concentration at which the confidence band around the predicted sample concentration just includes zero as a lower bound (with a given significance level). Comments that "with calibration over moderate-to-wide calibration ranges, the assumption of constant variance is almost always false." Suggests using weighted least-squares regression in situations of nonconstant variance.

Montgomery, D. C.; Johnson, L. A. (1976). *Forecasting and Time Series Analysis*. McGraw-Hill, New York.

Morrison, G. H. (1971). Evaluation of lunar elemental analyses. *Anal. Chem.*, **43**, 22A–31A.

Examination of chemical analysis data from numerous elements in moon rocks from the Lunar Analysis Program of the U.S. National Aeronautics and Space Administration. Comments on nonconstant variance: ". . . in the low-ppm range, the spread in values is somewhat greater than at higher concentrations."

Nalimov, V. V. (1963). *The Application of Mathematical Statistics to Chemical Analysis*. Pergamon Press, Oxford.

Referenced in Boumans (1978). Provides more elaborate statistical calculations to take into account the uncertainty of the calibration function.

Nalimov, V. V.; Nedler, V. V. (1961). *Zavodsk. Lab.*, **27**, 861–861.

Referenced in Liteanu and Rica (1975a). Study of the relationship between the background noise and the analytical signal in the frame of the problem of the detection limit.

Nalimov, V. V.; Nedler, V. V.; Men'shova, N. P. (1961). *Ind. Lab.*, **27**, 164–164.
Referenced in Boumans and Maessen (1966). Discussion of detection limits.

NBS (1961). *A Manual of Radioactivity Procedures*. Handbook 80. National Bureau of Standards, Washington, DC.
Uses the term *minimum detectable activity* (or *mass*) for the detection limit.

Neter, J.; Wasserman, W.; Kutner, M. (1990). *Applied Linear Regression Models*, 2nd ed. Homewood IL: Irwin.

Nicholson, W. L. (1971). In *Developments in Applied Spectroscopy*, Vol. 6. Grove, E. L.; Perkins, A. J., Eds. Plenum Press, New York.
Provides a discussion of detection limits.

O'Haver, T. C. (1976). In *Trace Analysis: Spectroscopic Methods for Elements*. Winefordner, J. D., Ed. Wiley, New York.
Referenced in Kirchmer (1983). The limit of detection should apply to a complete analytical procedure and not to a given instrument or instrumental method. Also, an analyte cannot be measured quantitatively at the detection limit.

Omenetto, N.; Winefordner, J. D. (1979). Atomic fluorescence spectrometry basic principles and applications. *Prog. Anal. At. Spectrosc.* **2**, 1–183.
Extensive calculations of detection limits for atomic spectrometric methods.

Oppenheimer, L.; Capizzi, T. P.; Weppelman, R. M.; Mehta, H. (1983). Determining the lowest limit of reliable assay measurement. *Anal. Chem.*, **55**, 638–643.
Terminology: "Lowest Limit of Reliable Assay Measurement-LLORAM". Unweighted and weighted least-squares prediction intervals (Lc, Ld, and Lq). Ivermectin (an antiparasitic agent) data. Discusses modeling the weights (power and quadratic functions) but approximate weights at the limits. Tries also using a log transformation.

Oresic, L. S.; Grdinic, V. (1990). Kaiser's 3-sigma criterion: a review of the limit of detection. *Acta Pharm. Jugosl.*, **40**, 21–61.
Referenced in Coleman et al. (1997). A general detection limit reference. Referenced in Clark and Whitfield (1994). A detection limit review with over 500 citations.

Ortiz, M. C.; Arcos, J.; Juarros, J. V.; Lopez-Palacios, J.; Sarabia, L. A. (1993). Robust procedure for calibration and calculation of the detection limit of trimipramine by adsorptive stripping voltammetry at a carbon paste electrode. *Anal. Chem.*, **65**, 678–682.
Uses a least-median-squares approach for calibration and for calculating the limit of detection. Also uses the noncentrality parameter instead of Student's t.

Ott, W. R. (1990). A physical explanation of the lognormality of pollutant concentrations. *J. Air Waste Manage. Assoc.*, **40**, 1378–1383.
Provides a physical justification for the assumption of lognormality of environmental data.

Owen, D. B. (1963). *Factors for One-Sided Tolerance Limits and for Variable Sampling Plans*. Monograph SCR-607, 19th ed. Sandia Corporation, Livermore, CA.
Original reference for tables of tolerance interval test statistics (K) presented in Guttman (1970).

Owen, D. B. (1965). The power of student's t-test. *J. Am. Stat. Assoc.*, **60**, 320–333.
Provides tables of delta values (include central t and noncentral t) used in Clayton et al. (1987).

Owen, W. J.; DeRouen, T. A. (1980). Estimation of the mean for log-normal data containing zeros and left-censored values, with applications to the measurement of worker exposure to air contaminants. *Biometrics*, **36**, 707–719.

Adapts Aitchison's result to the case of a censored lognormal distribution and applies it to environmental data.

Owens, K. G.; Bauer, C. F.; Grant, C. L. (1988). In *Detection in Analytical Chemistry: Importance, Theory, and Practice*. ACS Symposium Series 361. Currie, L. A., Ed. American Chemical Society, Washington, DC.

Discusses the use of calibration models for calculating detection limits—do not actually calculate detection limits. Experimental design considerations: concentration range, number of standards, distribution of standards, replication of standards, and preparation of standards. Discuss lack of fit (LOF) testing. Discusses heterogeneity of variance. Favors the use of weighted least squares over transformations. Good explanation of interval shapes with and without weighting (also figures).

Oxenford, J. L.; McGeorge, L. J.; Jenniss, S. W. (1989). Determination of practical quantitation levels for organic compounds in drinking water *J. Am. Water Works Assoc.*, **84**, 149–154.

Interlaboratory study conducted to determine practical quantitation levels (PQLs) for 22 organic compounds in drinking water. Data presented show a decrease in precision and accuracy with decreasing concentration.

Pallesen, L. (1985). The interpretation of analytical measurements made near the limit of detection. Unpublished work.

Referenced in Berthouex (1993). Describes a new approach to understanding the precision of measurements at low concentration—like the Rocke and Lorenzato (1995) two-component model. Error consists of background noise (b_i), which is constant, and analytical error (a_i), which is proportional to concentration. No data provided to demonstrate this approach.

Pankow, J. F. (1988). Detection limits and the environmental agenda. *Environ. Sci. Technol.*, **22**, 1372–1373.

Discussion of the implications of attaining detection limits far below risk levels.

Parsons, M. L. (1969). The definition of detection limits. *J. Chem. Educ.*, **46**, 290–292.

Calculates the standard deviation (s) as a function of the RMS noise (N_{rms}) as $s = N_{rms} (2/n)^{0.5}$. Defines the detection limit as that concentration that gives a signal-to-RMS noise (S/N_{rms}) of $(S/N_{rms}) = t(2/n)^{0.5}$. Finds the determination limit by choosing a relative standard deviation (RSD) and calculating the required S/N_{rms} as $(S/N_{rms}) = RSD \times 100 (2/n)^{0.5}$. Comments that the detection limit could be calculated from the determination limit more accurately than it could be measured experimentally because S/N_{rms} is a linear function of the concentration near the detection limit. Uses the equation $C_{detect} = [(S/N)_{detect} C_{determ}]/(S/N)_{determin}$.

Patterson, C. C.; Settle, D. M. (1976). *7th Materials Research Symposium*. National Bureau of Standards Special Publication. 422. U.S. Government Printing Office, Washington, DC.

Referenced in Currie (1985). Provides an illustration of assumed constancy of B shown seriously in error.

Pearson, E. S.; Hartley, H. O. (1976). *Biometrika Tables for Statisticians*. Biometrika Trust, London.

Source of many useful statistical tables.

Persson, T.; Rootzen, H. (1977). Simple and highly efficient estimators for a type I censored normal sample. *Biometrika*, **64**, 123–128.

Method for handling nondetects.

Peters, D. G.; Hayes, J. M.; Hieftje, G. M. (1974). In *Chemical Separation and Measurements*. W.B. Saunders, Philadelphia.

Referenced in Long and Winefordner (1983). Textbook containing methods for calculating the detection limit.

Pettinati, J. D.; Swift, C. E. (1977). Collaborative study of precision characteristics of the AOAC method for crude fat in meat and meat products. *J. Assoc. Off. Anal. Chem.*, **60**, 600–608.

Provides data (fat in meat) supporting the hypothesis that the coefficient of variation is related to concentration.

Porter, P. S.; Ward, R. C.; Bell, H. F. (1988). The detection limit. *Environ. Sci. Technol.*, **22**, 856–861.

Concerned primarily with data censoring issues, not detection limit calculation procedures.

Puschel, R. (1968). Zum Problem der "Genauigkeit" chemischer Analysen. *Mikrochim. Acta*, **4**, 783–801.

Not in English. Discusses the dependence of standard deviation on concentration (with figures). Equation: $\log(s \times 100/x) = 0.13 - 0.32 \log p$.

Ramirez-Munoz, J. (1966). Qualitative and quantitative sensitivity in flame photometry. *Talanta*, **13**, 87–101.

Defines *qualitative sensitivity* as the minimum amount (or concentration) of analyte that allows its identification, and *quantitative sensitivity* as the minimum amount (or concentration) of analyte that allows its determination.

Robinson, J. W. (1960). *Anal. Chem.*, **32**, 1A–17A.

Use the terms *detection limit* and *sensitivity* interchangeably.

Rocke, D. M.; Durbin, B. (1998), *Models and Estimators for Analytical Measurement Methods with Non-constant Variance.* Technical Report. Center for Image Processing and Integrated Computing, University of California at Davis, Davis, CA.

Describes alternative approaches to estimating the parameters of the Rocke and Lorenzato model.

Rocke, D. M.; Lorenzato, S. (1995). A two-component model for measurement error in analytical chemistry. *Technometrics*, **37**, 176–184.

Proposes a two-component model for standard deviation as a function of concentration—use maximum likelihood estimation. First component: standard deviation is constant at low concentrations (additive error). Second component: standard deviation is proportional to concentration at higher concentrations (multiplicative error). Approximate as $s_x = (a_0 + a_1 X^2)^{0.5}$. Detection limit equation is not very well explained (confidence interval–based critical level with $t = 3$). Cadmium (atomic absorption) and toluene (GC/MS) data. See Appendix A for a discussion of the Horwitz papers.

Rodbard, D. (1971). In *Competitive Protein Binding Assays.* Odell, W. D.; Daughaday, W. H., Eds. J.B. Lippincott, Philadelphia.

Referenced in Rodbard (1978). Most radioimmunoassasys show nonuniformity of variance.

Rodbard, D. (1978). Statistical estimation of the minimal detectable concentration ("sensitivity") for radioligand assays. *Anal. Biochem.*, **90**, 1–12.

Terminology: minimum detectable dose (MDD) and minimum detectable concentration (MDC). Apply to radioimmunoassay data. Acknowledge nonconstant variance but assume that variance is constant between zero and MDC. MDC (response units) = $Y_{blank} - ts(1/n_{blank} + 1/n_{MDC})^{0.5}$ (if the calibration curve has a positive slope, the minus sign before the t should be replaced by a plus sign). Suggests a quadratic model for variance as a function of concentration.

Rodbard, D.; Cooper, J. A. (1974). In *In Vitro Procedures with Radioisotopes in Medicine*. International Atomic Energy Agency, Vienna.

Referenced in Finney (1976). Discusses various sources of analysis error. Proposes using a quadratic or other polynomial function to represent the variance.

Rodbard, D.; Hutt, D. M. (1974). In *Radioimmunoassay and Related Procedures in Medicine*, Vol. 1. International Atomic Energy Agency, Vienna.

Referenced in Finney (1976). Expand on Rodbard and Cooper (1974) paper that uses a quadratic or other polynomial function to represent the analysis variance.

Rodbard, D.; Lenox, R. H.; Wray, H. L.; Ramseth, D. (1976). Statistical characterization of the random errors in the radioimmunoassay dose–response variable. *Clin. Chem.*, **22**, 350–358.

Discusses the presence of nonconstant variance in radioimmunoassay analyses. Suggests using weighted least-squares regression by modeling the variance to calculate the weights. Uses the following models: linear ($s^2 = a_0 + a_1 Y$), quadratic ($s^2 = a_0 + a_1 Y + a_2 Y^2$), log-log [$\log(s^2) = \log(a_0) + j \log(Y)$], proportional ($s^2 = a_1 Y$), and power ($s^2 = a_0 Y^j$). All the models provide satisfactory results for the data presented here; however, the authors recommend checking multiple models for each case.

Rogers, L. B. (1986). The inexact imprecise science of trace analysis. *J. Chem. Educ.*, **63**, 3–6.

Emphasizes the importance of increasing the reliability of analytical data through the use of quality assurance procedures.

Rogers, L. B. (1990). The new generation of measurement. *Anal. Chem.*, **62**, 703A–711A.

The coefficient of variation for replicates obtained for a concentration near 1 ppb can be 50% or more.

Roos, J. B. (1962). The limit of detection of analytical methods. *Analyst*, **87**, 832–833.

Points out that false negative error is often ignored.

Rosner, B. (1983). Percentage points for a generalized ESD many-outlier procedure. *Technometrics*, **25**, 165–172.

Widely used outlier test.

Ruessmann, H. H.; Brooks, K. R. (1964). *Acta Chim. Acad. Sci. Hung.*, **42**, 1–1.

Referenced in Gabriels (1970). Calculates the detection limit as a function of the standard deviation on the blank value.

Russell, B. J.; Shelton, J. P.; Walsh, A. (1957). An atomic-absorption spectrophotometer and its application to the analysis of solutions. *Spectrochimica Acta*, **8**, 317–328.

Uses the terms *detection limit* and *sensitivity* interchangeably.

Rutherford, E.; Geiger, H. (1910). The probability variations in the distribution of alpha particles. *Philos. Mag. 6th Ser.*, **20**, 698–704.

Early reference on the use of the Poisson distribution.

Ryan, T.; Joiner, B. (1973). *Normal Probability Plots and Tests for Normality*. Technical Report. Penn State University, University Park, PA.

S. A. B. (1986). Detection limits. *Anal. Chem.*, **58**, 986A–986A.

Summary of a detection limit symposium held at an ACS national meeting in New York City. Proceedings to appear in an ACS Symposium Series book, *Detection Limits: From Basic Concepts to Practical Application*, (1987).

Sanders, P. F.; Lippincott, R. L.; Eaton, A. (1996). Determining quantitation levels for regulatory purposes. *J. Am. Water Works Assoc.*, **88**, 104–114.

Defines the reliable detection level (RDL) as twice the MDL. Defines the quantitation level (QL) as the level that is greater than or equal to the RQLs for a majority of labs in a certified laboratory community. Survey of 51 certified drinking water laboratories to determine QLs for 36 compounds.

Sarhan, A. E.; Greenberg, B. G. (eds.) (1962). *Contributions to Order Statistics.* Wiley, New York.

Fundamental text on order statistics, useful in deriving nonparametric prediction limits.

Saw, J. G. (1961). Estimation of the normal population parameters given a type I censored sample. *Biometrika,* **48,** 367–377.

Scharf, F. (1964). *Experimentia,* **20,** 470–470.

Referenced in Gabriels (1970). Points out that if the difference between the blank and the smallest detectable quantity is of interest, the variability of both has to be taken into account.

Schmee, J.; Gladstein, D.; Nelson, W. (1985). Confidence limits of a normal distribution from singly censored samples using maximum likelihood. *Technometrics,* **27,** 119–128.

Direct estimation of the confidence interval for a censored normal random variable.

Schneider, H. (1986). *Truncated and Censored Samples from Normal Populations.* Marcel Dekker, New York.

Early book on analysis of censored data useful for the treatment of nondetects in environmental data.

Schuller, P. L.; Horwitz, W.; Stoloff, L. (1976). Review of aflatoxin methodology: a review of sampling plans and collaboratively studied methods of analysis for aflatoxins. *J. Assoc. Off. Anal. Chem.,* **59,** 1315–1343.

Provides data (aflatoxins) supporting the hypothesis that the coefficient of variation is related to concentration.

Schwartz, L. M. (1977). Nonlinear calibration. *Anal. Chem.,* **49,** 2062–2068.

Discusses nonconstant variance; mentions GC detector response explicitly as an example where variance is a function of concentration. Discusses the calculation of confidence (our prediction) intervals around nonlinear calibration curves.

Schwartz, L. M. (1979). Calibration curves with nonuniform variance. *Anal. Chem.,* **51,** 723–727.

Discusses the calculation of confidence (our prediction) intervals around linear and nonlinear calibration curves in situations of nonconstant variance using weighting factors. Use an example of benzene analyzed by GC. Mentions that there is agreement among statistics texts with regard to the minimal effect of weighteng factors on fitted regression curves. There is a significant effect on the confidence interval, however.

Scott, R. H.; Kokot, M. L. (1975). Application of inductively coupled plasmas to the analysis of geochemical samples. *Anal. Chim. Acta.,* **75,** 257–270.

The coefficient of variation decreases as concentration increases for ICP data.

Scott, J. W.; Whiddon, N. T.; Maddalone, R. F. (1993). *Compliance Monitoring Detection and Quantitation Levels for Utility Aqueous Discharges.* EPRI Final Report TR-103205. Electric Power Research Institute, Palo Alto, CA.

Can compute Y_c using the standard deviation at L_c as opposed to using the standard deviation at zero.

Searle, S. R. (1987). *Linear Models for Unbalanced Data.* Wiley, New York.

Book on variance components models.

Searle, S. R.; Casella, G.; McCulloch, C. E. (1992). *Variance Components Models.* Wiley, New York.

Referenced in Gibbons (1995). Replaces *s* with unbiased variance estimate based on the appropriate variance components model for the problem at hand.

Sen, P. K. (1968). Estimates of the regression coefficient based on Kendall's tau. *J. Am. Stat. Assoc.*, **63**, 1379–1389.

Source of Sen's test, which is a widely used nonparametric trend estimator.

Shapiro, S. S.; Francia, R. S. (1972). An approximate analysis of variance test for normality. *J. Am. Stat. Assoc.*, **67**, 215–216.

Extension of the Shapiro–Wilk test of normality for samples of 50 or more.

Shapiro, S. S.; Wilk, M. B. (1965). An analysis of variance test for normality (complete samples). *Biometrika*, **52**, 591–611.

Original source of the Shapiro–Wilk test of normality.

Shapiro, S. S.; Wilk, M. B.; Chen, H. J. (1968). A comparative study of various tests for normality. *J. Am. Stat. Assoc.*, **63**, 1343–1372.

Sharaf, M. A.; Illman, D. L.; Kowalski, B. R. (1986). *Chemometrics*. Wiley, New York.

Referenced in Analytical Methods Committee (1987). Terminology: limit of guaranteed detection.

Sillars, I. M.; Silver, R. S. (1944). The accuracy of measurements of dissolved oxygen in water. *J. Soc. Chem. Ind.*, **63**, 177–179.

Possibly the first authors to introduce statistical concepts to determine the smallest concentration that could be regarded as significantly different from zero, termed the *significance range*. Calculated as: $ts/n^{0.5}$.

Skogerboe, R. K. (1969). In *Flame Emission and Atomic Absorption Spectrometry*. Dean, J. A.; Rains, T. C., Eds. Marcel Dekker, New York.

Sensitivity should be a measure of the ability to determine a small change in concentration of analyte at any concentration level. It is limited by the slope of the analytical curve and the reproducibility with which the analyte signal can be measured.

Skogerboe, R. K.; Grant, C. L. (1970). Comments on the definitions of the terms sensitivity and detection limit. *Spectrosc. Lett.*, **3**, 215–220.

Attempts to resolve the confusion between the terms *detection limit* and *sensitivity*. Defines the sensitivity as the ability to determine a small change in analyte concentration (depends on the slope of the calibration curve and the reproducibility or precision), while the detection limit is based on the standard deviation of the blank. Suggests that the standard deviation of the blank (for calculating LODs) be determined over a period of days to months.

Slavin, W.; Sprague, S.; Manning, D. C. (1964). Detection limits in analytical atomic absorption spectroscopy. *At. Absorpt. Newsl.*, **18**, 1–12.

Referenced in Ramirez-Munoz (1966). Defines the sensitivity as the concentration that produces an absorption of 1%. Defines the relative detection limit as the concentration that produces absorption equivalent to twice the magnitude of the fluctuation in the background.

Smith, R. M.; Bain, L. J. (1976). Correlation type goodness-of-fit statistics with censored sampling. *Commun. Stat. Theor. Methods*, **A5**(2), 119–132.

Useful test of normality in censored samples (i.e., nondetects).

Smith, E. D.; Mathews, D. M. (1967). Least square regression lines. *J. Chem. Educ.*, **44**, 757–759.

Discussion of nonconstant variance in least-squares regression. Uses gas chromatographic data as an example.

Smith, R. A.; Hirsch, R. M.; Slack, J. R. (1982). A study of trends in total phosphorus measurements at NASQAN stations. USGS Water Supply Paper 2190, U.S. Geological Survey, Alexandria, VA.

Snedecor, G. W.; Cochran, W. G. (1980). *Statistical Methods.* Iowa State University Press, Ames, IA.

   One of the best applied statistics books.

Snelson, J. T. (1976). *Analysis of Organochlorine Residues in Butter Fat.* Document CX/PR 77/9. Food and Agriculture Organization, Rome.

   Referenced in Horwitz (1982). Provides data (pesticide residues) supporting the hypothesis that the coefficient of variation is related to concentration.

Specker, H. (1966). Methoden der Spurenanalyse und ihre Anwendungsgrenzen. *Z. Anal. Chem.*, **221**, 33–43.

   Not in English. Referenced in Liteanu and Rica (1975a). Study of the relationship between the background noise and the analytical signal in the frame of the problem of the detection limit.

Spectroscopy Group of the Institute of Physics and Physical Society (1964). *Limitations of Detection in Spectrochemical Analysis.*, Hilger & Watts, London.

   Conference in Exeter, England devoted solely to limitations of detection.

St. John, P. A.; McCarthy, W. J.; Winefordner, J. D. (1966). Applications of signal-to-noise theory in molecular luminescence spectrometry. *Anal. Chem.*, **38**, 1828–1835.

   Defines the theoretical minimum detectable sample concentration as the concentration that has a signal-to-noise ratio of 2.

St. John, P. A.; McCarthy, W. J.; Winefordner, J. D. (1967). A statistical method for evaluation of limiting detectable sample concentrations. *Anal. Chem.*, **39**, 1495–1497.

   Assume constant variance. Terminology: limiting detectable sample concentration (C_m). This is in S/N-based approach. C_m is defined as the concentration that results in $(S/N)_{Cm} = (t \times 2^{0.5})/n^{0.5})$. Calculate $(S/N)_{Cm}$, then determine whether the sample concentration yields a S/N above or below $(S/N)_{Cm}$.

Stevenson, C. L.; Winefordner, J. D. (1991). Estimating detection limits in ultratrace analysis: I. The variability of estimated detection limits. *Appl. Spectrosc.*, **45**, 1217–1224.

   Accepts limit of detection estimates as being equivalent if within a factor of 2 or 3. Discusses calculation of the variability and confidence intervals for the estimated LOD.

Sutherland, G. L. (1965). Residue analytical limit of detectability. *Residue Rev.*, **10**, 85–96.

   Terminology: limit of detectability. Use only the standard deviation of a blank and ignore false negative error. "Detectability limits cannot be defined in terms of recoveries." Limits of detection and quantification not distinguished.

Svoboda, V.; Gerbatsch, R. (1968). Zur Definition von Grenzwerten fur das Nachweisvermögen. *Z. Anal. Chem.*, **242**, 1–13.

   Not in English. Referenced in Liteanu and Rica (1975a). Study of the relationship between the background noise and the analytical signal in the frame of the problem of the detection limit.

Taylor, J. K. (1987). *Quality Assurance of Chemical Measurements.* Lewis Publishers, Chelsea, MI.

   Referenced in Sanders et al. (1996). Comments that precision of measurements at the MDL are generally poor.

Theil, H. (1950). A rank-invariant method of linear and polynomial regression analysis. *Proc. K. Ned. Akad. Wet., Part 3*, **A53**, 1397–1412.

Thompson, M.; Howarth, R. J. (1976). Duplicate analysis in geochemical practice. *Analyst*, **101**, 690–698.

Mentions situations where a non-Gaussian distributions can arise, for example, if the concentration levels are near the detection limit and subzero readings are set to zero. Mentions nonconstant variance; suggests modeling the standard deviation with a polynomial. Defines the detection limit as the concentration at which the precision is unity. Expands the *complete analytical procedure* of Kaiser to include a set of samples of the analyte in a specific matrix, an exactly defined analytical procedure, and the particular instrumentation used.

Tietjen, G. L.; Moore, R. H. (1972). Some Grubbs-type statistics for the detection of several outliers. *Technometrics*, **14**(3), 583–597.

Method for the sequential screening of outliers.

USATHAMA (1987). *USATHAMA QA Program*, 2nd ed. U.S. Army Toxic and Hazardous Materials Agency, Aberdeen Proving Ground, MD.

Referenced in Grant et al. (1991). Original reference for the certified reporting limit (CRL).

USEPA (1982). *Appendix A: Methods for Chemical Analysis of Wastewater.* U.S. Environmental Protection Agency, EMSL, Cincinnati, OH. Referenced in Kimbrough and Wakakuwa (1993). Description of MDL approach.

USEPA (1984a). Polynuclear aromatic hydrocarbons. *Fed. Reg.*, **49**, 43344–43352.

Defines the MDL as "the minimum concentration of a substance that can be measured and reported with 99% confidence that the value is above zero." Lists MDLs for various PAHs using HPLC.

USEPA (1984b). Definition and procedure for determination of the method detection limit: Revision 1.11. *Fed. Reg.*, **49**, 43430–43431.

Terminology: method detection limit. General procedure: (1) estimate the detection limit; (2) prepare reagent (blank) water; (3) make a standard with an analyte level 1 to 5 times the estimated detection limit (must be less than 10 times the MDL in reagent water); (4) take 7 aliquots of the standard and process through the entire analytical procedure; (5) calculate the standard deviation (s); (6) calculate the MDL as MDL = $3.143s$. Optional iterative process to check the reasonableness of the estimated MDL.

USEPA (1985). National primary drinking water regulations; volatile synthetic organic chemicals: proposed rule. *Fed. Reg.*, **50**, 46902–46933.

Practical quantitation level (PQL) definition: The PQL is defined as the lowest achievable level of analytical quantitation during routine laboratory operating conditions within specified limits of precision and accuracy. The PQL can be estimated as 5 to 10 times the MDL.

USEPA (1987). List (phase 1) of hazardous constituents for ground-water monitoring data from hazardous waste facilities: final rule. *Fed. Reg.*, **52**, 25942–25953.

Provides estimates of practical quantitation limits (PQL's) for analytes analyzed by particular methods.

USEPA (1988). Statistical methods for evaluating ground-water monitoring data from hazardous waste facilities: final rule. 40CFR Part 264. *Fed. Reg.*, **53**(196), 39720–39731 (Oct. 11).

Revised Subtitle C rule.

USEPA (1989). Method detection limits and practical quantitiation levels: proposed rule. *Fed. Reg.*, **54**, 22100–22104.

"The practical quantitation level (PQL) is the lowest concentration that can be reliably measured within specified limits of precision and accuracy during routine laboratory operating conditions." "The PQL is generally 5 to 10 times the MDL for relatively clean matrices such as finished drinking water." Estimates MDLs and PQLs for inorganics and sythetic organics. Referenced in Rocke and Lorenzato (1995). Defines the PQL as the concentration in which the coefficient of variation falls to 0.2.

USEPA (1991). Solid waste disposal facility criteria: final rule. *Fed. Reg.*, **56**(196), 50978–51119 (Oct. 9).

Subtitle D rule.

USEPA (1992). *Statistical Analysis of Ground-Water Monitoring Data at RCRA Facilities.* Addendum to Interim Final Guidance. Office of Solid Waste, July 1992.

Revised guidance for Subtitle C and D statistical applications.

USEPA (1993). *Guidance on Evaluation, Resolution, and Documentation of Analytical Problems Associated with Compliance Monitoring.* EPA/821-B-93-001. U.S. Environmental Protection Agency, Washington, DC.

USEPA (1996). *Section 7, U.S. EPA DBP/ICR Analytical Methods Manual, Office of Water.* EPA 814-B-96-002. U.S. Government Printing Office, Washington, D.C.

Referenced in Kimbrough (1997). Reference for the minimum reporting level (MRL), the lowest concentration that meets the data quality objectives of the information collection rule (ICR).

Versieck, J.; Cornelis, R. (1980). Normal levels of trace elements in human blood plasma or serum. *Anal. Chim. Acta*, **116**, 217–254.

Provides data (trace elements) supporting the hypothesis that the coefficient of variation is related to concentration.

Watson, J. E. (1980). *Upgrading Environmental Data.* EPA 520/1-80-012. U.S. Environmental Protection Agency, Office of Radiation Programs, Washington, DC.

Referenced in Chambless et al. (1992). Reports on the widespread misuse of the estimated MDC as the basis for data censoring in the fields of radiation detection and analytical chemistry.

Watt, D. E.; Ramsden, D. (1964). *High Sensitivity Counting Techniques.* Macmillan, New York.

Terminology: minimum detectable activity (or mass).

Watts, R. R. (1980). In *Optimizing Chemical Laboratory Performance Through the Application of Quality Assurance Principles.* Garfield, F. M., Ed. Association of Official Analytical Chemists, Arlington, VA.

Provides data supporting the hypothesis that the coefficient of variation is related to concentration.

Wernimont, G. T. (1985). *Use of Statistics to Develop and Evaluate Analytical Methods.* Association of Official Analytical Chemists, Arlington, VA.

Referenced in Grant et al. (1991). Mentions the need to protect against false negatives.

West, T. S. (1969). Comparison of sensitivity of atomic-absorption and atomic fluorescence spectroscopy. *Spectrosc. Lett.*, **2**, 179–183.

Response to criticisms by Willis (1969) that West's laboratory used *sensitivity* and *detection limit* interchangeably without exact definition. Comments that the detection limit has some merit in absorption studies; however, sensitivity is a more realistic and meaningful term.

West, T. S.; Williams, X. K. (1968a). Atomic fluorescence spectroscopy of silver using a high-intensity hollow-cathode lamp as source. *Anal. Chem.*, **40**, 335–339.

Compares detection limits for silver in atomic fluorescence and absorption spectroscopy.

West, T. S.; Williams, X. K. (1968b). Atomic-fluorescence spectroscopy of magnesium with a high-intensity hollow-cathode lamp as line source. *Anal. Chim. Acta.*, **42**, 29–37.

Compares detection limits for magnesium in atomic fluorescence and absorption spectroscopy.

Wilk, M. B.; Shapiro, S. S. (1968). The joint assessment of normality of several independent samples. *Technometrics*, **10**, 825–839.

Generalization of the Shapiro–Wilk test of normality for the case of several groups, (e.g., multiple upgradient wells).

Wilks, S. S. (1941). Determination of sample sizes for setting tolerance limits. *Ann. Math. Stat.*, **12**, 91–96.

Willis, J. B. (1969). Comparison of detection limits in atomic fluorescence and absorption spectroscopy. *Spectrosc. Lett.*, **2**, 191–196.

Comments that various authors used the terms *detection limit* and *sensitivity* interchangeably.

Willits, N. (1993). Personal communication, University of California at Davis, Davis, CA.

Wilson, A. L. (1961). The precision and limit of detection of analytical methods. *Analyst*, **86**, 72–74.

Proposes using s_c to calculate the detection limit, where $s_c = (s_a^2 + s_b^2)^{0.5}$, s_a is the standard deviation of the sample, s_b is the standard deviation of the blank, and $c = a - b$. Quote: "The probability of detection is reasonably high when the limit of detection is set at values $\geq 2s_b$." This method ignores false negative error.

Wilson, A. L. (1973). The performance characteristics of analytical methods: III. *Talanta.*, **20**, 725–732.

Terminology: criterion of detection, corresponds to Currie's critical level; and limit of detection, corresponds to Currie's detection limit. Comment that there is "no valid reason why any one confidence level should be used universally."

Winefordner, J. D. (1976). In *Trace Analysis: Spectroscopic Methods for Elements*. Wiley, New York.

Review of detection limit calculations.

Winefordner, J. D.; Overfield, C. V. (1967). Estimation of limits of detection for the flame emission gas chromatographic detector. *J. Chromatogr.*, **31**, 1–8.

Evaluates a specific expression that can be used to calculate the limit of detectability for atomic emission flame spectrometry. Shows that the signal-to-noise ratio is a linear function of the analyte concentration near the detection limit.

Winefordner, J. D.; Rutledge, M. (1985). Comparison of calculated detection limits in molecular absorption, molecular luminescence, Raman, molecular ionization, and photothermal spectrometry. *Appl. Spectrosc.*, **39**, 377–391.

Calculates detection limits for absorption, luminescence, photoionization, Raman, and photothermal molecular spectrometric methods. The detection limit concept was developed for the specific purpose of permitting scientists to compare analytical instruments and methodologies to assist in developing more powerful techniques.

Winefordner, J. D.; Vickers, T. J. (1964a). Calculation of the limit of detectability in atomic absorption flame spectrometry. *Anal. Chem.*, **36**, 1947–1954.

Derives a specific expression that can be used to calculate the limit of detectability for atomic absorption flame spectrometry.

Winefordner, J. D.; Vickers, T. J. (1964b). Calculation of the limit of detectability in atomic emission flame spectrometry. *Anal. Chem.*, **36**, 1939–1946.

Derives a specific expression that can be used to calculate the limit of detectability for atomic emission flame spectrometry.

Winefordner, J. D.; Ward, J. L. (1980). The reliability of detection limits in analytical chemistry. *Anal. Lett.*, **13**, 1293–1297.

Suggests five guidelines to improve the reliability of detection limits. Stresses the importance of "following rigid statistical rules and documentation of the experimental system, the experimental parameters and the entire analytical procedure."

Winefordner, J. D.; Parsons, M. L.; Mansfield, J. M.; McCarthy, W. J. (1967a). Derivation of expressions for calculation of limiting detectable atomic concentration in atomic fluorescence flame spectrometry. *Anal. Chem.*, **39**, 436–442.

Derives a specific expression that can be used to calculate the limit of detectability for atomic fluorescence flame spectrometry.

Winefordner, J. D.; McCarthy, W. J.; St. John, P. A. (1967b). The selection of optimum conditions for spectrochemical methods. *J. Chem. Educ.*, **44**, 80–83.

Uses signal-to-noise theory to define the minimum detectable sample concentration (C_{min}).

Winefordner, J. D.; Cetorelli, J. J.; McCarthy, W. J. (1968). Estimation of minimum detectable sample concentrations obtained with near and middle infrared detectors for infrared absorption spectrophotometry. *Talanta*, **15**, 207–212.

Terminology: minimum detectable sample concentration, C_{mp}, defined as that concentration resulting in a fixed minimal detectable absorbance, A_{min}. Equation for absorption spectrophotometry: $C_{mp} = A_{min}/eb$, where e is the molar absorptivity and b is the sample cell thickness. A_{min} is typically taken to be about 0.002 for ultraviolet or visible absorption spectrophotometry of condensed substances, and about 0.02 for infrared absorption spectrophotometry of condensed substances. Have shown that the signal-to-noise (RMS) ratio is a linear function of the analyte concentration near the detection limit.

Winefordner, J. D.; Schulman, S. G.; O'ttaver, T. C. (1972). *Luminescence Spectrometry in Analytical Chemistry.* Wiley-Interscience, New York.

Discussion of detection limits.

Wing, J.; Wahlgren, M. A. (1967). Detection sensitivities in nuclear activation with an isotopic fast-neutron source. *Anal. Chem.*, **39**, 85–89.

Terminology: relative detection sensitivities. determines relative detection sensitivities in nuclear activation of 73 chemical elements.

Yohe, T. L.; Hertz, C. D. (1991). In *Proceedings of the Water Quality and Technology Conference*, Orlando, FL.

Referenced in Kimbrough (1997).

Youden, W. J.; Steiner, E. H. (1975). *Statistical Manual of the Association of Official Analytical Chemists.* Association of Official Analytical Chemists, Washington, DC.

Zil'Berstein, K. H. (1977). *Spectrochemical Analysis of Pure Substances.* Adam Hilger, Bristol, Gloucestershire, England.

Referenced in Boumans (1978). Key reference for detection limits.

Zitter, H.; God, C. (1971). Ermittlung, Auswertung und Ursachen von Fehlern bei Betriebsanalysen. *Fresenius Z. Anal. Chem.*, **255**, 1–9.

Not in English. Referenced in Analytical Methods Committee (1987). Suggests a linear function to model the standard deviation as a function of concentration.

Zorn, M. E.; Gibbons, R. D.; Sonzogni, W. C. (1997). Weighted least-squares approach to calculating limits of detection and quantification by modeling variability as a function of concentration. *Anal. Chem.*, **69**, 3069–3075.

Proposes a weighted tolerance interval equation. Calculates unweighted and weighted prediction and tolerance interval-based L_c, L_d, L_q, and AML for 16 polychlorinated biphenyl congeners. Three models of standard deviation as a function of concentration (quadratic, exponential, and two-component Rocke and Lorenzato) used to calculate weights at the limits.

Zorn, M. E.; Gibbons, R. D.; Sonzogni, W. C. (1999). Evaluation of approximate methods for calculating the limit of detection and limit of quantification. *Environ. Sci. Technol.*, **33**, 2291–2295.

Evaluates "approximate" methods for calculating limits of detection and quantification that are less computationally complex than statistically rigorous methods proposed previously. The approximate methods use data at multiple spiking concentrations, are iterative, can be derived from either prediction intervals or statistical tolerance intervals, and require at a minimum ordinary least-squares regression for calculating the intercept and slope.

INDEX

Accepted value, 330
Accuracy, 12–13, 323
ACS Limit of Quantitation (LOQ), 58
Alternative minimum level (AML), 79–82
AML software, 280
Analysis of Variance (ANOVA), 108–111, 213–215. *See also* Variance, components
Anthropogenic compounds, 2. *See also* VOCs
Assessment monitoring, 1, 218–228, 323
ASTM, 72–73, 89

Between-laboratory, *see* interlaboratory
Bias, 14, 17–18, 324
Blank samples, 115–116, 330
   calibration blanks, 115
   field blanks, 115
   instrument blanks, 115
   laboratory blanks, 115
   matrix blanks, 115
   method blanks, 115
   reagent blanks, 115
   solvent blanks, 115
   system blanks, 115
   trip blanks, 115
Blind studies, 116, 330
   double-blind studies, 116
BLUE, *see* Censored data, linear estimators
Bonferroni:
   adjustment, 207
   inequality, 207

CABF t-test, 215–216
Calibrate, 324
Calibration:
   based detection limits, 61–78
   based quantification limits, 79–89
   designs, 34
   function, 11
   intercept, 11, 332

iteratively reweighted least-squares, 39–40, 68–69, 120–121
linear model, 11, 39
ordinary least squares (OLS), 34, 35
slope, 11, 337
unweighted least squares, *see* Calibration, OLS
weighted least squares (WLS), 35–39, 66–68
CARStat, 280
Censoring:
   mechanism, 137
   point, 137–138
   Type I, 145
   Type II, 145
Censored data, 136–163
   alternate linear estimators, 145–153
   comparison of estimators, 158–161
   conceptual foundation, 137–138
   Delta distributions, 153–155
   imputation, 139
   linear estimators, 144–145
   MLE, 139–143
   regression methods, 155–157
   RMLE, 143–144
   substitution of order statistics, 157–158
Central limit theorem, 31
Certified reference material, 330
Coefficient of variation (COV, CV), 330
Comparison to background, 252–272
   lognormal prediction limits for individual samples, 262–264
   lognormal prediction limits for the mean, 265–267
   lognormal prediction limits for the median, 264–265
   nonparametric prediction limits for individual samples, 267
   nonparametric prediction limits for the median, 267, 272

Comparison to background (*Cont.*)
  normal prediction limits for individual samples, 253–258
  normal prediction limits for the mean, 258–262
Comparison to standards, 131–135
  confidence interval for true concentration, 134
  L_C, 131–133
  L_D, 131–133
  L_Q, 132–133
Compliance monitoring, 1, 218–227, 324
Confidence level, 23
Confidence limits, 27–30, 229–251
  for a mean, 1, 231–232
  for a percentile, 243–251
  lognormal mean, 232–233, 238–241
  lognormal median, 232
  lower, 229–231, 333
  nonparametric, 242–243
  normal, 1, 231–232
  upper, 230–231, 337
Control charts, 211–213
Corrective action monitoring, 1, 218–228, 325
Critical level (L_C, or LC), 14–15, 17, 331. *See also* L_C

Data quality objectives (DQOs), 325
Decision limit, *see* Critical level
DETECT software, 280
Detection, 14
Detection limit, 2, 15, 17, 325. *See also* Calibration and L_D.
Detection monitoring, 1, 194–217, 325. *See also* Prediction intervals
  Intrawell comparisons, 211–213
  landfill, 205
  leachate, 205
  plume, 206
Determination limit, 58. *See also* Estimators, L_Q
DL, *see* Detection limit
Drift, 331
DUMPStat, 280
Dunnett's test, 256–258
Dynamic range, 331

EPA method, *see* Method detection limit
Error:
  measurement, 325
  random, *see* random error
  sampling, 336
  standard deviation, 326
  systematic, 33

Estimates:
  empirical Bayes (EB), 119, 125
  expected a posteriori (EAP), 119
  marginal maximum likelihood (MMLE), 119–122
  restricted maximum likelihood (RMLE), 111
Estimators:
  L_C, 14–15, 17, 48
  L_D, 2, 14–15, 17, 49
  L_Q, 2, 15, 16, 17
  least squares, *see* Calibration, OLS
  linear, *see* Censored data, linear estimators
  maximum likelihood, *see* Censored data, MLE
  weighted least squares, *see* Calibration, WLS
Evaluation function, 12
Experimental design, 104–116
  between-laboratory, 105–108
  equispaced design, 105
  maximum spiking concentration, 105
  selection of calibration standards, 107
  semigeometric design, 105
  within-laboratory, 105–108
  Youden pair design, 106, 111–113
Exponential model, 38–39, 80

False negative rate, 53–54, 331
False positive rate, 53–54, 331
Filtering, 331

Gain, 331
Gauge repeatability, 336
Gaurdband, 332
GC/MS, 43
Geometric mean, 240–241, 276–277
Groundwater sampling and analysis, 224–228
  groundwater-surface water interface, 226
  long-term monitoring, 226–227
  natural attenuation, 227–228
  site-wide evaluation, 277–279

Homoscedasticity, 35
Hubaux-Vos model, 62–63
Hybrid model, 87. *See also* Rocke and Lorenzato model
Hypothesis testing, 21–27
  alternative, 21
  null, 21
Hysteresis, 332

Interference, 332. *See also* Matrix effects
Inter-laboratory studies, 17–20, 117–127
  calibration, 117–121
  copper data, 124

detection limit, 121
method validation study, 332
quantification limit, 123
random-effects model, 119–122. *See also* Variance, components
round-robin study, 332
Interlaboratory detection estimate (IDE), 70–78, 118
Interlaboratory quantitation estimate (IQE), 85–89, 118
Interlaboratory standard deviation (ILSD) model, 77–78
Inter-well comparisons, 206–212
Intra-well comparisons, 211–213
Inverse normal, 155
Inverse prediction, 333
Iteratively reweighted least-squares, *see* Calibration

Kaiser-Currie method, 49–50
Kendall test, *see* Mann-Kendall test
Kurtosis, *see* Normality tests and Outlier tests

Land's coefficients:
  90% confidence, 233–234
  95% confidence, 235–236
  99% confidence, 236–237
  tabled values, 283–321
Limit:
  confidence, *see* Confidence limit
  of detection, *see* Detection limit
  prediction, *see* Prediction intervals
  of quantitation, *see* Quantification limit
  tolerance, *see* Tolerance intervals
Lognormal:
  confidence limits, 333. *See also* Confidence limits
  distribution, 333
  prediction limits, *see* Prediction limits and Comparison to background
  tolerance limits, *see* Calibration
Long term monitoring, 273–274

Mann-Kendall test, 197–200
Matrix effects, 10, 113–114
Maximum contaminant level (MCL), 59, 220
Mean:
  arithmetic, 333
  geometric, *see* Geometric mean
  recovery curve, 77
  response curve, 77
Measurand, 333
Median, 334
Metrology, 334

Measurement:
  assurance, 333
  assurance program, 333
  capability study (MCS), 326
  error, 19
  process, 9
  routine, 9–10
  uncertainty, 90–92, 326
Method detection limit (MDL), 15, 48, 50–52, 56, 328
Minimum level, 57, 59
Mixed-effects model, 17
Monitoring well network, 205
Multivariate:
  hypergeometric distribution, 211
  normality, 256–258
  t-distribution, 257

Noise, 328
Noncentrality parameter, 54, 64
Noncentral t-distribution, 26, 54, 64
Nonparametric:
  confidence limits, *see* Confidence limits, nonparametric
  prediction limits, 210–211
  trend tests, *see* Trend detection
Normal:
  confidence limits, *see* confidence limits, normal distribution, 334
  prediction limits, *see* prediction limits, normal
  tolerance limits, *see* tolerance limits
Normality tests, 164–182
  censored data, 177–181
  D'Agostino's test, 170–172
  graphical approaches, 165–166
  Kolmogorov-Smirnov test, 181–182
  moment tests, 173–175
  multiple group tests, 175–177
  Shapiro Francia test, 170
  Shapiro-Wilk test, 166–170

Order statistics, 157–158, 210–211
Outlier tests, 183–193, 334
  comparison of tests, 191–193
  Dixon's test, 190–191
  E_m-test, 188–190
  Kurtosis test, 188
  Rosner's test, 184–187
  Shapiro-Wilk test, 188
  Skewness test, 187–188

Phase I study, 334
Phase II study, 335
Phase III evaluation, 274–277

INDEX

Potential area of concern (PAOC), 220, 335
Practical quantification limit (PQL), 58–59
Precision, 328
    function, 335
    tolerance ratio (P/T Ratio), 328
Prediction intervals, 30–32, 52, 335
    censored data, 209
    for the mean, 258–261, 265
    for the median, 264, 267, 272
    for a single value, *see* Comparison to standards
    lower prediction limit, 333
    multiple constituents, 209
    multiple locations, 207–208
    single well and constituent, 206–207
    OLS, 40
    verification resampling, 208–209
    WLS, 41–42

Quality assurance (QA), 9–11, 328
Quality control (QC), 9–11, 328
Quantification, 14, 16
Quantification limit, 2, 15, 17, 57–60, 329.
    See also Calibration and L_Q
Quantitation, *see* Quantification

Random, 329
    error, 329
    sampling, 335
RCRA, 215
Reference method, 335
Relative measurement uncertainty, 16, 92, 95
Relative standard deviation, 16, 80 335
Repeatability, 19, 329
Replicate sample, 335
Reproducibility, 329
Rocke and Lorenzato model, 36–38, 55, 82, 121

S-Plus, 280
Sample, 329
    blank, *see* Blank samples
    blind, *see* Blind studies
    check, 330
    duplicate, 331
    replicate, 335
    representative, 335
    subsample, 335
    verification, *see* Verification sampling
Sanitas, 280
Seasonality, *see* Trend detection
Sen's test, *see* Trend detection
Selectivity, 336
Sensitivity, 12, 336

Signal-to-noise ratio (SNR), 336
Significance level, 23, 336
Significant digits, 90–103, 329
    fractions of digits, 93–94
    MRSD format, 102–103
    reporting a measurement with uncertainty, 101–102
    relationship to detection, 96–98
    relationship to quantification, 98–101
Significant figures, *see* Significant digits
Single-operator precision, 336
Skewness, *see* Normality tests and Outlier tests
Soil sampling, 223–224
Spatial variability, *see* Detection monitoring
Specificity, 337
Spiking concentration, 62
Standard, 329
    operating procedure, 327
    Reference material (SRM), 10, 329
Standard deviation, 16, 337
Statistical power, 23
Statistical significance, 26–27

Tolerance intervals, 32–33, 52–53
    OLS, 40–41, 65
    WLS, 42–43
Traceability, 330
Trend detection, 194–203
    Mann-Kendall test, 197–200
    seasonal Mann-Kendall test, 200–203
    Sen's test, 195–197
    statistical properties, 203
True value, 1, 330
Type I error, 23, 48
Type II error, 23, 49

Variance:
    components, 108–111, 337
    function, 36–39
Verification sampling, 337. *See also* Detection monitoring and Prediction intervals
VOCs, 2

Waste stream sampling, 228
Weighted least squares—WLS, 55, 66–68, 337.
    See also Calibration, weighted least squares
Within-laboratory study, 19
W-statistic, *see* Normality tests, Shapiro-Wilk test

Youden:
    pair design, 106, 111–113, 337
    plot, 112, 337